Lecture Notes in Mathematics

2237

More information about this series at http://www.springer.com/series/304

CEMPI CENTRE EUROPÉEN
POUR LES MATHÉMATIQUES, LA PHYSIQUE ET
LEURS INTERACTIONS

David Coupier

Editor

Stochastic Geometry

Modern Research Frontiers

 Springer

Editor
David Coupier
LAMAV
Université Polytechnique des Hauts de
France (UPHF)
Valenciennes, France

ISSN 0075-8434   ISSN 1617-9692   (electronic)
Lecture Notes in Mathematics
ISBN 978-3-030-13546-1   ISBN 978-3-030-13547-8   (eBook)
https://doi.org/10.1007/978-3-030-13547-8

Mathematics Subject Classification (2010): Primary: 60

This Springer imprint is published by the registered company Springer Nature Switzerland AG.
The registered company address is: Gewerbestrasse 11, 6330 Cham, Switzerland

Preface

This manuscript is the third volume of the "CEMPI subseries" common to Lecture Notes in Mathematics and Lecture Notes in Physics.

Stochastic geometry can be succinctly described as the study of random spatial patterns. It is traditionally considered to have been born in the eighteenth century with the well-known Buffon's needle problem, and it has received growing attention from the mathematical community during the second part of the twentieth century. Nowadays, stochastic geometry draws on a large variety of subdomains of mathematics such as point processes, geometric random graphs, percolation, convex geometry, random fields and spatial statistics and admits a wide range of applications. Let us cite for example astronomy, computational geometry, telecommunication networks, image analysis and stereology, and material science.

The GDR GeoSto (http://gdr-geostoch.math.cnrs.fr/) is a national research structure funded by the CNRS and created in 2012, which aims to federate the French community working on stochastic geometry. Since 2014, its yearly meeting has been preceded by two introductory courses. This volume contains five of these introductory lectures.

The first chapter is a historically motivated introduction to stochastic geometry, whereas each of the other four gives an introduction to one important branch of contemporary stochastic geometry, which we have called "Highlights in Stochastic Geometry". The chapters also have in common their proximity to applications. Let us describe them briefly.

The first chapter, entitled "Some classical problems in random geometry", is based on lectures by Pierre Calka (Université de Rouen, France) given in Lille in 2014. It presents four historical questions on geometric probabilities, namely the Buffon's needle problem, the Bertrand paradox, the Sylvester four-point problem and the bicycle wheel problem. Through each of these classical problems, the author highlights the topics currently most active in stochastic geometry.

The second chapter entitled "Understanding spatial point patterns through intensity and conditional intensities", written by Jean-François Coeurjolly (Université du Québec, Canada) and Frédéric Lavancier (Université de Nantes, France), is based on lectures by Jean-François Coeurjolly given in Lille in 2014. This chapter presents point process statistics from an original point of view, namely through intensities and conditional intensities, in order to capture interactions between points of the process. With several references to implemented packages, the authors are concerned with remaining close to applications: their exposition should be very useful for the working spatial statistician.

The third chapter, entitled "Stochastic methods for image analysis", is based on lectures by Agnès Desolneux (CMLA and ENS Paris-Saclay, France) given in Poitiers in 2015. It is about stochastic methods for computer vision and image analysis. It starts with a very nice introduction to the Gestalt theory that is a psychophysiological theory of human visual perception, which can be translated into a mathematical framework thanks to the non-accidentalness principle, and is illustrated by many convincing pictures.

While fractal analysis of one-dimensional signals is mainly based on the use of fractional Brownian motion, its generalization to higher dimensions, motivated in particular by medical imaging questions, requires a deep understanding of random fields. The fourth chapter, entitled "Introduction to random fields and scale invariance", based on lectures by Hermine Biermé (Université de Poitiers, France) given in Nantes in 2016, appears as a useful survey and an overview of recent achievements for the worker in this field.

The fifth chapter, entitled "Introduction to the theory of Gibbs point processes", is based on lectures by David Dereudre (Université de Lille, France) given in Nantes in 2016. This last chapter is a very clear and stimulating review of Gibbs point processes, an area of research with a long tradition in statistical physics and novel impulses from stochastic geometry and spatial statistics. After a careful introduction, the author summarizes recent results on the main challenging question in the topic: the uniqueness of Gibbs point processes in infinite volume.

In conclusion, this volume offers a unique and accessible overview (up to the frontiers of recent research) of the most active fields in stochastic geometry. We hope that it will make the reader want to go further.

Valenciennes, France   David Coupier
November 2018

Acknowledgements

On behalf of GDR GeoSto, we sincerely acknowledge the authors of the five chapters contained in this book for their precious work and all the participants of the annual meetings of the GDR without whom this book would not exist.

Since its creation in 2012, the GDR GeoSto is totally funded by the French National Center for Scientific Research (CNRS).

The edition of this manuscrit was made possible by the Labex CEMPI (ANR-11-LABX-0007-01).

Contents

Contributors

Hermine Biermé LMA, UMR CNRS 7348, Université de Poitiers, Chasseneuil, France

Pierre Calka University of Rouen, LMRS, Saint-Étienne-du-Rouvray, France

Jean-François Coeurjolly Université du Québec à Montréal, Département de Mathématiques, Montréal, QC, Canada

David Dereudre University Lille, Villeneuve-d'Ascq, France

Agnès Desolneux CNRS, CMLA and ENS Paris-Saclay, Paris, France

Frédéric Lavancier Université de Nantes, Laboratoire de Mathématiques Jean Leray, Nantes, France

Chapter 1
Some Classical Problems in Random Geometry

Pierre Calka

Abstract This chapter is intended as a first introduction to selected topics in random geometry. It aims at showing how classical questions from recreational mathematics can lead to the modern theory of a mathematical domain at the interface of probability and geometry. Indeed, in each of the four sections, the starting point is a historical practical problem from geometric probability. We show that the solution of the problem, if any, and the underlying discussion are the gateway to the very rich and active domain of integral and stochastic geometry, which we describe at a basic level. In particular, we explain how to connect Buffon's needle problem to integral geometry, Bertrand's paradox to random tessellations, Sylvester's four-point problem to random polytopes and Jeffrey's bicycle wheel problem to random coverings. The results and proofs selected here have been especially chosen for non-specialist readers. They do not require much prerequisite knowledge on stochastic geometry but nevertheless comprise many of the main results on these models.

1.1 Introduction: Geometric Probability, Integral Geometry, Stochastic Geometry

Geometric probability is the study of geometric figures, usually from the Euclidean space, which have been randomly generated. The variables coming from these random spatial models can be classical objects from Euclidean geometry, such as a point, a line, a subspace, a ball, a convex polytope and so on.

It is commonly accepted that geometric probability was born in 1733 with Buffon's original investigation of the falling needle. Subsequently, several open questions appeared including Sylvester's four-point problem in 1864, Bertrand's paradox related to a random chord in the circle in 1888 and Jeffreys's bicycle wheel problem in 1946. Until the beginning of the twentieth century, these questions were

P. Calka (✉)
University of Rouen, LMRS, Saint-Étienne-du-Rouvray, France
e-mail: pierre.calka@univ-rouen.fr

© Springer Nature Switzerland AG 2019
D. Coupier (ed.), *Stochastic Geometry*, Lecture Notes in Mathematics 2237,
https://doi.org/10.1007/978-3-030-13547-8_1

1

all considered as recreational mathematics and there was a very thin theoretical background involved which may explain why the several answers to Bertrand's question were regarded as a paradox.

After a course by G. Herglotz in 1933, W. Blaschke developed a new domain called *integral geometry* in his papers *Integralgeometrie* in 1935–1937, see e.g. [17]. It relies on the key idea that the mathematically natural probability models are those that are invariant under certain transformation groups and it provides mainly formulas for calculating expected values, i.e. integrals with respect to rotations or translations of random objects. Simultaneously, the modern theory of probability based on measure theory and Lebesgue's integral was introduced by S.N. Kolmogorov in [72].

During and after the Second World War, people with an interest in applications in experimental science—material physics, geology, telecommunications, etc.-realized the significance of random spatial models. For instance, in the famous foreword to the first edition of the reference book [31], D.G. Kendall narrates his own experience during the War and how his Superintendent asked him about the strength of a sheet of paper. This question was in fact equivalent to the study of a random set of lines in the plane. Similarly, J.L. Meijering published a first work on the study of crystal aggregates with random tessellations while he was working for the Philips company in 1953 [82]. In the same way, C. Palm who was working on telecommunications at Ericsson Technics proved a fundamental result in the one-dimensional case about what is nowadays called the Palm measure associated with a stationary point process [96]. All of these examples illustrate the general need to rigorously define and study random spatial models.

We traditionally consider that the expression *stochastic geometry* dates back to 1969 and was due to D.G. Kendall and K. Krickeberg at the occasion of the first conference devoted to that topic in Oberwolfach. In fact, I. Molchanov and W.S. Kendall note in the preface of [95] that H.L. Frisch and J.M. Hammersley had already written the following lines in 1963 in a paper on percolation: *Nearly all extant percolation theory deals with regular interconnecting structures, for lack of knowledge of how to define randomly irregular structures. Adventurous readers may care to rectify this deficiency by pioneering branches of mathematics that might be called stochastic geometry or statistical topology.*

For more than 50 years, a theory of stochastic geometry has been built in conjunction with several domains, including

- the theory of point processes and queuing theory, see notably the work of Mecke [79], Stoyan [122], Neveu [93], Daley [34] and [41],
- convex and integral geometry, see e.g. the work of Schneider [111] and Weil [127] as well as their common reference book [113],
- the theory of random sets, mathematical morphology and image analysis, see the work of Kendall [67], Matheron [78] and Serra [115],
- combinatorial geometry, see the work of Ambartzumian [3].

It is worth noting that this development has been simultaneous with the research on spatial statistics and analysis of real spatial data coming from experimental science, for instance the work of B. Matérn in forestry [77] or the numerous papers in geostatistics, see e.g. [94].

In this introductory lecture, our aim is to describe some of the best-known historical problems in geometric probability and explain how solving these problems and their numerous extensions has induced a whole branch of the modern theory of stochastic geometry. We have chosen to embrace the collection of questions and results presented in this lecture under the general denomination of *random geometry*. In Sect. 1.2, Buffon's needle problem is used to introduce a few basic formulas from integral geometry. Section 1.3 contains a discussion around Bertrand's paradox which leads us to the construction of random lines and the first results on selected models of random tessellations. In Sect. 1.4, we present some partial answers to Sylvester's four-point problem and then derive from it the classical models of random polytopes. Finally, in Sect. 1.5, Jeffrey's bicycle wheel problem is solved and is the front door to more general random covering and continuum percolation.

We have made the choice to keep the discussion as non-technical as possible and to concentrate on the basic results and detailed proofs which do not require much prerequisite knowledge on the classical tools used in stochastic geometry. Each topic is illustrated by simulations which are done using *Scilab 5.5*. This chapter is intended as a foretaste of some of the topics currently most active in stochastic geometry and naturally encourages the reader to go beyond it and carry on learning with reference to books such as [31, 95, 113].

Notation and Convention The Euclidean space $\mathbb{R}^d$ of dimension $d \geq 1$ and with origin denoted by o is endowed with the standard scalar product $\langle \cdot, \cdot \rangle$, the Euclidean norm $\|\cdot\|$ and the Lebesgue measure V_d. The set $B_r(x)$ is the Euclidean ball centered at $x \in \mathbb{R}^d$ and of radius $r > 0$. We denote by $\mathbb{B}^d$ (resp. $\mathbb{S}^{d-1}$, $\mathbb{S}_+^{d-1}$) the unit ball (resp. the unit sphere, the unit upper half-sphere). The Lebesgue measure on $\mathbb{S}^{d-1}$ will be denoted by σ_d. We will use the constant $\kappa_d = V_d(\mathbb{B}^d) = \frac{1}{d}\sigma_d(\mathbb{S}^{d-1}) = \frac{\pi^{\frac{d}{2}}}{\Gamma(\frac{d}{2}+1)}$. Finally, a convex compact set of $\mathbb{R}^d$ (resp. a compact intersection of a finite number of closed half-spaces of $\mathbb{R}^d$) will be called a d-dimensional *convex body* (resp. *convex polytope*).

1.2 From Buffon's Needle to Integral Geometry

In this section, we describe and solve the four century-old needle problem due to Buffon and which is commonly considered as the very first problem in geometric probability. We then show how the solution to Buffon's original problem and to one of its extensions constitutes a premise to the modern theory of integral geometry. In particular, the notion of intrinsic volumes is introduced and two classical integral formulas involving them are discussed.

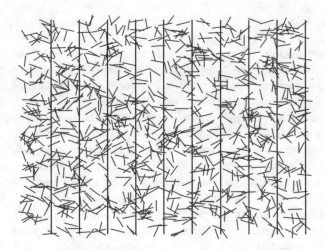

Fig. 1.1 Simulation of Buffon's needle problem with the particular choice $\ell/D = 1/2$: over 1000 samples, 316 were successful (red), 684 were not (blue)

1.2.1 Starting from Buffon's Needle

In 1733, Georges-Louis Leclerc, comte de Buffon, raised a question which is nowadays better known as Buffon's needle problem. The solution, published in 1777 [20], is certainly a good candidate for the first-ever use of an integral calculation in probability theory. First and foremost, its popularity since then comes from being the first random experiment which provides an approximation of π.

Buffon's needle problem can be described in modern words in the following way: a needle is dropped *at random* onto a parquet floor which is made of parallel strips of wood, each of same width. What is the probability that it falls across a vertical line between two strips (Fig. 1.1)?

Let us denote by D the width of each strip and by ℓ the length of the needle. We assume for the time being that $\ell \leq D$, i.e. that only one crossing is possible. The randomness of the experiment is described by a couple of real random variables, namely the distance R from the needle's mid-point to the closest vertical line and the angle Θ between a horizontal line and the needle.

The chosen probabilistic model corresponds to our intuition of a *random* drop: the variables R and Θ are assumed to be independent and both uniformly distributed on $(0, D/2)$ and $(-\pi/2, \pi/2)$ respectively.

Now there is intersection if and only if $2R \leq \ell \cos(\Theta)$. Consequently, we get

$$p = \frac{2}{\pi D} \int_{-\frac{\pi}{2}}^{\frac{\pi}{2}} \int_0^{\frac{1}{2}\ell \cos(\theta)} \mathrm{d}r \mathrm{d}\theta = \frac{2\ell}{\pi D}.$$

This remarkable identity leads to a numerical method for calculating an approximate value of π. Indeed, repeating the experiment n times and denoting by S_n the number of hits, we can apply Kolmogorov's law of large numbers to show that $\frac{2\ell n}{DS_n}$ converges almost surely to π with an error estimate provided by the classical central limit theorem.

In 1860, Joseph-Émile Barbier provided an alternative solution for Buffon's needle problem, see [6] and [71, Chapter 1]. We describe it below as it solves at the same time the so-called *Buffon's noodle problem*, i.e. the extension of Buffon's needle problem when the needle is replaced by any planar curve of class C^1.

Let us denote by p_k, $k \geq 0$, the probability of exactly k crossings between the vertical lines and the needle. Henceforth, the condition $\ell \leq D$ is not assumed to be fulfilled any longer as it would imply trivially that $p = p_1$ and $p_k = 0$ for every $k \geq 2$. We denote by $f(\ell) = \sum_{k \geq 1} k p_k$ the mean number of crossings. The function f has the interesting property of being additive, i.e. if two needles of respective lengths ℓ_1 and ℓ_2 are pasted together at one of their endpoints and in the same direction, then the total number of crossing is obviously the sum of the numbers of crossings of the first needle and of the second one. This means that $f(\ell_1 + \ell_2) = f(\ell_1) + f(\ell_2)$. Since the function f is increasing, we deduce from its additivity that there exists a positive constant α such that $f(\ell) = \alpha \ell$.

More remarkably, the additivity property still holds when the two needles are not in the same direction. This implies that for any finite polygonal line $\mathscr{C}$, the mean number of crossings with the vertical lines of a rigid noodle with same shape as $\mathscr{C}$, denoted by $f(\mathscr{C})$ with a slight abuse of notation, satisfies

$$f(\mathscr{C}) = \alpha \mathscr{L}(\mathscr{C}) \tag{1.1}$$

where $\mathscr{L}(\cdot)$ denotes the arc length. Using both the density of polygonal lines in the space of piecewise C^1 planar curves endowed with the topology of uniform convergence and the continuity of the functions f and $\mathscr{L}$ on this space, we deduce that the formula (1.1) holds for any piecewise C^1 planar curve.

It remains to make the constant α explicit, which we do when replacing $\mathscr{C}$ by the circle of diameter D. Indeed, almost surely, the number of crossings of this noodle with the vertical lines is 2, which shows that $\alpha = \frac{2}{\pi D}$. In particular, when K is a convex body of $\mathbb{R}^2$ with diameter less than D and $p(K)$ denotes the probability that K intersects one of the vertical lines, we get that

$$p(K) = \frac{1}{2} f(\partial K) = \frac{\mathscr{L}(\partial K)}{\pi D}.$$

Further extensions of Buffon's needle problem with more general needles and lattices can be found in [18]. In the next subsection, we are going to show how to derive the classical Cauchy-Crofton's formula from similar ideas.

1.2.2 Cauchy-Crofton formula

We do now the opposite of Buffon's experiment, that is we fix a noodle which has the shape of a convex body K of $\mathbb{R}^2$ and let a *random* line fall onto the plane. We then count how many times in mean the line intersects K.

This new experiment requires to define what a random line is, which means introducing a measure on the set of all lines of $\mathbb{R}^2$. We do so by using the polar equation of such a line, i.e. for any $\rho \in \mathbb{R}$ and $\theta \in [0, \pi)$, we denote by $L_{\rho,\theta}$ the line

$$L_{\rho,\theta} = \rho(\cos(\theta), \sin(\theta)) + \mathbb{R}(-\sin(\theta), \cos(\theta)).$$

Noticing that there is no natural way of constructing a probability measure on the set of random lines which would satisfy translation and rotation invariance, we endow the set $\mathbb{R} \times [0, \pi)$ with its Lebesgue measure. The integrated number of crossings of a line with a C^1 planar curve $\mathscr{C}$ is then represented by the function

$$g(\mathscr{C}) = \int_{-\infty}^{\infty} \int_0^{\pi} \#(L_{\rho,\theta} \cap \mathscr{C}) \mathrm{d}\theta \mathrm{d}\rho.$$

The function g is again additive under any concatenation of two curves so it is proportional to the arc length. A direct calculation when $\mathscr{C}$ is the unit circle then shows that

$$g(\mathscr{C}) = 2\mathscr{L}(\mathscr{C}). \tag{1.2}$$

This result is classically known in the literature as the Cauchy-Crofton formula [28, 33]. Going back to the initial question related to a convex body K, we apply (1.2) to $\mathscr{C} = \partial K$ and notice that $\#(L_{\rho,\theta} \cap \mathscr{C})$ is equal to $2\mathbf{1}_{\{L_{\rho,\theta} \cap K \neq \emptyset\}}$. We deduce that

$$\mathscr{L}(\partial K) = \int_{-\infty}^{\infty} \int_0^{\pi} \mathbf{1}_{\{L_{\rho,\theta} \cap K \neq \emptyset\}} \mathrm{d}\theta \mathrm{d}\rho. \tag{1.3}$$

1.2.3 Extension to Higher Dimension

We aim now at extending (1.3) to higher dimension, that is we consider the set $\mathscr{K}^d$ of convex bodies of $\mathbb{R}^d$, $d \geq 2$, and for any element K of $\mathscr{K}^d$, we plan to calculate integrals over all possible k-dimensional affine subspaces L_k of the content of $L_k \cap K$. This requires to introduce a set of fundamental functionals called *intrinsic volumes* on the space of convex bodies of $\mathbb{R}^d$. This is done through the rewriting of the volume of the parallel set $(K + B_\rho(o))$ as a polynomial in $\rho > 0$. Indeed, we are going to prove that there exists a unique set of d functions $V_0, \cdots, V_{d-1}$ such that

$$V_d(K + B_\rho(o)) = \sum_{k=0}^{d} \kappa_{d-k} \rho^{d-k} V_k(K). \tag{1.4}$$

The identity (1.4) is known under the name of *Steiner formula* and was proved by Steiner for $d = 2$ and 3 in 1840 [120]. In particular, the renormalization with the multiplicative constant κ_{d-k} guarantees that the quantity $V_k(K)$ is really intrinsic to K, i.e. that it does not depend on the dimension of the underlying space. We explain below how to prove (1.4), following closely [112, Section 1].

   In the first step, we start by treating the case when K is a convex polytope P. We denote by $\mathscr{F}$ the set of all faces of P and by $\mathscr{F}_k$, $0 \le k \le d$, the subset of $\mathscr{F}$ consisting of the k-dimensional faces of P. For any face $F \in \mathscr{F} \setminus \{P\}$, the open *outer normal cone* of F denoted by $N_P(F)$ is the set of $x \in \mathbb{R}^d \setminus \{o\}$ such that there exists an affine hyperplane H with normal vector x satisfying $H \cap P = F$ and $\langle x, y - h \rangle \le 0$ for every $y \in P$ and $h \in H$. In particular, when $F \in \mathscr{F}_k$, $0 \le k \le (d - 1)$, $N_P(F)$ is a $(d - k)$-dimensional cone. Moreover, $(F + N_P(F))$ is the set of points outside P whose nearest point in P lies in F and no other lower-dimensional face. The set of all sets $(F + N_P(F))$, often called the normal fan of P, is a partition of $\mathbb{R}^d \setminus P$, see Fig. 1.2. Consequently, from the decomposition of

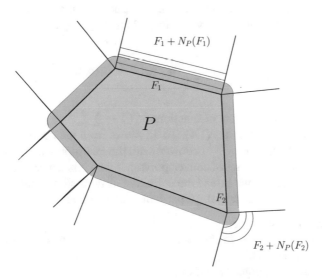

Fig. 1.2 A two-dimensional convex polygon P (gray), the region $(P + B_r(o)) \setminus P$ (pink), the sets $(F + N_P(F))$ with two examples for the edge F_1 and the vertex F_2 (striped regions)

$(P + B_\rho(o)) \setminus P$ combined with Fubini's theorem, we get

$$V_d(P + B_\rho(o)) = V_d(P) + \sum_{F \in \mathscr{F} \setminus \{P\}} V_d(F + (N_P(F) \cap B_\rho(o)))$$

$$= V_d(P) + \sum_{k=0}^{d-1} \sum_{F \in \mathscr{F}_k} V_k(F) \gamma(F, P) \rho^{d-k} \kappa_{d-k} \qquad (1.5)$$

where $\gamma(F, P)$ is the normalized area measure on the unit sphere $\mathbb{S}^{d-k-1}$ of $N_P(F) \cap \mathbb{S}^{d-k-1}$. In particular, we deduce from (1.5) that (1.4) holds for $K = P$ as soon as

$$V_k(P) = \sum_{F \in \mathscr{F}_k} V_k(F) \gamma(F, P).$$

In the second step, we derive the Steiner formula for any convex body K. In order to define $V_k(K)$, we use the trick to rewrite (1.4) for a polytope P and for several values of ρ, namely $\rho = 1, \cdots, (d + 1)$. The $(d + 1)$ equalities constitute a Vandermonde system of $(d + 1)$ equations in $(\kappa_d V_0(P), \cdots, \kappa_0 V_d(P))$. These equations are linearly independent because the Vandermonde determinant of n pairwise distinct real numbers is different from zero. When solving the system by inverting the Vandermonde matrix, we construct a sequence $\alpha_{k,l}$, $0 \leq k \leq d$, $1 \leq l \leq (d + 1)$ such that for every polytope P and $0 \leq k \leq d$,

$$V_k(P) = \sum_{l=1}^{d+1} \alpha_{k,l} V_d(P + B_l(o)).$$

It remains to check that the set of functions $V_k(\cdot) = \sum_{l=1}^{d+1} \alpha_{k,l} V_d(\cdot + B_l(o))$, $0 \leq k \leq d$, defined on $\mathscr{K}^d$ satisfies (1.4). This follows from the continuity of V_d, and hence of all V_k on the space $\mathscr{K}^d$ endowed with the Hausdorff metric and from the fact that (1.4) holds on the set of convex polytopes which is dense in $\mathscr{K}^d$.

For practical reasons, we extend the definition of intrinsic volumes to $K = \emptyset$ by taking $V_k(\emptyset) = 0$ for every $k \geq 0$. Of particular interest are:

– the functional V_0 equal to $\mathbf{1}_{\{K \neq \emptyset\}}$,
– the functional V_1 equal to the so-called *mean width* up to the multiplicative constant $d\kappa_d/(2\kappa_{d-1})$,
– the functional V_{d-1} equal to half of the Hausdorff measure of ∂K.

Furthermore, Hadwiger's theorem, which asserts that any additive, continuous and motion-invariant function on $\mathscr{K}^d$ is a linear combination of the V_k's, provides an alternative way of characterizing intrinsic volumes [49].

We now go back to our initial purpose, i.e. extending the Cauchy-Crofton formula to higher dimension. We do so in two different ways:

First, when $K \in \mathscr{K}^2$, an integration over ρ in (1.3) shows that

$$\mathscr{L}(\partial K) = 2V_1(K) = \int_0^\pi V_1(K|L_{0,\theta}) d\theta$$

where $(K|L_{0,\theta})$ is the one-dimensional orthogonal projection of K onto $L_{0,\theta}$. When $K \in \mathscr{K}^d$, $d \geq 2$, Kubota's formula asserts that any intrinsic volume V_k can be recovered up to a multiplicative constant as the mean of the Lebesgue measure of the projection of K onto a uniformly distributed random k-dimensional linear subspace. In other words, there exists an explicit positive constant c depending on d and k but not on K such that for every $K \in \mathscr{K}^d$,

$$V_k(K) = c \int_{\mathrm{SO}_d} V_k(K|\mathscr{R}(L)) d\nu_d(\mathscr{R}) \qquad (1.6)$$

where L is a fixed k-dimensional linear subspace of $\mathbb{R}^d$, SO_d is the usual special orthogonal group of $\mathbb{R}^d$ and ν_d is its associated normalized Haar measure.

Secondly, with a slight rewriting of (1.3), we get for any $K \in \mathscr{K}^2$

$$\mathscr{L}(\partial K) = \iint_{\mathbb{R} \times (0,\pi)} V_0(K \cap (\mathrm{rot}_\theta(L_{0,0}) + \rho(\cos(\theta), \sin(\theta)))) d\rho d\theta,$$

where rot_θ is the rotation around o and of angle θ. When $K \in \mathscr{K}^d$, $d \geq 2$, for any $0 \leq l \leq k \leq d$ and any fixed $(d-k+l)$-dimensional linear subspace L_{d-k+l}, the Crofton formula states that the k-th intrinsic volume of K is proportional to the mean of the l-th intrinsic volume of the intersection of K with a uniformly distributed random $(d-k+l)$-dimensional affine subspace, i.e. there exists an explicit positive constant c' depending on d, k and l but not on K such that

$$V_k(K) = c' \int_{\mathrm{SO}_d} \int_{L_{d-k+l}^\perp} V_l(K \cap (\mathscr{R}L_{d-k+l} + t)) dt d\nu_d(\mathscr{R}), \qquad (1.7)$$

where L_{d-k+l} is a fixed $(d-k+l)$-dimensional affine subspace of $\mathbb{R}^d$.

For the proofs of (1.6) and (1.7) with the proper explicit constants and for a more extensive account on integral geometry and its links to stochastic geometry, we refer the reader to the reference books [110, Chapters 13–14], [109, Chapters 4–5] and [113, Chapters 5–6]. We will show in the next section how some of the formulas from integral geometry are essential to derive explicit probabilities related to the Poisson hyperplane tessellation.

1.3   From Bertrand's Paradox to Random Tessellations

In this section, we recall Bertrand's problem which leads to three different and perfectly correct answers. This famous paradox questions the several potential models for constructing random lines in the Euclidean plane. The fundamental choice of the translation-invariance leads us to the definition of the stationary Poisson line process. After that, by extension, we survey a few basic facts on two examples of stationary random tessellations of $\mathbb{R}^d$, namely the Poisson hyperplane tessellation and the Poisson-Voronoi tessellation.

1.3.1   Starting from Bertrand's Paradox

In the book entitled *Calcul des Probabilités* and published in 1889 [12], J. Bertrand asks for the following question: a chord of the unit circle is chosen at random. What is the probability that it is longer than $\sqrt{3}$, i.e. the edge of an equilateral triangle inscribed in the circle?

The paradox comes from the fact that there are several ways of choosing a chord *at random*. Depending on the model that is selected, the question has several possible answers. In particular, there are three correct calculations which show that the required probability is equal to either $1/2$, or $1/3$ or $1/4$. Still a celebrated and well-known mathematical brain-teaser, Bertrand's problem questions the foundations of the modern probability theory when the considered variables are not discrete. We describe below the three different models and solutions (Fig. 1.3).

Solution 1 (Random Radius) We define a random chord through the polar coordinates (R, Θ) of the orthogonal projection of the origin onto it. The variable Θ is assumed to be uniformly distributed on $(0, 2\pi)$ because of the rotation-invariance of the problem while R is taken independent of Θ and uniformly distributed in $(0, 1)$. The length of the associated chord is $2\sqrt{1 - R^2}$. Consequently, the required

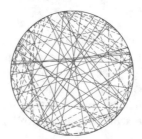

Fig. 1.3 Simulation of Bertrand's problem with 100 chords: (**a**) Solution 1 (left): 54 successful (plain line, red), 46 unsuccessful (dotted line, blue). (**b**) Solution 2 (middle): 30 successful. (**c**) Solution 3 (right): 21 successful

probability is

$$p_1 = P(2\sqrt{1 - R^2} \geq \sqrt{3}) = P(R \leq 1/2) = 1/2.$$

Solution 2 (Random Endpoints) We define a random chord through the position Θ of its starting point in the anticlockwise direction and the circular length Θ' to its endpoint. Again, Θ is uniformly distributed on $(0, 2\pi)$ while Θ' is chosen independent of Θ and also uniformly distributed in $(0, 2\pi)$. The length of the associated chord is $2\sin(\Theta'/2)$. Consequently, the required probability is

$$p_2 = P(2\sin(\Theta'/2) \geq \sqrt{3}) = P(\Theta'/2 \in (\pi/3, 2\pi/3)) = \frac{\frac{4\pi}{3} - \frac{2\pi}{3}}{2\pi} = \frac{1}{3}.$$

Solution 3 (Random Midpoint) We define a random chord through its midpoint X. The random point X is assumed to be uniformly distributed in the unit disk. The length of the associated chord is $2\sqrt{1 - \|X\|^2}$. Consequently, the required probability is

$$p_3 = P(\|X\| \leq 1/2) = \frac{V_2(B_o(1/2))}{V_2(B_o(1))} = \frac{1}{4}.$$

In conclusion, as soon as the model, i.e. the meaning that is given to the word *random*, is fixed, all three solutions look perfectly correct. J. Bertrand considers the problem as ill-posed, that is he does not decide in favor of any of the three. Neither does H. Poincaré in his treatment of Bertrand's paradox in his own *Calcul des probabilités* in 1912 [100]. Actually, they build on it a tentative formalized probability theory in a continuous space. Many years later, in his 1973 paper *The well-posed problem* [66], E.T. Jaynes explains that a natural way for discriminating between the three solutions consists in favoring the one which has the most invariance properties with respect to transformation groups. All three are rotation invariant but only one is translation invariant and that is Solution 1. And in fact, in a paper from 1868 [33], long before Bertrand's book, M.W. Crofton had already proposed a way to construct random lines which guarantees that the mean number of lines intersecting a fixed closed convex set is proportional to the arc length of its boundary.

Identifying the set of lines $L_{\rho,\theta}$ with $\mathbb{R} \times [0, \pi)$, we observe that the previous discussion means that the Lebesgue measure $d\rho d\theta$ on $\mathbb{R} \times [0, \pi)$ plays a special role when generating random lines. Actually, it is the only rotation and translation invariant measure up to a multiplicative constant. The construction of a natural random set of lines in $\mathbb{R}^2$ will rely heavily on it.

1.3.2   Random Sets of Points, Random Sets of Lines and Extensions

Generating random geometric shapes in the plane requires to generate random sets of points and random sets of lines. Under the restriction to a fixed convex body K, the most natural way to generate random points consists in constructing a sequence of independent points which are uniformly distributed in K. Similarly, in view of the conclusion on Bertrand's paradox, random lines can be naturally taken as independent and identically distributed lines with common distribution

$$\frac{1}{\mu_2(K)} \mathbf{1}_{\{L_{\rho,\theta} \cap K \neq \emptyset\}} \mathrm{d}\rho \mathrm{d}\theta$$

where

$$\mu_2(\cdot) = \iint_{\mathbb{R}\times(0,\pi)} \mathbf{1}_{\{L_{\rho,\theta} \cap \cdot \neq \emptyset\}} \mathrm{d}\rho \mathrm{d}\theta.$$

These constructions present two drawbacks: first, they are only defined inside K and not in the whole space and secondly, they lead to undesired dependencies. Indeed, when fixing the total number of points or lines thrown in K, the joint distribution of the numbers of points or lines falling into several disjoint Borel subsets of K is multinomial. Actually, there is a way of defining a more satisfying distribution on the space of locally finite sets of points (resp. lines) in the plane endowed with the σ-algebra generated by the set of functions which to any set of points (resp. lines) associates the number of points falling into a fixed Borel set of $\mathbb{R}^2$ (resp. the number of lines intersecting a fixed Borel set of $\mathbb{R}^2$). Indeed, for any fixed $\lambda > 0$, there exists a random set of points (resp. lines) in the plane such that:

- for every Borel set B with finite Lebesgue measure, the number of points falling into B (resp. the number of lines intersecting B) is Poisson distributed with mean $\lambda V_2(B)$ (resp. $\lambda \mu_2(B)$)
- for every finite collection of disjoint Borel sets $B_1, \cdots, B_k, k \geq 1$, the numbers of points falling into B_i (resp. lines intersecting B_i) are mutually independent.

This random set is unique in distribution and both translation and rotation invariant. It is called a *homogeneous Poisson point process* (resp. *isotropic and stationary Poisson line process*) of intensity λ. For a detailed construction of both processes, we refer the reader to e.g. [113, Section 3.2].

The Poisson point process can be naturally extended to $\mathbb{R}^d$, $d \geq 3$, by replacing V_2 with V_d. Similarly, we define the isotropic and stationary Poisson hyperplane process in $\mathbb{R}^d$ by replacing μ_2 by a measure μ_d which is defined in the following way. For any $\rho \in \mathbb{R}$ and $\mathbf{u} \in \mathbb{S}^{d-1}_+$, we denote by $H_{\rho,\mathbf{u}}$ the hyperplane containing the point $\rho\mathbf{u}$ and orthogonal to $\mathbf{u}$. Let μ_d be the measure on $\mathbb{R}^d$ such that for any

Borel set B of $\mathbb{R}^d$,

$$\mu_d'(B) = \iint_{\mathbb{R} \times \mathbb{S}_+^{d-1}} \mathbf{1}_{\{H_{\rho,\mathbf{u}} \cap B \neq \emptyset\}} d\rho d\sigma_d(\mathbf{u}). \tag{1.8}$$

Another possible extension of these models consists in replacing the Lebesgue measure V_d (resp. the measure μ_d) by any locally finite measure which is not a multiple of V_d (resp. of μ_d). This automatically removes the translation invariance in the case of the Poisson point process while the translation invariance is preserved in the case of the Poisson hyperplane process only if the new measure is of the form $d\rho d\nu_d(\mathbf{u})$ where ν_d is a measure on $\mathbb{S}_+^{d-1}$. For more information on this and also on non-Poisson point processes, we refer the reader to the reference books [31, 35, 70, 93]. In what follows, we only consider homogeneous Poisson point processes and isotropic and stationary Poisson hyperplane processes, denoted respectively by $\mathscr{P}_\lambda$ and $\widehat{\mathscr{P}}_\lambda$. These processes will constitute the basis for constructing stationary random tessellations of the Euclidean space.

1.3.3 On Two Examples of Random Convex Tessellations

The Poisson hyperplane process $\widehat{\mathscr{P}}_\lambda$ induces naturally a tessellation of $\mathbb{R}^d$ into convex polytopes called *cells* which are the closures of the connected components of the set $\mathbb{R}^d \setminus \bigcup_{H \in \widehat{\mathscr{P}}_\lambda} H$. This tessellation is called the (isotropic and stationary) *Poisson hyperplane tessellation* of intensity λ. In dimension two, with probability one, any crossing between two lines is an X-crossing, i.e. any vertex of a cell belongs to exactly four different cells while in dimension d, any k-dimensional face of a cell, $0 \leq k \leq d$, is included in the intersection of $(d-k)$ different hyperplanes and belongs to exactly 2^{d-k} different cells almost surely.

Similarly, the Poisson point process $\mathscr{P}_\lambda$ also generates a tessellation in the following way: any point x of $\mathscr{P}_\lambda$, called a *nucleus*, gives birth to its associated *cell* $C(x|\mathscr{P}_\lambda)$ defined as the set of points of $\mathbb{R}^d$ which are closer to x than to any other point of $\mathscr{P}_\lambda$ for the Euclidean distance, i.e.

$$C(x|\mathscr{P}_\lambda) = \{y \in \mathbb{R}^d : \|y - x\| \leq \|y - x'\| \forall x' \in \mathscr{P}_\lambda\}.$$

This tessellation is called the *Voronoi tessellation* generated by $\mathscr{P}_\lambda$ or the *Poisson-Voronoi tessellation* in short. In particular, the cell $C(x|\mathscr{P}_\lambda)$ with nucleus x is bounded by portions of bisecting hyperplanes of segments $[x, x']$, $x' \in \mathscr{P}_\lambda \setminus \{x\}$. The set of vertices and edges of these cells constitute a graph which is random and embedded in $\mathbb{R}^d$, sometimes referred to as the *Poisson-Voronoi skeleton*. In dimension two, with probability one, any vertex of this graph belongs to exactly three different edges and three different cells while in dimension d, any k-dimensional face of a cell, $0 \leq k \leq d$, is included in the intersection of

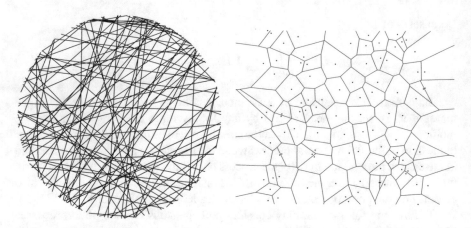

Fig. 1.4 Simulation of the Poisson line tessellation (left) and the Poisson-Voronoi tessellation (right) in the square

$(d+1-k)(d-k)/2$ different bisecting hyperplanes and belongs to exactly $(d+1-k)$ different cells almost surely (Fig. 1.4).

The Poisson hyperplane tessellation has been used as a natural model for the trajectories of particles inside bubble chambers [45], the fibrous structure of paper [83] and the road map of a city [54]. The Voronoi construction was introduced in the first place by R. Descartes as a possible model for the shape of the galaxies in the Universe [36]. The Poisson-Voronoi tessellation has appeared since then in numerous applied domains, including telecommunication networks [5] and materials science [74, 102].

In both cases, stationarity of the underlying Poisson process makes it possible to do a statistical study of the tessellation. Indeed, let f be a translation-invariant, measurable and non-negative real-valued function defined on the set $\mathscr{P}^d$ of convex polytopes of $\mathbb{R}^d$ endowed with the topology of the Hausdorff distance. For $r > 0$, let $\mathbf{C}_r$ and N_r be respectively the set of cells included in $B_r(o)$ and its cardinality. Then, following for instance [32], we can apply Wiener's ergodic theorem to get that, when $r \to \infty$,

$$\frac{1}{N_r} \sum_{C \in \mathbf{C}_r} f(C) \to \frac{1}{E(V_d(C_o)^{-1})} E\left(\frac{f(C_o)}{V_d(C_o)}\right) \quad \text{almost surely} \qquad (1.9)$$

where C_o is the almost-sure unique cell containing the origin o in its interior.

This implies that two different cells are of particular interest: the cell C_o, often called the *zero-cell*, and the cell $\mathscr{C}$ defined in distribution by the identity

$$E(f(\mathscr{C})) = \frac{1}{E(V_d(C_o)^{-1})} E\left(\frac{f(C_o)}{V_d(C_o)}\right). \qquad (1.10)$$

The convergence at (1.9) suggests that $\mathscr{C}$ has the law of a cell chosen uniformly at random in the whole tessellation, though such a procedure would not have any clear mathematical meaning. That is why $\mathscr{C}$ is called the *typical cell* of the tessellation even if it is not defined almost surely and it does not belong to the tessellation either. In particular, the typical cell is not equal in distribution to the zero-cell and is actually stochastically smaller since it has a density proportional to $V_d(C_o)^{-1}$ with respect to the distribution of C_o. Actually, it is possible in the case of the Poisson hyperplane tessellation to construct a realization of $\mathscr{C}$ which is almost surely strictly included in C_o [80]. This fact can be reinterpreted as a multidimensional version of the classical *bus waiting time paradox*, which says the following: if an individual arrives at time t at a bus stop, the time between the last bus he/she missed and the bus he/she will take is larger than the typical interarrival time between two consecutive busses.

Relying on either (1.9) or (1.10) may not be easy when calculating explicit mean values or distributions of geometric characteristics of $\mathscr{C}$. Another equivalent way of defining the distribution of $\mathscr{C}$ is provided by the use of a so-called Palm distribution, see e.g. [96], [90, Section 3.2], [113, Sections 3.3,3.4] and [73, Section 9]. For sake of simplicity, we explain the procedure in the case of the Poisson-Voronoi tessellation only. Let f be a measurable and non-negative real-valued function defined on $\mathscr{P}^d$. Then, for any Borel set B such that $0 < V_d(B) < \infty$, we get

$$E(f(\mathscr{C})) = \frac{1}{\lambda V_d(B)} E \left(\sum_{x \in \mathscr{P}_\lambda \cap B} f(C(x \mid \mathscr{P}_\lambda) - x) \right). \tag{1.11}$$

where $C(x \mid \mathscr{P}_\lambda) - x$ is the set $C(x \mid \mathscr{P}_\lambda)$ translated by $-x$. The fact that the quantity on the right-hand side of the identity (1.11) does not depend on the Borel set B comes from the translation invariance of the Poisson point process $\mathscr{P}_\lambda$. It is also remarkable that the right-hand side of (1.11) is again a mean over the cells with nucleus in B but contrary to (1.9) there is no limit involved, which means in particular that B can be as small as one likes, provided that $V_d(B) > 0$.

Following for instance [90, Proposition 3.3.2], we describe below the proof of (1.11), i.e. that for any translation-invariant, measurable and non-negative function f, the Palm distribution defined at (1.11) is the same as the limit of the means in the law of large numbers at (1.9). Indeed, let us assume that $\mathscr{C}$ be defined in law by the identity (1.11) and let us prove that (1.10) is satisfied, i.e. that $\mathscr{C}$ has a density proportional to $V_d(\cdot)^{-1}$ with respect to C_o. By classical arguments from measure theory, (1.11) implies that for any non-negative measurable function F defined on the product space $\mathscr{P}^d \times \mathbb{R}^d$,

$$\lambda \int E(F(\mathscr{C}, x)) dx = E \left(\sum_{x \in \mathscr{P}_\lambda} F(C(x \mid \mathscr{P}_\lambda) - x, x) \right).$$

Applying this to $F(C, x) = \frac{f(C)}{V_d(C)} \mathbf{1}_{\{-x \in C\}}$ and using the translation invariance of f, we get

$$\lambda E(f(\mathscr{C})) = E\left(\sum_{x \in \mathscr{P}_\lambda} \frac{f(C(x|\mathscr{P}_\lambda))}{V_d(C(x|\mathscr{P}_\lambda))} \mathbf{1}_{o \in C(x|\mathscr{P}_\lambda)}\right) = E\left(\frac{f(C_o)}{V_d(C_o)}\right). \qquad (1.12)$$

Applying (1.12) to $f = 1$ and $f = V_d$ successively, we get

$$E(V_d(\mathscr{C})) = \frac{1}{E(V_d(C_o)^{-1})} = \frac{1}{\lambda}. \qquad (1.13)$$

Combining (1.12) and (1.13), we obtain that the typical cell $\mathscr{C}$ defined at (1.11) satisfies (1.10) so it is equal in distribution to the typical cell defined earlier through the law of large numbers at (1.9).

In addition to these two characterizations of the typical cell, the Palm definition (1.11) of $\mathscr{C}$ in the case of the Poisson-Voronoi tessellation provides a very simple realization of the typical cell $\mathscr{C}$: it is equal in distribution to the cell $C(o|\mathscr{P}_\lambda \cup \{o\})$, i.e. the Voronoi cell associated with the nucleus o when the origin is added to the set of nuclei of the tessellation. This result is often called Slivnyak's theorem, see e.g. [113, Theorem 3.3.5].

The classical problems related to stationary tessellations are mainly the following:

(a) making a global study of the tessellation, for instance on the set of vertices or edges: calculation of mean global topological characteristics per unit volume, proof of limit theorems, etc;

(b) calculating mean values and whenever possible, moments and distributions of geometric characteristics of the zero-cell or the typical cell or a typical face;

(c) studying rare events, i.e. estimating distribution tails of the characteristics of the zero-cell or the typical cell and proving the existence of limit shapes in some asymptotic context.

Problem (a) As mentioned earlier, Cowan [32] showed several laws of large numbers by ergodic methods which were followed by second-order results from the seminal work due to F. Avram and Bertsimas on central limit theorems [4] to the more recent additions [53] and [55]. Topological relationships have been recently established in [128] for a general class of stationary tessellations.

Problem (b) The question of determining mean values of the volume or any combinatorial quantity of a particular cell was tackled early. The main contributions are due notably to Matheron [78, Chapter 6] and Miles [85, 86] in the case of the Poisson hyperplane tessellation and to Møller [88] in the case of the Poisson-Voronoi tessellation. Still, to the best of our knowledge, some mean values are unknown like for instance the mean number of k-dimensional faces of the Poisson-Voronoi typical cell for $1 \leq k \leq (d-1)$ and $d \geq 3$. Regarding explicit distributions, several works are noticeable [11, 22, 23] but in the end, it seems that very few have been computable up to now.

Problem (c) The most significant works related to this topic have been rather recent: distribution tails and large-deviations type results [43, 56], extreme values [29], high dimension [57]... Nevertheless, many questions, for instance regarding precise estimates of distribution tails, remain open to this day. One of the most celebrated questions concerns the shape of large cells. A famous conjecture stated by D.G. Kendall in the forties asserts that large cells from a stationary and isotropic Poisson line tessellation are close to the circular shape, see e.g. the foreword to [31]. This remarkable feature is in fact common to the Crofton cell and the typical cell of both the Poisson hyperplane tessellation and the Poisson-Voronoi tessellation in any dimension. It was formalized for different meanings of *large* cells and proved, with an explicit probability estimate of the deviation to the limit shape, by D. Hug, M. Reitzner and R. Schneider in a series of breakthrough papers, see e.g. [58, 61, 62].

Intentionally, we have chosen to skip numerous other models of tessellations. Noteworthy among these are those generated by non-Poisson point processes [44], Johnson-Mehl tessellations [89] and especially STIT tessellations [92].

In the next subsection, we collect a few explicit first results related to Problem (b), i.e. the mean value and distribution of several geometric characteristics of either the zero-cell C_o or the typical cell $\mathscr{C}$.

1.3.4   Mean Values and Distributional Properties of the Zero-Cell and of the Typical Cell

This subsection is not designed as an exhaustive account on the many existing results related to zero and typical cells from random tessellations. Instead, we focus here on the basic first calculations which do not require much knowledge on Poisson point processes. For more information and details, we refer the reader to the reference books [90], [113, Section 10.4] and to the survey [23].

1.3.4.1   The Zero-Cell of a Poisson Hyperplane Tessellation

We start with the calculation of the probability for a certain convex body K to be included in C_o the zero-cell or the typical cell. In the case of the planar Poisson line tessellation, let K be a convex body containing the origin—if not, K can be replaced by the convex hull of $K \cup \{o\}$. Since the number of lines from the Poisson line process intersecting K is a Poisson variable with mean $\lambda \mu_2(K)$, we get

$$P(K \subset C_o) = P(\{L_{\rho,\theta} \cap K = \emptyset \ \forall L_{\rho,\theta} \in \mathscr{P}_\lambda)$$
$$= \exp(-\mu_2(K))$$
$$= \exp(-\mathscr{L}(\partial K)),$$

where the last equality comes from the Cauchy-Crofton formula (1.3) and is obviously reminiscent of Buffon's needle problem. For $d \geq 3$, we get similarly, thanks to (1.7) applied to $k = 1$ and $l = 0$,

$$P(K \subset C_o) = \exp(-\mu_d(K)) = \exp(-\kappa_{d-1}V_1(K)). \qquad (1.14)$$

The use of Crofton formula for deriving the probability $P(K \subset C_o)$ may explain why the zero-cell of the isotropic and stationary Poisson hyperplane tessellation is often referred to as the *Crofton cell*.

Applying (1.14) to $K = B_r(o)$, $r > 0$, and using the equality $V_1(\mathbb{B}^d) = \frac{d\kappa_d}{\kappa_{d-1}}$, we obtain that the radius of the largest ball centered at o and included in C_o is exponentially distributed with mean $(d\kappa_d)^{-1}$.

1.3.4.2 The Typical Cell of a Poisson Hyperplane Tessellation

Let us denote by $f_0(\cdot)$ the number of vertices of a convex polytope. In the planar case, some general considerations show without much calculation that $E(f_0(\mathscr{C}))$ is equal to 4. Indeed, we have already noted that with probability one, any vertex of a cell belongs to exactly 4 cells and is the highest point of a unique cell. Consequently, there are as many cells as vertices. In particular, the mean $\frac{1}{N_r}\sum_{C \in \mathbf{C}_r} f_0(C)$ is equal, up to boundary effects going to zero when $r \rightarrow \infty$, to 4 times the ratio of the total number of vertices in $B_r(o)$ over the total number of cells included in $B_r(o)$, i.e. converges to 4. Other calculations of means and further moments of geometric characteristics can be found notably in [78, Chapter 6] and in [83–85].

One of the very few explicit distributions is the law of the inradius of $\mathscr{C}$, i.e. the radius of the largest ball included in $\mathscr{C}$. Remarkably, this radius is equal in distribution to the radius of the largest ball centered at o and included in C_o, i.e. is exponentially distributed with mean $(d\kappa_d)^{-1}$. This result is due to R.E. Miles in dimension two and is part of an explicit construction of the typical cell $\mathscr{C}$ based on its inball [86], which has been extended to higher dimension since then, see e.g. [24].

1.3.4.3 The Typical Cell of a Poisson-Voronoi Tessellation

In this subsection, we use the realization of the typical cell $\mathscr{C}$ of a Poisson-Voronoi tessellation as the cell $C(o|\mathscr{P}_\lambda \cup \{o\})$, as explained at the end of Sect. 1.3.3. We follow the same strategy as for the zero-cell of a Poisson hyperplane tessellation, i.e. calculating the probability for a convex body K to be contained in $\mathscr{C}$ and deducing from it the distribution of the inradius of $\mathscr{C}$.

Fig. 1.5 Voronoi Flower
(red) of the convex body K
(black) with respect to o

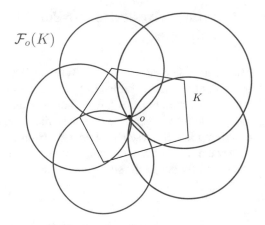

$\mathcal{F}_o(K)$

K

o

Let K be a convex body containing the origin. The set K is contained in $C(o|\mathscr{P}_\lambda \cup \{o\})$ if and only if o is the nearest nucleus to any point in K, which means that for every $x \in K$, the ball $B_{\|x\|}(x)$ does not intersect $\mathscr{P}_\lambda$. Let us consider the set

$$\mathscr{F}_o(K) = \bigcup_{x \in K} B_{\|x\|}(x)$$

that we call the *Voronoi flower* of K with respect to o, see Fig. 1.5.

Using the fact that the number of points of $\mathscr{P}_\lambda$ in $\mathscr{F}_o(K)$ is Poisson distributed with mean $\lambda V_d(\mathscr{F}_o(K))$, we get

$$P(K \subset C(o|\mathscr{P}_\lambda \cup \{o\})) = \exp(-\lambda V_d(\mathscr{F}_o(K))).$$

Applying this to $K = B_o(r)$, $r > 0$, we deduce that the radius of the largest ball centered at o and included in $C(o|\mathscr{P}_\lambda \cup \{o\})$ is Weibull distributed with tail probability equal to $\exp(-\lambda 2^d \kappa_d r^d)$, $r > 0$.

Similarly to the case of the typical cell of a Poisson line tessellation, some direct arguments lead us to the calculation of $E(f_0(\mathscr{C}))$ in dimension two: any vertex of a cell belongs to exactly 3 cells and with probability one, is the either highest or lowest point of a unique cell. Consequently, there are as twice as many vertices as cells. In particular, the mean $\frac{1}{N_r} \sum_{C \in \mathbf{C}_r} f_0(C)$ is equal, up to boundary effects going to zero when $r \to \infty$, to 3 times the ratio of the total number of vertices in $B_r(o)$ over the total number of cells included in $B_r(o)$, i.e. equal to 6. This means that $E(f_0(\mathscr{C})) = 6$.

In the next section, we describe a different way of generating random polytopes: they are indeed constructed as convex hulls of random sets of points.

1.4   From Sylvester's Four-Point Problem to Random
Polytopes

This section is centered around Sylvester's four-point problem, another historical problem of geometric probability which seemingly falls under the denomination of recreational mathematics but in fact lays the foundations of an active domain of today's stochastic geometry, namely the theory of random polytopes. We aim at describing first Sylvester's original question and some partial answers to it. We then survey the topic of random polytopes which has been largely investigated since the sixties, partly due to the simultaneous development of computational geometry.

1.4.1   Starting from Sylvester's Four-Point Problem

In 1864, J.J. Sylvester published in *The Educational Times* [123] a problem which is nowadays known under the name of *Sylvester's four-point problem* and can be rephrased in the following way: given a convex body K in the plane, what is the probability that 4 random points inside K are the vertices of a convex quadrilateral (Fig. 1.6)?

Let us denote by $p_4(K)$ the probability that 4 random points which are independent and uniformly distributed in K are the vertices of a convex quadrilateral. If not, one of the 4 points is included in the triangle whose vertices are the 3 other points. Denoting by $\overline{A}(K)$ the mean area of a random triangle whose vertices are 3 i.i.d. uniform points in K, we get the identity

$$p_4(K) = 1 - 4\frac{\overline{A}(K)}{V_2(K)}. \qquad (1.15)$$

Fig. 1.6 Simulation of Sylvester's four-point problem in the triangle: the convex hull of the 4 uniform points (red) is either a convex quadrilateral (left) or a triangle (right)

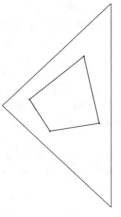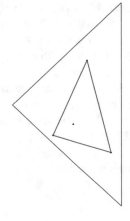

Solving Sylvester's problem is then equivalent to calculating the mean value $\overline{A}(K)$. In particular, the quantity $\frac{\overline{A}(K)}{V_2(K)}$ is scaling invariant and also invariant under any area-preserving affine transformation.

1.4.1.1 Calculation of Sylvester's Probability in the Case of the Disk

In this subsection, we provide an explicit calculation of $p_4(\mathbb{D})$ where $\mathbb{D}$ is the unit disk. As in the next subsection, we follow closely the method contained in [68, pages 42–46].

Step 1 We can assume that one of the points is on the boundary of the disk. Indeed, isolating the farthest point from the origin, we get

$$V_2(\mathbb{D})^3 \overline{A}(\mathbb{D}) = 3 \int_{\mathbb{D}} \left[\iint_{\mathbb{D}^2} \mathbb{1}_{\{\|x_1\|, \|x_2\| \in (0, \|x_3\|)\}} V_2(\mathrm{Conv}(\{x_1, x_2, x_3\}))) \mathrm{d}z_1 \mathrm{d}z_2 \right] \mathrm{d}z_3.$$

Now for a fixed $z_3 \in \mathbb{D} \setminus \{o\}$, we apply the change of variables $z_i' = \frac{z_i}{\|z_3\|}$, $i = 1, 2$, in the double integral and we deduce that

$$\overline{A}(\mathbb{D}) = \frac{3}{\pi^3} \int_{\mathbb{D}} \|z_3\|^6 \left[\iint_{\mathbb{D}^2} V_2(\mathrm{Conv}(\{z_1', z_2', \frac{z_3}{\|z_3\|}\}))) \mathrm{d}z_1 \mathrm{d}z_2' \right] \mathrm{d}z_3.$$

$$(1.16)$$

Since the double integral above does not depend on z_3, we get

$$\overline{A}(\mathbb{D}) = \frac{3}{4\pi^2} I \tag{1.17}$$

where

$$I = \iint_{\mathbb{D}^2} V_2(\mathrm{Conv}(\{z_0, z_1, z_2\})) \mathrm{d}z_1 \mathrm{d}z_2,$$

z_0 being a fixed point on the boundary of $\mathbb{D}$.

Step 2 Let us now calculate I, i.e. π^2 times the mean area of a random triangle with one deterministic vertex on the unit circle and two random vertices independent and uniformly distributed in $\mathbb{D}$.

For sake of simplicity, we replace the unit disk $\mathbb{D}$ by its translate $\mathbb{D} + (0, 1)$ and the fixed point on the boundary of $\mathbb{D} + (0, 1)$ is chosen to be equal to o. This does not modify the integral I. The polar coordinates (ρ_i, θ_i) of z_i, $i = 1, 2$, satisfy $\rho_i \in (0, 2\sin(\theta_i))$ and

$$V_2(\mathrm{Conv}(\{o, z_1, z_2\})) = \frac{1}{2} \rho_1 \rho_2 |\sin(\theta_2 - \theta_1)|. \tag{1.18}$$

Consequently, we obtain

$$
I = \iint_{0<\theta_1<\theta_2<\pi} \left[\int_0^{2\sin(\theta_1))} \rho_1^2 d\rho_1 \int_0^{2\sin(\theta_2)} \rho_2^2 d\rho_2 \right] \sin(\theta_2 - \theta_1) d\theta_1 d\theta_2
$$

$$
= \frac{64}{9} \iint_{0<\theta_1<\theta_2<\pi} \sin^3(\theta_1) \sin^3(\theta_2) \sin(\theta_2 - \theta_1) d\theta_1 d\theta_2
$$

$$
= \frac{35\pi}{36}.
$$

Conclusion Combining this result with (1.17) and (1.15), we get

$$
p_4(\mathbb{D}) = 1 - \frac{35}{12\pi^2} \approx 0.70448\ldots
$$

1.4.1.2 Calculation of Sylvester's Probability in the Case of the Triangle

In this subsection, we calculate the probability $p_4(\mathbb{T})$ where the triangle $\mathbb{T}$ is the convex hull of the three points o, $(1, 1)$ and $(1, -1)$. We recall that the calculation of $p_4(K)$ is invariant under any area-preserving affine transformation.

Step 1 We can assume that one of the points is on the edge facing o. Indeed, denoting by (x_i, y_i) the Cartesian coordinates of z_i, $i = 1, 2, 3$, we get

$$
V_2(\mathbb{T})^3 \overline{A}(\mathbb{T}) = 3 \int_{\mathbb{T}} \left[\iint_{\mathbb{T}^2} \mathbf{1}_{\{x_1,x_2\in(0,x_3)\}} V_2(\text{Conv}(\{z_1, z_2, z_3\})) dz_1 dz_2 \right] dz_3.
$$

Now for a fixed $z_3 \in \mathbb{T} \setminus \{o\}$, we apply the change of variables $z_i' = \frac{z_i}{x_3}$, $i = 1, 2$, in the double integral. We get

$$
\overline{A}(\mathbb{T}) = 3 \int_{x_3=0}^1 \int_{y_3=-x_3}^{x_3} \left[\iint_{\mathbb{T}^2} x_3^6 V_2(\text{Conv}(\{z_1', z_2', \frac{z_3}{x_3}\})) dz_1' dz_2' \right] dz_3
$$

$$
= 6 \int_{x_3=0}^1 \int_{y_3=0}^{x_3} x_3^6 \left[\iint_{\mathbb{T}^2} V_2(\text{Conv}(\{z_1', z_2', \frac{x_3}{x_3}\})) dz_1' dz_2' \right] dz_3.
$$

Finally, for fixed x_3, we apply the change of variables $h = \frac{y_3}{x_3}$. Since the double integral in square brackets above does not depend on x_3, we get

$$
\overline{A}(\mathbb{T}) = 6 \int_0^1 x_3^7 dx_3 \int_{h=0}^1 \left[\iint_{\mathbb{T}^2} V_2(\text{Conv}(\{z_1, z_2, (1, h)\})) dz_1 dz_2 \right] dh
$$

$$
= \frac{3}{4} \int_{h=0}^1 I'(h) dh \tag{1.19}
$$

where $I'(h) = \iint_{\mathbb{T}^2} V_2(\text{Conv}(\{z_1, z_2, (1, h)\})) dz_1 dz_2$.

Step 2 Let us now calculate $I'(h)$, $0 < h < 1$, i.e. the mean area of a random triangle with one deterministic vertex at $(1, h)$ on the vertical edge and two random vertices independent and uniformly distributed in $\mathbb{T}$. The point $(1, h)$ divides $\mathbb{T}$ into two subtriangles, the upper triangle $T_+(h) = \mathrm{Conv}(\{o, (1, h), (1, 1)\})$ and the lower triangle $T_-(h) = \mathrm{Conv}(\{o, (1, h), (1, -1)\})$. Let us rewrite $I'(h)$ as

$$I'(h) = 2I'_{+,-}(h) + V_2(T_+(h))^2 \widehat{A}(T_+(h)) + V_2(T_-(h))^2 \widehat{A}(T_-(h)) \qquad (1.20)$$

where

$$I'_{+,-}(h) = \int_{z_1 \in T_+(h), z_2 \in T_-(h)} V_2(\mathrm{Conv}(\{z_1, z_2, (1, h)\})))dz_1 dz_2 \qquad (1.21)$$

and $\widehat{A}(T)$, for a triangle T, is the mean area of a random triangle which shares a common vertex with T and has two independent vertices uniformly distributed in T.

We start by making explicit the quantity $\widehat{A}(T)$ for any triangle T. It is invariant under any area-preserving affine transformation and is multiplied by λ^2 when T is rescaled by λ^{-1}. Consequently, it is proportional to $V_2(T)$, i.e.

$$\widehat{A}(T) = \widehat{A}(\mathbb{T}) V_2(T). \qquad (1.22)$$

We now calculate

$$\widehat{A}(\mathbb{T}) = \int_{\mathbb{T}^2} V_2(\mathrm{Conv}(\{o, z_1, z_2\}))dz_1 dz_2.$$

The polar coordinates (ρ_i, θ_i) of x_i, $i = 1, 2$, satisfy $\rho_i \in (0, \cos^{-1}(\theta_i))$ and equality (1.18). Consequently, we obtain

$$
\begin{aligned}
\widehat{A}(\mathbb{T}) &= \int_{-\frac{\pi}{4}}^{\frac{\pi}{4}} \int_{-\frac{\pi}{4}}^{\frac{\pi}{4}} \left[\int_0^{\cos^{-1}(\theta_1)} \rho_1^2 d\rho_1 \int_0^{\cos^{-1}(\theta_2)} \rho_2^2 d\rho_2 \right] \sin(\theta_2 - \theta_1) d\theta_1 d\theta_2 \\
&= \frac{1}{9} \iint_{-\frac{\pi}{4} < \theta_1 < \theta_2 < \frac{\pi}{4}} \frac{\sin(\theta_2 - \theta_1)}{\cos^3(\theta_1) \cos^3(\theta_2)} d\theta_1 d\theta_2 \\
&= \frac{4}{27}. \qquad (1.23)
\end{aligned}
$$

Combining (1.22) and (1.23), we obtain in particular that

$$\widehat{A}(T_+(h)) = \frac{2(1-h)}{27}, \quad \text{and} \quad \widehat{A}(T_-(h)) = \frac{2(1+h)}{27}. \qquad (1.24)$$

We turn now our attention to the quantity $I'_{+,-}(h)$ defined at (1.21). Using the rewriting of the area of a triangle as half of the non-negative determinant of two vectors and introducing $g_+(h)$ (resp. $g_-(h)$) as the center of mass of $T_+(h)$ (resp. $T_-(h)$), we obtain

$$
\begin{aligned}
I'_{+,-}(h) &= \frac{1}{2} \int_{z_1 \in T_+(h), z_2 \in T_-(h)} \det(z_1 - (1, h), z_2 - (1, h)) dz_1 dz_2 \\
&= \frac{1}{2} \int_{z_1 \in T_+(h)} \det\left((z_1 - (1, h)), \int_{z_2 \in T_-(h)} (z_2 - (1, h))\right) dz_1 \\
&= \frac{V_2(T_-(h))}{2} \det\left(\int_{z_1 \in T_+(h)} (z_1 - (1, h)) dz_1, (g_-(h) - (1, h))\right) \\
&= V_2(T_-(h)) V_2(T_+(h)) V_2(\text{Conv}(\{g_+(h), g_-(h), (1, h)\})) \\
&= \frac{(1-h)(1+h)}{4} \frac{V_2(\mathbb{T})}{9} = \frac{1 - h^2}{36}.
\end{aligned}
$$
(1.25)

Inserting (1.24) and (1.25) into (1.20), we get

$$
I'(h) = \frac{1 - h^2}{18} + \frac{(1 - h)^3}{54} + \frac{(1 + h)^3}{54} = \frac{5 + 3h^2}{54}.
$$
(1.26)

Conclusion Combining (1.26) with (1.19) and (1.15), we get $\overline{A}(\mathbb{T}) = \frac{1}{12}$ and

$$
p_4(T) = \frac{2}{3} = 0.66666\ldots
$$

1.4.1.3 Extremes of $p_4(K)$

In 1917, W. Blaschke proved a monotonicity result for Sylvester's four-point problem, namely that the probability $p_4(K)$ is maximal when K is a disk and minimal when K is a triangle, see [15] and [16, §24, §25]. Because of (1.15), this amounts to saying that the mean area of a random triangle in a unit area convex body K is minimal when K is a disk and maximal when K is a triangle.

This assertion is due to the use of symmetrization techniques which have become classical since then in convex and integral geometry. We describe below the main arguments developed by W. Blaschke and also rephrased in a nice way in the historical note [99].

Let K be a unit area convex body of $\mathbb{R}^2$ and let us denote by $x_{\min}$ and $x_{\max}$ the minimal and maximal projection on the x-axis of a point of K. The boundary of K is parametrized by two functions $f_+, f_- : [x_{\min}, x_{\max}] \longrightarrow \mathbb{R}$ such that $f_+ \geq f_-$ with equality at $x_{\min}$ and $x_{\max}$. We use the term *Steiner symmetrization* of K with

Fig. 1.7 Two symmetrizations (red) of a convex body delimited by translates of the curves $f_+(x) = \sqrt{x}$ and $f_-(x) = x(x - 2 + \frac{\sqrt{2}}{2})$ on the interval $[0, 2]$ (black) and the image of a triangle by symmetrization (blue): the Steiner symmetrization (left) and the shaking (right)

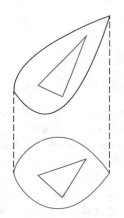
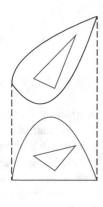

respect to the x-axis for the transformation

$$\mathscr{S} : \begin{cases} K & \longrightarrow \mathbb{R}^2 \\ (x, y) & \longmapsto \left(x, y - \frac{f_+(x) + f_-(x)}{2}\right) \end{cases}$$

In other words, $\mathscr{S}$ sends any segment which is the intersection of K with a vertical line to its unique vertical translate which is symmetric with respect to the x-axis; see Fig. 1.7. In particular, $\mathscr{S}$ is area-preserving and the image of K is a convex body which is symmetric with respect to the x-axis, see e.g. [109, Section 10.3]. Let us show that

$$\overline{A}(K) \geq \overline{A}(\mathscr{S}(K)) \tag{1.27}$$

Let $z_i = (x_i, y_i)$, $i = 1, 2, 3$, be three points of K. In particular, we start by noticing that the area of the parallelogram spanned by the two vectors $(z_2 - z_1)$ and $(z_3 - z_1)$ is twice the area of the triangle $\mathrm{Conv}(\{z_1, z_2, z_3\})$. Consequently, we get

$$V_2(\mathrm{Conv}(\{z_1, z_2, z_3\})) = \frac{1}{2}\left| \begin{matrix} x_2 - x_1 & x_3 - x_1 \\ y_2 - y_1 & y_3 - y_1 \end{matrix} \right| = \frac{1}{2}\left| \begin{matrix} 1 & 1 & 1 \\ x_1 & x_2 & x_3 \\ y_1 & y_2 & y_3 \end{matrix} \right|. \tag{1.28}$$

We will also use the points $z_i^* = \mathscr{S}(z_i) = (x_i, y_i^*)$, $\overline{z_i^*} = (x_i, -y_i^*)$ and $w_i = (x_i, y_i - 2y_i^*)$, $i = 1, 2, 3$. In particular, the identity $\mathscr{S}(w_i) = \overline{z_i^*}$ is satisfied and the two triangles $\mathrm{Conv}(\{z_1, z_2, z_3\})$ and $\mathrm{Conv}(\{w_1, w_2, w_3\})$ have same area. Consequently,

$$V_2(\mathrm{Conv}(\{z_1, z_2, z_3\})) + V_2(\mathrm{Conv}(\{w_1, w_2, w_3\}))$$

$$\geq \frac{1}{2}\left| \begin{matrix} 1 & 1 & 1 \\ x_1 & x_2 & x_3 \\ y_1 & y_2 & y_3 \end{matrix} \right| - \left| \begin{matrix} 1 & 1 & 1 \\ x_1 & x_2 & x_3 \\ y_1 - 2y_1^* & y_2 - 2y_2^* & y_3 - 2y_3^* \end{matrix} \right| \Bigg|$$

$$\geq \frac{1}{2} \left| \begin{matrix} 1 & 1 & 1 \\ x_1 & x_2 & x_3 \\ 2y_1^* & 2y_2^* & 2y_3^* \end{matrix} \right| \, |$$

$$= 2V_2(\text{Conv}(\{z_1^*, z_2^*, z_3^*\})). \tag{1.29}$$

Integrating (1.29) with respect to $z_1, z_2, z_3 \in K$ and using the fact that both the transformation $\mathscr{S}$ and the reflection with respect to the x-axis preserve the Lebesgue measure, we obtain (1.27). It remains to use the fact that the equality in (1.27) is satisfied only when K is an ellipse. In fact, for any convex body K, there exists a sequence of lines such that the image of K under consecutive applications of Steiner symmetrizations with respect to the lines of that sequence converges to a disk [26]. The function $\overline{A}(\cdot)$ being continuous on the set of convex bodies, we obtain $\overline{A}(K) \geq \overline{A}(\frac{1}{\sqrt{\pi}}\mathbb{D})$ and therefore $p_4(K) \leq p_4(\mathbb{D})$.

We turn now our attention to the proof of $p_4(K) \leq p_4(\mathbb{T})$. The method relies on a transformation $\mathscr{T}$ in the same spirit as the Steiner symmetrization, called *Schüttelung* or *shaking* and defined as follows:

$$\mathscr{T} :: \begin{cases} K & \longrightarrow \mathbb{R}^2 \\ (x, y) & \longmapsto (x, y - f_-(x)) \end{cases}$$

In other words, $\mathscr{T}$ sends any segment which is the intersection of K with a vertical line to its unique vertical translate with a lower-end on the x-axis, see Fig. 1.7. In particular, $\mathscr{T}$ is area-preserving and preserves the convexity.

Using both (1.28) and the fact that K is a unit-area convex body which satisfies the equality

$$K = \{(x, y) : x \in [x_{\min}, x_{\max}], f_-(x) \leq y \leq f_+(x)\},$$

we get that

$$\overline{A}(K) = \iiint_{[x_{\min}, x_{\max}]} I(x_1, x_2, x_3) \mathrm{d}x_1 \mathrm{d}x_2 \mathrm{d}x_3 \tag{1.30}$$

where

$$I(x_1, x_2, x_3)$$

$$= \int_{f_-(x_1)}^{f_+(x_1)} \int_{f_-(x_2)}^{f_+(x_2)} \int_{f_-(x_3)}^{f_+(x_3)} \frac{1}{2} \left| \begin{matrix} 1 & 1 & 1 \\ x_1 & x_2 & x_3 \\ y_1 & y_2 & y_3 \end{matrix} \right| \mathrm{d}y_3 \mathrm{d}y_2 \mathrm{d}y_1$$

$$= \int_{f_-(x_1)}^{f_+(x_1)} \int_{f_-(x_2)}^{f_+(x_2)} \int_{f_-(x_3)}^{f_+(x_3)} \frac{1}{2} |a_1 y_1 + a_2 y_2 + a_3 y_3| \mathrm{d}y_3 \mathrm{d}y_2 \mathrm{d}y_1.$$

and with $a_1 = a_1(x_1, x_2, x_3) = (x_3 - x_2)$, $a_2 = a_2(x_1, x_2, x_3) = (x_1 - x_3)$ and $a_3 = a_3(x_1, x_2, x_3) = (x_2 - x_1)$. For sake of simplicity, the dependency of the coefficients a_i on the coordinates x_i is omitted. When x_1, x_2, x_3 are fixed, the function I calculates the 4-dimensional volume of the parallelepiped region delimited by the rectangular basis $[f_-(x_1), f_+(x_1)] \times [f_-(x_2), f_+(x_2)] \times [f_-(x_3), f_+(x_3)] \times \{0\}$ and the surface of equation $y_4 = \frac{1}{2}|a_1 y_1 + a_2 y_2 + a_3 y_3|$. In particular, if we allow the rectangular basis to be translated, the integral I only depends on the distance $\mathscr{D}$ in $\mathbb{R}^3$ from the midpoint $(\frac{f_-(x_1)+f_+(x_1)}{2}, \frac{f_-(x_2)+f_+(x_2)}{2}, \frac{f_-(x_3)+f_+(x_3)}{2})$ of the rectangular basis to the set $\{(y_1, y_2, y_3) : a_1 y_1 + a_2 y_2 + a_3 y_3 = 0\}$ and is even an increasing function of $\mathscr{D}$. We notice that $\mathscr{D}$ satisfies

$$\mathscr{D} = \frac{|a_1(f_-(x_1) + f_+(x_1)) + a_2(f_-(x_2) + f_+(x_2)) + a_3(f_-(x_3) + f_+(x_3))|}{2\sqrt{a_1^2 + a_2^2 + a_3^2}}$$

$$= \frac{1}{2\sqrt{a_1^2 + a_2^2 + a_3^2}} \left| \begin{matrix} 1 & 1 & 1 \\ x_1 & x_2 & x_3 \\ f_-(x_1) & f_-(x_2) & f_-(x_3) \end{matrix} \right| + \left| \begin{matrix} 1 & 1 & 1 \\ x_1 & x_2 & x_3 \\ f_+(x_1) & f_+(x_2) & f_+(x_3) \end{matrix} \right| \,.$$

Now, the two determinants in the last equality above are two times the algebraic areas of two triangles whose vertices are on ∂K and have respective x-coordinates x_1, x_2 and x_3. Because of the convexity of K, they must have opposite signs. Consequently, we get from the triangular inequality $||a| - |b|| \le |a - b|$ that

$$\mathscr{D} \le \frac{1}{2\sqrt{a_1^2 + a_2^2 + a_3^2}} \left| \begin{matrix} 1 & 1 & 1 \\ x_1 & x_2 & x_3 \\ f_+(x_1) - f_-(x_1) & f_+(x_2) - f_-(x_2) & f_+(x_3) - f_-(x_3) \end{matrix} \right| \,.$$

In particular, after application of the transformation $\mathscr{T}$, the distance is equal to the right-hand side of the inequality above. The integral $I(x_1, x_2, x_3)$ being an increasing function of $\mathscr{D}$, it is greater, which implies thanks to (1.30) that

$$\overline{A}(K) \le \overline{A}(\mathscr{T}(K)).$$

It remains to use the fact that there exists a sequence of lines such that the image of K under consecutive applications of the *Schüttelung* operations with respect to these lines converges to a triangle [13]. Therefore, we get the inequality $\overline{A}(K) \le \overline{A}(\mathbb{T})$ and thanks to (1.15), the required inequality $p_4(K) \ge p_4(\mathbb{T})$.

1.4.2   Random Polytopes

There are several ways of extending Sylvester's initial question:

(a) increasing the number of random points inside K and ask for the probability that n i.i.d. points uniformly distributed in a two-dimensional convex body K are *in convex position*, i.e. are extreme points of their convex hull;
(b) increasing the dimension and ask for the probability that $(d + 2)$ or more i.i.d. points uniformly distributed in a d-dimensional convex body K are the vertices of a convex polytope;
(c) increasing the number of random points in any dimension and ask more general questions, such as the distribution and mean value of the number of extreme points and of other characteristics of the convex hull;
(d) replacing the uniform distribution in K by another probability distribution in $\mathbb{R}^d$.

The topic of random polytopes has become more popular in the last 50 years and this is undoubtedly due in part to the birth of computational geometry and the need to get quantitative information on the efficiency of algorithms in discrete geometry, in particular algorithms designed for the construction of the convex hull of multivariate data. We describe below the state of the art on each of the problems above.

Problem (a)   Let us denote by $p_n(K)$ the probability that n i.i.d. points uniformly distributed in K are in convex position. Using combinatorial arguments, P. Valtr obtained explicit calculations for $p_n([0, 1]^2)$ [124] and $p_n(\mathbb{T})$ [125]. More recently, J.-F. Marckert provided a recursive formula for $p_n(\mathbb{D})$, which he implements to derive explicit values up to $p_8(\mathbb{D})$ [75]. Though there is no general formula for every K, the sequences $p_n(K)$ for all convex bodies K share a common asymptotic behavior when $n \to \infty$. In a breakthrough paper [8] in 1999, I. Bárány showed that for every K with area 1, when $n \to \infty$,

$$\log p_n(K) = -2n \log(n) + n \log(\frac{1}{4}e^2 \mathrm{pa}(K)) + o(n)$$

where $f(n) = o(g(n))$ means that $\lim_{n \to \infty} f(n)/g(n) = 0$ and $\mathrm{pa}(K)$ is the supremum of the so-called affine perimeter of all convex bodies included in K, see Sect. 1.4.2.2.

Problem (b)   Let us denote by $p_n^{(d)}(K)$ the probability that n i.i.d. points uniformly distributed in a convex body K of $\mathbb{R}^d$ are in convex position. J.F.C. Kingman calculated $p_{d+2}^{(d)}(\mathbb{B}^d)$ in 1969 [69] and it was shown by H. Groemer in 1973 that $p_{d+2}^{(d)}(K)$ is minimal when K is the unit ball (or an ellipsoid) [48]. It is still undecided whether the d-dimensional simplex should maximize $p_{d+2}^{(d)}(K)$ though W. Blaschke had claimed that his proof in the case $d = 2$ could be directly extended [15]. Regarding the asymptotic behavior of $p_n^{(d)}(K)$ when $n \to \infty$, I. Bárány conjectured a two-term expansion in the spirit of the two-dimensional case and showed the following one-term expansion [9]:

$$\log p_n(K) = -\frac{2}{d-1} n \log(n) + O(n)$$

where $f(n) = O(g(n))$ means that f/g is bounded.

Problem (c) Let K be a convex body of $\mathbb{R}^d$ and let K_n be the convex hull of n i.i.d. points uniformly distributed in K. The natural questions on this model have to do with the shape of K_n and the distributions of the geometric characteristics of K_n such as the number of vertices, number of faces or the volume. They can be treated in the two different contexts of fixed n and n large. This will be the focus of the end of the section.

Problem (d) There have been several works related to Problem (d). Wendel's result described below is one of them in the case of a symmetric distribution with respect to o. Other papers have focused on isotropic distributions [27] and especially the Gaussian distribution, see e.g. [10, 107].

For a more detailed account on the topic of random polytopes, we refer the reader to [113, Chapter 8], the lecture [58] and the very exhaustive survey [106]. In the rest of the section, we will present a few results related to Problem (c) in both the non-asymptotic and asymptotic regimes.

1.4.2.1 Non-asymptotic Results

In this subsection, we describe two of the non-asymptotic results on the convex hull K_n of n i.i.d. points uniformly distributed in a convex body K of $\mathbb{R}^d$ (Fig. 1.8): the Efron identity relating first moments of functionals of K_n and Wendel's calculation of the probability that the origin o belongs to K_n. In the sequel, $f_k(\cdot)$ is the number of k-dimensional faces of a convex polytope. In particular, $f_0(\cdot)$ denotes the number of vertices.

In 1965, B. Efron proved an extension of (1.15), i.e. he provided an identity which connects in a very simple way the mean number of vertices of the convex hull of n points to the mean volume of the convex hull of $(n-1)$ points [39]. The calculation, which has been extended since then by several identities for higher moments due to Buchta [19], goes as follows. Let $X_1, \cdots, X_n$ be the n i.i.d. uniform points in K. Then almost surely,

$$f_0(K_n) = \sum_{k=1}^{n} \mathbf{1}_{\{X_k \notin \mathrm{Conv}(X_1, \cdots, X_{k-1}, X_{k+1}, \cdots, X_n)\}}.$$

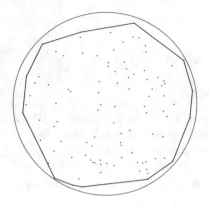 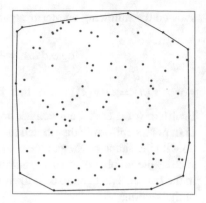

Fig. 1.8 Simulations of the random polytope K_{100} (black) when K is a disk (left) and K is a square (right)

Taking the expectation of this equality, we obtain

$$
\begin{aligned}
E(f_0(K_n)) &= n P(X_n \notin \mathrm{Conv}(X_1, \cdots, X_{n-1})) \\
&= n E(E(\mathbf{1}_{\{X_n \notin \mathrm{Conv}(X_1, \cdots, X_{n-1})\}} | X_1, \cdots, X_{n-1}) \\
&= n \left(1 - \frac{E(V_d(K_{n-1}))}{V_d(K)}\right).
\end{aligned}
$$

In 1962, J.G. Wendel showed an explicit formula for the probability that the origin o lies inside the convex hull of n i.i.d. points with a symmetric distribution with respect to o [129]. Let $X_1, \cdots, X_n$, $n \geq 1$, be the random points and we assume additionally that their common distribution is such that with probability one, all subsets of size d are linearly independent. We first notice that o is not in the convex hull if and only if there exists a half-space containing all the points, i.e. there exists $y \in \mathbb{R}^d$ such that $\langle y, X_k \rangle > 0$ for every $1 \leq k \leq n$. This implies in particular that the probability $P(o \notin \mathrm{Conv}(\{X_1, \cdots, X_n\}))$ equals 1 as soon as $n \leq d$ and $2^{-(n-1)}$ when $d = 1$. Now the calculation for $n \geq (d+1) \geq 3$ is done by purely combinatorial arguments.

Indeed, each X_k defines a set of authorized y which is a half-space bounded by the linear hyperplane H_k with normal vector X_k. Consequently, each connected component of the complement of $\bigcup_{k=1}^n H_k$ can be coded by a sequence in $\{-1, 1\}^n$ where $+1$ at the k-th position means that the connected component lies in the authorized half-space bounded by H_k. There are 2^n possible codes and we denote by $N_{d,n}$ the total number of connected components. The variable $N_{d,n}$ is almost surely constant, as we shall see later on. Recalling from the earlier discussion that a necessary and sufficient condition to have the origin outside of the convex hull is that the intersection of all authorized half-spaces bounded by the hyperplanes H_k is not empty, we obtain the following equivalence: o is not in the convex hull of $\{X_1, \cdots, X_n\}$ if and only if one of the connected components is coded by

$(1, \cdots, 1)$. This happens with probability

$$P(o \notin \mathrm{Conv}(\{X_1, \cdots, X_n\})) = \frac{N_{d,n}}{2^n}. \tag{1.31}$$

As announced earlier, the variable $N_{d,n}$ is constant on the event of probability one that any subset of size d of the n points is linearly independent. We calculate $N_{d,n}$ on this event by proving a recurrence relation. For fixed $n \geq 2$, the n-th hyperplane H_n separates into two subparts each connected component of $\mathbb{R}^d \setminus \bigcup_{k=1}^{n-1} H_k$ that it meets and leaves unchanged the remaining connected components of $\mathbb{R}^d \setminus \bigcup_{k=1}^{n-1} H_k$. The number of connected components of $\mathbb{R}^d \setminus \bigcup_{k=1}^{n-1} H_k$ that H_n meets is equal to the number of connected components of $H_n \setminus \bigcup_{k=1}^{n-1} H_k$, i.e. $N_{d-1,n-1}$ while the number of untouched connected components of $\mathbb{R}^d \setminus \bigcup_{k=1}^{n-1} H_k$ is $N_{d,n-1} - N_{d-1,n-1}$. Consequently, we get the relation

$$N_{d,n} = N_{d-1,n-1} + N_{d,n-1}.$$

Using that $N_{d,1} = 2$ for every $d \geq 1$, we deduce that $N_{d,n} = 2 \sum_{k=0}^{d-1} \binom{n-1}{k}$ thanks to Pascal's triangle. This last equality combined with (1.31) leads to

$$P(o \notin \mathrm{Conv}(\{X_1, \cdots, X_n\})) = 2^{-(n-1)} \sum_{k=0}^{d-1} \binom{n-1}{k}.$$

In particular, when $n \to \infty$, this probability goes to zero exponentially fast. In the next section, we investigate the general asymptotic behavior of K_n.

1.4.2.2   Asymptotic Results

When $n \to \infty$, K_n converges to K itself and studying the asymptotic behavior of K_n means being able to quantify the quality of the approximation of K by K_n. The first breakthrough is due to A. Rényi and R. Sulanke in 1963 and 1964 [107, 108]. They showed in particular that in the planar case, the mean number of vertices of K_n has a behavior which is highly dependent on the regularity of the boundary of K. Indeed, when ∂K is of class $\mathscr{C}^2$,

$$E(f_0(K_n)) \underset{n \to \infty}{\sim} 2^{\frac{1}{3}} 3^{-\frac{1}{3}} \Gamma\left(\frac{5}{3}\right) V_d(K)^{-\frac{1}{3}} \int_{\partial K} r_s^{-\frac{1}{3}} \, ds \, n^{\frac{1}{3}}. \tag{1.32}$$

where r_s is the radius of curvature of ∂K at s and $f(n) \underset{n \to \infty}{\sim} g(n)$ means that f/g has limit 1. The quantity $\int_{\partial K} r_s^{-\frac{1}{3}} \, ds$ is called the *affine perimeter* of K.

When K is a convex polygon itself, the number of vertices of K_n is expected to be *smaller* in mean as, roughly speaking, it does not require many edges to approximate the flat parts of the boundary of K. While the extreme points of K_n are more or less homogeneously spread along the curve ∂K when it is smooth, they are concentrated *in the corners*, i.e. around the vertices of K when K is a convex polygon. Consequently, the growth rate of $E(f_0(K_n))$ becomes logarithmic, as opposed to the polynomial rate from (1.32). Denoting by r the number of vertices of the convex polygon K, we get

$$E(f_0(K_n)) \underset{n \to \infty}{\sim} \frac{2r}{3} \log n. \tag{1.33}$$

The two estimates (1.32) and (1.33) have been extended in many directions: asymptotic means of $f_k(K_n)$, $1 \le k \le d$ and of $V_d(K_n)$ in any dimension and convergences [7, 105], same for the intrinsic volumes in the smooth case [103, 114], concentration estimates [126], second-order results [25, 104] and so on. Many basic questions remain unanswered, like for instance the asymptotic behavior of the mean intrinsic volumes of K_n when K is a polytope.

In the next section, we generate for the first time non-convex random sets as unions of translates of a so-called *grain*.

1.5   From the Bicycle Wheel Problem to Random Coverings and Continuum Percolation

In this section, we start with a practical problem which can be reinterpreted as a random covering problem on the circle by random arcs with fixed length. We solve it and discuss possible extensions. This leads us to a classical model of random covering of the Euclidean space called the Boolean model and we present briefly some of the questions related to it: covering of a particular set, continuum percolation and shapes of the connected components of the two phases.

1.5.1   Starting from the Bicycle Wheel Problem and Random Covering of the Circle

In 1989, C. Domb tells how his work during the second World War in radar research for the Admiralty led him to ask H. Jeffreys in 1946 about a covering problem [37]. H. Jeffreys related his question to his own *bicycle wheel problem* which he had formulated several years before in the following way: *a man is cycling along a road and passes through a region threwn with tacks. He wishes to know whether one has entered his tire. Because of the traffic, he can only snatch glances at random times. At each glance he has covered a fraction x of the wheel. What is the probability that after n glances, he has covered the whole wheel?* It turns out that the problem had

Fig. 1.9 Simulations of the bicycle wheel problem for $x = 0.01$: the random intervals in the cases $n = 20$ (left), $n = 50$ (middle) and $n = 100$ (right)

been solved by W.L. Stevens in 1939 [121], see also [119, Chapter 4]. Surprisingly, his method that we describe below relies exclusively on combinatorial arguments (Fig. 1.9).

We start by rephrasing the problem in mathematical terms in the following way: a set of n intervals of length x are placed randomly on a circle of length one and we aim at calculating the probability $q_n(x)$ that the circle is fully covered. The endpoints in the anticlockwise direction of the n random intervals of length x are denoted by $U_1, \cdots, U_n$ and are assumed to be n i.i.d. random variables uniformly distributed in $(0, 1)$. For every $1 \leq i \leq n$, we consider the event denoted by A_i that the endpoint of the i-th arc is not covered by the other $(n-1)$ random intervals. The key idea consists in noticing that the circle is fully covered by the n random intervals if and only if all endpoints are covered. Consequently, using the inclusion-exclusion principle, we obtain that the probability $p_{\text{cov}}(n, x)$ satisfies

$$1 - p_{\text{cov}}(n, x) = P\left(\cup_{i=1}^n A_i\right)$$

$$= \sum_{k=1}^n (-1)^{k+1} \sum_{1 \leq i_1 < i_2 < \cdots < i_k \leq n} P\left(A_{i_1} \cap A_{i_2} \cap \cdots \cap A_{i_k}\right)$$

$$= \sum_{k=1}^n (-1)^{k+1} \binom{n}{k} P(A_1 \cap A_2 \cap \cdots A_k), \qquad (1.34)$$

where the last equality comes from the fact that the variables $U_1, \cdots, U_n$ are exchangeable. It remains to calculate $P(A_1 \cap A_2 \cap \cdots A_k)$, $1 \leq k \leq n$, i.e. the probability that the endpoint of each of the first k random intervals is not covered, not only by the $(k-1)$ other intervals from the first bunch but also by the $(n-k)$ remaining intervals. This means that the event $A_1 \cap A_2 \cap \cdots A_k$ can be rewritten as

$$A_1 \cap A_2 \cap \cdots A_k = B_k \bigcap \left(\bigcap_{i=k+1}^n C_{i,k}\right) \qquad (1.35)$$

where B_k is the event that the endpoint of each of the first k random intervals is not covered by the $(k-1)$ other intervals from the first bunch and $C_{i,k}$, $k+1 \leq i \leq n$, is the event that the i-th random arc does not cover any of the endpoints from the first k random intervals.

On the event B_k, the k endpoints are at distance at least x from each other. Consequently, $P(B_k)$ is the probability that a random division of the circle into k parts produces parts of lengths larger than x. This is in particular the exact problem 666 that W.A. Whitworth solves in his book *Choice and Chance* published in 1870 [130]. By a direct integral calculation, we get

$$P(B_k) = (1 - kx)_+^{k-1}. \tag{1.36}$$

Conditional on the positions of the first k random intervals which satisfy the condition of the event B_k, the events $C_{i,k}$ are independent. The set of allowed positions on the circle for the endpoint of the i-th random arc is then the complement of a union of k disjoint intervals of length x. Consequently,

$$P\left(\bigcap_{i=k+1}^{n} C_{i,k} \Big| B_k \right) = (1 - kx)_+^{n-k}. \tag{1.37}$$

Combining (1.36), (1.37) with (1.35) and (1.34) shows that

$$p_{\text{cov}}(n, x)(x) = \sum_{k=0}^{n} (-1)^k \binom{n}{k} (1 - kx)_+^{n-1}.$$

In 1982, the formula was extended by A.F. Siegel and L. Holst to the explicit calculation of $p_{\text{cov}}(n, \mu)$, i.e. the probabiity to cover the circle with i.i.d. intervals which have random lengths such that these lengths are i.i.d. μ-distributed variables which are independent of the positions of the intervals on the circle [118]. They even provided the distribution of the number of uncovered gaps on the circle in this context. Following their work, T. Huillet obtained the joint distribution of the lengths of the connected components [63]. In [117], A.F. Siegel conjectured that $p_{\text{cov}}(n, \mu)$ satisfies a monotonicity result which is proved in [22] and is the following: if two probability distributions μ, ν on $(0, 1)$ are such that $\mu \leq \nu$ for the convex order, see e.g. [91, Chapter 1], then $p_{\text{cov}}(n, \mu) \leq p_{\text{cov}}(n, \mu)$. In particular, thanks to Jensen's inequality, this implies that $p_{\text{cov}}(n, x) \leq p_{\text{cov}}(n, \mu)$ where x is the mean of μ.

To the best of our knowledge, the most recent contribution in higher dimension is due to Bürgisser et al. [21] and contains on one hand an exact formula for the probability to cover the sphere with n spherical caps of fixed angular radius when this radius is larger than $\pi/2$ and on the other hand an upper bound for this probability when the angular radius is less than $\pi/2$.

Finally, a related question introduced by A. Dvoretzky in 1956 concerns the covering of the circle by an infinite number of intervals $(I_n)_n$ with deterministic lengths $(\ell_n)_n$ such that the sequence $(\ell_n)_n$ is non-increasing [38]. In 1972, L.A. Shepp showed that the circle is covered infinitely often with probability 1 if and only if the series $\sum_{n=1}^{\infty} n^{-2} \exp(\ell_1 + \cdots + \ell_n)$ is divergent [116].

1.5.2   A Few Basics on the Boolean Model

A natural extension of the bicycle wheel problem consists in considering random coverings of the Euclidean space by so-called *grains* with random positions and possibly random shapes or sizes. Considering the discussion on the translation invariance in Sect. 1.3, we construct directly such a model in $\mathbb{R}^d$. Let $\mathscr{P}_\lambda$ be a homogeneous Poisson point process of intensity λ and let K be a fixed non-empty compact set of $\mathbb{R}^d$, called the *grain*. Then the associated *Boolean model* is defined as the random set $\bigcup_{x \in \mathscr{P}_\lambda} (x + K)$, sometimes also called the *occupied phase* of the Boolean model. In the case when $K = B_r(o)$, $r > 0$, it was introduced by E.N. Gilbert in 1961 as a simplified approximation of the coverage of a radio transmission network where each individual can send a signal up to distance r [42]. When K is a random grain, for instance $K = B_R(o)$ with R a non-negative random variable, the model can be extended in the following way: the occupied phase is $\bigcup_{x \in \mathscr{P}_\lambda} (x + K_x)$ where $(K_x)_x$ is a collection of i.i.d. copies of K and independent of the Poisson point process $\mathscr{P}_\lambda$ (Fig. 1.10).

The Boolean model is well-adapted in a series of applied situations including flow in porous media [64], conduction in dispersions [76] and the elastic behavior of composites [131]. In practice, there are of course lots of fundamental statistical issues related in particular to the estimation of the intensity or of the grain distribution from the observation of the intersection of the Boolean model with a window or from sections or projections of this intersection on lower-dimensional subspaces. We shall omit this aspect and describe only the following probabilistic questions related to the model:

(a) estimating the covering probability of a particular set;
(b) concentrating on percolation, i.e. looking for the existence of an unbounded connected component of either the occupied or vacant phase;
(c) studying the geometry of the occupied or vacant phase or of their connected components.

Fig. 1.10 Simulation of the Boolean model in the unit square in the case $\lambda = 100$ and $K = B_R(o)$ where R is uniform on $(0, 0.1)$

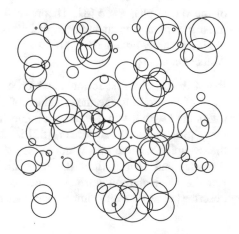

Problem (a) In the eighties, L. Flatto and D.J. Newman followed by S. Janson investigated the distribution of the number of random balls with fixed radius necessary to cover a bounded subset of $\mathbb{R}^d$ or a Riemannian manifold in two seminal works [40, 65]. Though this distribution is not explicit, S. Janson showed in particular a convergence in distribution, when the radius goes to 0, of the renormalized number to a Gumbel law. Regarding the covering of the whole space $\mathbb{R}^d$, it is shown in [51, Theorem 3.1] and [81, Proposition 7.3] that $\mathbb{R}^d = \bigcup_{x \in \mathcal{P}_\lambda} B_R(x)$ occurs almost surely when R^d is a non-integrable random variable and if not, the vacant set has infinite Lebesgue measure almost surely. P. Hall obtained upper and lower bounds for the probability of not covering $\mathbb{R}^d$ when R is deterministic, see [51, Theorem 3.11]. These results have been recently extended by a study of the covering of $\mathbb{R}^d$ by unions of balls $B_r(x)$ when the couples (x, r) belong to a Poisson point process in $\mathbb{R}^d \times (0, \infty)$ [14].

Problem (b) This question has been treated mostly in the case of i.i.d. grains of type $B_R(o)$ where R is a non-negative random variable. In [50], P. Hall shows notably that if $E(R^{2d-1})$ is finite, then there exists a critical intensity $\lambda_c \in (0, \infty)$ such that if $\lambda < \lambda_c$, all the connected components of the occupied phase are bounded almost surely. A breakthrough due to Gouéré [46] extends P. Hall's result in the following way: there exists a positive critical intensity λ_c under which all connected components of the occupied phase are bounded if and only if $E(R^d)$ is finite. We also refer to [47] for the extension of that result for unions of balls $B_r(x)$ when the couples (x, r) belong to a Poisson point process in $\mathbb{R}^d \times (0, \infty)$.

In the reference book [81] by R. Meester and R. Roy, it is shown that for any such Poisson Boolean model, the number of unbounded connected components of the occupied (resp. vacant) phase is either 0 or 1 almost surely, see Theorems 3.6 and 4.6 therein. Moreover, in the particular case of a two-dimensional Boolean model with almost surely bounded radii, criticality of the occupied and vacant phases coincide, i.e. there exists $\lambda_c \in (0, \infty)$ such that for $\lambda < \lambda_c$, there is possible percolation of the vacant phase and no unbounded component of the occupied phase almost surely, for $\lambda = \lambda_c$, neither the occupied phase nor the vacant phase percolates and for $\lambda > \lambda_c$, there is possible percolation of the occupied phase and no unbounded component of the vacant phase, see [81, Theorems 4.4, 4.5]. The equality of the two critical intensities of the vacant and occupied phases has been proved without the condition of almost surely bounded radii in two very recent works due to Penrose [98] and to Ahlberg et al. [1].

Problem (c) The first formulas which connect the mean values of the characteristics of the grain to the mean values of the characteristics of the Boolean model intersected with a window are available in seminal papers due to Miles [87] and Davy [30]. A few decades later, [52] investigates large deviation probabilities for the occupied volume in a window. More recently, in [59], asymptotic covariance formulas and central limit theorems are derived for a large set of geometric functionals, including intrinsic volumes and so-called Minkowski tensors. We describe below two very simple examples of questions related to (c).

1.5.2.1 The Spherical Contact Distribution of the Vacant Phase

We aim at determining the so-called *spherical contact distribution of the vacant phase*, i.e. the distribution of the radius R_c of the largest ball centered at o and included in the vacant phase, conditional on the event $\{o \notin \bigcup_{x \in \mathscr{P}_\lambda} (x + K)\}$. We observe that $R_c \geq r$ means that there is no point of $\mathscr{P}_\lambda$ at distance less than r from $-K$. Consequently, we get

$$P(R \geq r) = P(\mathscr{P}_\lambda \cap (-K + B_r(o)) = \emptyset)$$
$$= \exp(-\lambda V_d(K + B_r(o))).$$

When K is a convex body, we can use the Steiner formula (1.4) and get that

$$P(R \geq r) = \exp\left(-\lambda \sum_{k=0}^{d} \kappa_{d-k} V_k(K) r^{d-k}\right).$$

This calculation shows that when the grain K is convex, the quantity $\log(P(R \geq r))$ is a polynomial in r. In [60], it is shown that the converse is not true in general, unless the spherical contact distribution is replaced by another contact distribution.

1.5.2.2 The Number of Grains in a Typical Connected Component of the Occupied Phase

Similarly to the construction in Sect. 1.3 of the typical cell of a stationary tessellation, there is a way to define a typical connected component of the occupied phase: we add a deterministic point at the origin to the homogeneous Poisson point process and consider the connected component containing the origin of $\bigcup_{x \in \mathscr{P}_\lambda \cup \{o\}} (x + K)$. The aim of the calculation below is to derive a general formula for the distribution of the number $\mathscr{N}_o$ of grains contained in that connected component. The method below follows the work by Penrose [97], see also [101]. For any $n \geq 0$, we get

$$P(\mathscr{N}_o = n + 1) = E\left(\sum_{\{x_1, \cdots, x_n\} \in \mathscr{P}_\lambda^{(n)}} F(\{x_1, \cdots, x_n\}, \mathscr{P}_\lambda)\right) \tag{1.38}$$

where $\mathscr{P}_\lambda^{(n)}$ is the set of finite subsets of $\mathscr{P}_\lambda$ with exactly n elements and the functional $F(\{x_1, \cdots, x_n\}, \mathscr{P}_\lambda)$ is the indicator function of the event that the union $(o + K) \cup (x_1 + K) \cup \cdots \cup (x_n + K)$ is connected and all the remaining grains $(x + K)$, $x \in \mathscr{P}_\lambda \setminus \{x_1, \cdots, x_n\}$ are disconnected from that union. The expectation in (1.38) can be made explicit thanks to Mecke's formula for Poisson point processes, see e.g.

[113, Corollary 3.2.3]. We get indeed, for $n \geq 1$,

$$P(\mathcal{N}_o = n + 1) = \frac{\lambda^n}{n!} \int E(F(\{x_1, \cdots, x_n\}, \mathcal{P}_\lambda \cup \{x_1, \cdots, x_n\})) dx_1 \cdots dx_n$$

$$= \frac{\lambda^n}{n!} \int \mathbf{1}_{\{\bigcup_{0 \leq k \leq n} (x_i + K) \text{ connected}\}}$$

$$P(\forall x \in \mathcal{P}_\lambda, (x + K) \cap \bigcup_{0 \leq k \leq n} (x_i + K) = \emptyset) dx_1 \cdots dx_n$$

where for sake of simplicity, the origin o has been denoted by x_0. Using again the fact that $\mathcal{P}_\lambda$ is a Poisson point process, we deduce that

$$P(\mathcal{N}_o = n + 1)$$

$$= \frac{\lambda^n}{n!} \int \mathbf{1}_{\{\bigcup_{0 \leq k \leq n} (x_i + K) \text{ connected}\}} e^{-\lambda V_d((\bigcup_{0 \leq k \leq n} (x_i + K)) + (-K))} dx_1 \cdots dx_n.$$

When $K = B_r(o)$ for fixed $r > 0$, the previous formula becomes

$$P(\mathcal{N}_o = n + 1) = \frac{\lambda^n}{n!} \int \mathbf{1}_{\{\bigcup_{0 \leq k \leq n} B_r(x_i) \text{ connected}\}} e^{-\lambda V_d(\bigcup_{0 \leq k \leq n} B_{2r}(x_i))} dx_1 \cdots dx_n.$$

Alexander [2] showed that when $\lambda \to \infty$, this probability satisfies

$$\log P(\mathcal{N}_o = n + 1) = -\lambda \kappa_d r^d + (d - 1)n \log(\frac{\lambda}{n}) + O(1)$$

where $f(\lambda) = O(g(\lambda))$ means that the function f/g is bounded for large λ. A side result is the so-called phenomenon of *compression* which says roughly that in a high-density Boolean model, the density inside a bounded connected component is larger than the ambient density.

Acknowledgements The author warmly thanks two anonymous referees for their careful reading of the original manuscript, resulting in an improved and more accurate exposition.

References

1. D. Ahlberg, V. Tassion, A. Teixeira, Existence of an unbounded vacant set for subcritical continuum percolation (2017). https://arxiv.org/abs/1706.03053
2. K.S. Alexander, Finite clusters in high-density continuous percolation: compression and sphericality. Probab. Theory Relat. Fields **97**, 35–63 (1993)
3. R.V. Ambartzumian, A synopsis of combinatorial integral geometry. Adv. Math. **37**, 1–15 (1980)
4. F. Avram, D. Bertsimas, On central limit theorems in geometrical probability. Ann. Appl. Probab. **3**(4), 1033–1046 (1993)

5. F. Baccelli, S. Zuyev, Poisson-Voronoi spanning trees with applications to the optimization of communication networks. Oper. Res. **47**, 619–631 (1999)
6. J.-E. Barbier, Note sur Ie probleme de l'aiguille et Ie jeu du joint couvert. J. Math. Pures Appl. **5**, 273–286 (1860)
7. I. Bárány, Random polytopes in smooth convex bodies. Mathematika **39**, 81–92 (1992)
8. I. Bárány, Sylvester's question: the probability that n points are in convex position. Ann. Probab. **27**, 2020–2034 (1999)
9. I. Bárány, A note on Sylvester's four-point problem. Studia Sci. Math. Hungar. **38**, 733–77 (2001)
10. Y.M. Baryshnikov, R.A. Vitale, Regular simplices and Gaussian samples. Discret. Comput. Geom. **11**, 141–147 (1994)
11. V. Baumstark, G. Last, Gamma distributions for stationary Poisson flat processes. Adv. Appl. Probab. **41**, 911–939 (2009)
12. J. Bertrand, *Calcul des probabilités* (Gauthier-Villars, Paris, 1889)
13. T. Biehl, Über Affine Geometrie XXXVIII, Über die Schüttlung von Eikörpern. Abh. Math. Semin. Hamburg Univ. **2**, 69–70 (1923)
14. H. Biermé, A. Estrade, Covering the whole space with Poisson random balls. ALEA Lat. Am. J. Probab. Math. Stat. **9**, 213–229 (2012)
15. W. Blaschke, Lösung des "Vierpunktproblems" von Sylvester aus der Theorie der geometrischen Wahrscheinlichkeiten. Leipziger Berichte **69**, 436–453 (1917)
16. W. Blaschke, *Vorlesungen über Differentialgeometrie II: Affine Differentialgeometrie* (Springer, Berlin, 1923)
17. W. Blaschke, Integralgeometrie 2: Zu Ergebnissen von M.W. Crofton. Bull. Math. Soc. Roum. Sci. **37**, 3–11 (1935)
18. D. Bosq, G. Caristi, P. Deheuvels, A. Duma P. Gruber, D. Lo Bosco, V. Pipitone, *Marius Stoka: Ricerca Scientifica dal 1951 al 2013*, vol. III (Edizioni SGB, Messina, 2014)
19. C. Buchta, An identity relating moments of functionals of convex hulls. Discret. Comput. Geom. **33**, 125–142 (2005)
20. G.-L.L. Comte de Buffon, *Histoire naturelle, générale et particulière, avec la description du cabinet du Roy*. Tome Quatrième (Imprimerie Royale, Paris, 1777)
21. P. Bürgisser, F. Cucker, M. Lotz, Coverage processes on spheres and condition numbers for linear programming. Ann. Probab. **38**, 570–604 (2010)
22. P. Calka, The distributions of the smallest disks conta ining the Poisson-Voronoi typical cell and the Crofton cell in the plane. Adv. Appl. Probab. **34**, 702–717 (2002)
23. P. Calka, Tessellations, in *New Perspectives in Stochastic Geometry*, ed. by W.S. Kendall, I. Molchanov (Oxford University Press, Oxford, 2010), pp. 145–169
24. P. Calka, Asymptotic methods for random tessellations, in *Stochastic Geometry, Spatial Statistics and Random Fields*, ed. by E. Spodarev. Lecture Notes in Mathematics, vol. 2068 (Springer, Heidelberg, 2013), pp. 183–204
25. P. Calka, T. Schreiber, J.E. Yukich, Brownian limits, local limits and variance asymptotics for convex hulls in the ball. Ann. Probab. **41**, 50–108 (2013)
26. C. Carathéodory, E. Study, Zwei Beweise des Satzes daß der Kreis unter allen Figuren gleichen Umfanges den größten Inhalt hat. Math. Ann. **68**, 133–140 (1910)
27. H. Carnal, Die konvexe Hülle von *n* rotationssymmetrisch verteilten Punkten. Z. Wahrscheinlichkeit. und verw. Gebiete **15**, 168–176 (1970)
28. A. Cauchy, Notes sur divers théorèmes relatifs à la rectification des courbes, et à la quadrature des surfaces. C. R. Acad. Sci. Paris **13**, 1060–1063 (1841)
29. N. Chenavier, A general study of extremes of stationary tessellations with examples. Stochastic Process. Appl. **124**, 2917–2953 (2014)
30. P. Davy, Projected thick sections through multi-dimensional particle aggregates. J. Appl. Probab. **13**, 714–722 (1976)
31. S.N. Chiu, D. Stoyan, W.S. Kendall, J. Mecke, *Stochastic Geometry and its Applications*, 3rd edn. Wiley Series in Probability and Statistics (Wiley, Chichester, 2013)

32. R. Cowan, The use of ergodic theorems in random geometry. Adv. Appl. Probab. **10**, 47–57 (1978)
33. M.W. Crofton, On the theory of local probability, applied to straight lines drawn at random in a plane; the methods used being also extended to the proof of certain new theorems in the integral calculus. Philos. Trans. R. Soc. Lond. **156**, 181–199 (1868)
34. D.J. Daley, Asymptotic properties of stationary point processes with generalized clusters. Z. Wahrscheinlichkeitstheorie und Verw. Gebiete **21**, 65–76 (1972)
35. D.J. Daley, D. Vere-Jones, *An Introduction to the Theory of Point Processes*. Springer Series in Statistics (Springer, New York, 1988)
36. R. Descartes, *Principia Philosophiae* (Louis Elzevir, Amsterdam, 1644)
37. C. Domb, Covering by random intervals and one-dimensional continuum percolation. J. Stat. Phys. **55**, 441–460 (1989)
38. A. Dvoretzky, On covering a circle by randomly placed arcs. Proc. Natl. Acad. Sci. U S A **42**, 199–203 (1956)
39. B. Efron, The convex hull of a random set of points. Biometrika **52**, 331–343 (1965)
40. L. Flatto, D.J. Newman Random coverings. Acta Math. **138**, 241–264 (1977)
41. P. Franken, D. König, U. Arndt, V. Schmidt, *Queues and Point Processes* (Akademie-Verlag, Berlin, 1981)
42. E.N. Gilbert, Random plane networks. J. Soc. Ind. Appl. Math. **9**, 533–543 (1961)
43. A. Goldman, Sur une conjecture de D.G. Kendall concernant la cellule de Crofton du plan et sur sa contrepartie brownienne. Ann. Probab. **26**, 1727–1750 (1998)
44. A. Goldman, The Palm measure and the Voronoi tessellation for the Ginibre process. Ann. Appl. Probab. **20**, 90–128 (2010)
45. S. Goudsmit, Random distribution of lines in a plane. Rev. Mod. Phys. **17**, 321–322 (1945)
46. J.-B. Gouéré, Subcritical regimes in the Poisson Boolean model of continuum percolation. Ann. Probab. **36**, 1209–1220 (2008)
47. J.-B. Gouéré, Subcritical regimes in some models of continuum percolation. Ann. Appl. Probab. **19**, 1292–1318 (2009)
48. H. Groemer, On some mean values associated with a randomly selected simplex in a convex set. Pac. J. Math. **45**, 525–533 (1973)
49. H. Hadwiger, *Vorlesungen Über Inhalt, Oberfläche und Isoperimetrie* (Springer, Berlin, 1957)
50. P. Hall, On continuum percolation. Ann. Probab. **13**, 1250–1266 (1985)
51. P. Hall, *Introduction to the Theory of Coverage Processes* (Wiley, New York, 1988)
52. L. Heinrich, Large deviations of the empirical volume fraction for stationary Poisson grain models. Ann. Appl. Probab. **15**, 392–420 (2005)
53. L. Heinrich, L. Muche, Second-order properties of the point process of nodes in a stationary Voronoi tessellation. Math. Nachr. **281**, 350–375 (2008). Erratum Math. Nachr. **283**, 1674–1676 (2010)
54. L. Heinrich, H. Schmidt, V. Schmidt, Limit theorems for stationary tessellations with random inner cell structures. Adv. Appl. Probab. **37**, 25–47 (2005)
55. L. Heinrich, H. Schmidt, V. Schmidt, Central limit theorems for Poisson hyperplane tessellations. Ann. Appl. Probab. **16**, 919–950 (2006)
56. H.J. Hilhorst, Asymptotic statistics of the n-sided planar Poisson–Voronoi cell. I. Exact results. J. Stat. Mech. Theory Exp. **9**, P09005 (2005)
57. J. Hörrmann, D. Hug, M. Reitzner, C. Thäle, Poisson polyhedra in high dimensions. Adv. Math. **281**, 1–39 (2015)
58. D. Hug, Random polytopes, in *Stochastic Geometry, Spatial Statistics and Random Fields*, ed. by E. Spodarev. Lecture Notes in Mathematics, vol. 2068 (Springer, Heidelberg, 2013), pp. 205–238
59. D. Hug, G. Last, M. Schulte, Second order properties and central limit theorems for geometric functionals of Boolean models. Ann. Appl. Probab. **26**, 73–135 (2016)
60. D. Hug, G. Last, W. Weil, Polynomial parallel volume, convexity and contact distributions of random sets. Probab. Theory Relat. Fields **135**, 169–200 (2006)

61. D. Hug, M. Reitzner, R. Schneider, The limit shape of the zero cell in a stationary Poisson hyperplane tessellation. Ann. Probab. **32**, 1140–1167 (2004)
62. D. Hug, R. Schneider, Asymptotic shapes of large cells in random tessellations. Geom. Funct. Anal. **17**, 156–191 (2007)
63. T. Huiller, Random covering of the circle: the size of the connected components. Adv. Appl. Probab. **35**, 563–582 (2003)
64. A. Hunt, R. Ewing, B. Ghanbarian, *Percolation theory for flow in porous media*, 3rd edn. Lecture Notes in Physics, vol. 880 (Springer, Cham, 2014)
65. S. Janson, Random coverings in several dimensions. Acta Math. **156**, 83–118 (1986)
66. E.T. Jaynes, The well-posed problem. Found. Phys. **3**, 477–492 (1973)
67. D.G. Kendall, *Foundations of a Theory of Random Sets. Stochastic Geometry (A Tribute to the Memory of Rollo Davidson)* (Wiley, London, 1974), pp. 322–376
68. M.G. Kendall, P.A.P. Moran, *Geometrical Probability* (Charles Griffin, London, 1963)
69. J.F.C. Kingman, Random secants of a convex body. J. Appl. Probab. **6**, 660–672 (1969)
70. J.F.C. Kingman, *Poisson Processes* (Clarendon Press, Oxford, 1993)
71. D.A. Klain, G.-C. Rota, *Introduction to Geometric Probability* (Cambridge University Press, Cambridge, 1997)
72. A.N. Kolmogorov, *Grundbegriffe der Wahrscheinlichkeitsrechnung* (Springer, Berlin, 1933)
73. G. Last, M. Penrose, *Lectures of the Poisson Process* (Cambridge University Press, Cambridge, 2017)
74. W. Lefebvre, T. Philippe, F. Vurpillot, Application of Delaunay tessellation for the characterization of solute-rich clusters in atom probe tomography. Ultramicroscopy **111**, 200–206 (2011)
75. J.-F. Marckert, The probability that n random points in a disk are in convex position. Braz. J. Probab. Stat. **31**(2), 320–337 (2017)
76. K.Z. Markov, C.I. Christov, On the problem of heat conduction for random dispersions of spheres allowed to overlap. Math. Models Methods Appl. Sci. **2**, 249–269 (1992)
77. B. Matérn, Spatial variation: Stochastic models and their application to some problems in forest surveys and other sampling investigations. Meddelanden Fran Statens Skogsforskningsinstitut, vol. 49, Stockholm (1960)
78. Matheron, G.: *Random Sets and Integral Geometry*. Wiley Series in Probability and Mathematical Statistics (Wiley, New York, 1975)
79. J. Mecke, Stationäre zufällige Masse auf lokalkompakten Abelschen Gruppen. Z. Wahrscheinlichkeitstheorie und Verw. Gebiete **9**, 36–58 (1967)
80. J. Mecke, On the relationship between the 0-cell and the typical cell of a stationary random tessellation. Pattern Recogn. **32**, 1645–1648 (1999)
81. R. Meesters, R. Roy, *Continuum Percolation* (Cambridge University Press, New York, 1996)
82. J.L. Meijering, Interface area, edge length and number of vertices in crystal aggregates with random nucleation. Philips Res. Rep. **8**, 270–90 (1953)
83. R.E. Miles, Random polygons determined by random lines in a plane I. Proc. Natl. Acad. Sci. U S A **52**, 901–907 (1964)
84. R.E. Miles, Random polygons determined by random lines in a plane II. Proc. Natl. Acad. Sci. U S A **52**, 1157–1160 (1964)
85. R.E. Miles, The random division of space. Suppl. Adv. Appl. Probab. **4**, 243–266 (1972)
86. R.E. Miles, The various aggregates of random polygons determined by random lines in a plane. Adv. Math. **10**, 256–290 (1973)
87. R.E. Miles, Estimating aggregate and overall characteristics from thich sections by transmission microscopy. J. Microsc. **107**, 227–233 (1976)
88. J. Møller, Random tessellations in $\mathbb{R}^d$. Adv. Appl. Probab. **21**, 37–73 (1989)
89. J. Møller, Random Johnson–Mehl tessellations. Adv. Appl. Probab. **24**, 814–844 (1992)
90. J. Møller, *Lectures on Random Voronoi Tessellations*. Lecture Notes in Statistics, vol. 87 (Springer, New York, 1994)
91. A. Müller, D. Stoyan, *Comparison Methods for Stochastic Models and Risks*. Wiley Series in Probability and Statistics (Wiley, Chichester, 2002)

92. W. Nagel, V. Weiss, Crack STIT tessellations: characterization of stationary random tessellations stable with respect to iteration. Adv. Appl. Probab. **37**, 859–883 (2005)
93. J. Neveu, Processus ponctuels, in *École d'été de Probabilités de Saint-Flour*. Lecture Notes in Mathematics, vol. 598 (Springer, Berlin, 1977), pp. 249–445
94. *New Advances in Geostatistics. Papers from Session Three of the 1987 MGUS Conference held in Redwood City, California, April 13–15, 1987*. Mathematical Geology, vol. 20 (Kluwer Academic/Plenum Publishers, Dordrecht, 1988), pp. 285–475
95. *New Perspectives in Stochastic Geometry*, ed. by W.S. Kendall, I. Molchanov (Oxford University Press, Oxford, 2010)
96. C. Palm, Intensitätsschwankungen im Fernsprechverkehr. Ericsson Technics **44**, 1–189 (1943)
97. M. Penrose, On a continuum percolation model. Adv. Appl. Probab. **23**, 536–556 (1991)
98. M. Penrose, Non-triviality of the vacancy phase transition for the Boolean model (2017). https://arxiv.org/abs/1706.02197
99. R.E. Pfiefer, The historical development of J. J. Sylvester's four point problem. Math. Mag. **62**, 309–317 (1989)
100. H. Poincaré, *Calcul des probabilités* (Gauthier-Villars, Paris, 1912)
101. J. Quintanilla, S. Torquato, Clustering in a continuum percolation model. Adv. Appl. Probab. **29**, 327–336 (1997)
102. C. Redenbach, On the dilated facets of a Poisson-Voronoi tessellation. Image Anal. Stereol. **30**, 31–38 (2011)
103. M. Reitzner, Stochastical approximation of smooth convex bodies. Mathematika **51**, 11–29 (2004)
104. M. Reitzner, Central limit theorems for random polytopes. Probab. Theory Relat. Fields **133**, 483–507 (2005)
105. M. Reitzner, The combinatorial structure of random polytopes. Adv. Math. **191**, 178–208 (2005)
106. M. Reitzner, Random polytopes, in *New Perspectives in Stochastic Geometry*, ed. by W.S. Kendall, I. Molchanov (Oxford University Press, Oxford, 2010), pp. 45–76
107. A. Rényi, R. Sulanke, Über die konvexe Hülle von n zufällig gewählten Punkten. Z. Wahrscheinlichkeitsth. verw. Geb. **2**, 75–84 (1963)
108. A. Rényi, R. Sulanke, Über die konvexe Hülle von n zufällig gewählten Punkten. II. Z. Wahrscheinlichkeitsth. verw. Geb. **3**, 138–147 (1964)
109. R. Schneider, *Convex Bodies: The Brunn-Minkowski Theory* (Cambridge University Press, Cambridge, 1993)
110. L.A. Santaló, *Integral Geometry and Geometric Probability*. Encyclopedia of Mathematics and its Applications, vol. 1 (Addison-Wesley, Reading, 1976)
111. R. Schneider, Random hyperplanes meeting a convex body. Z. Wahrsch. Verw. Gebiete **61**, 379–387 (1982)
112. R. Schneider, Integral geometric tools for stochastic geometry, in *Stochastic Geometry*, ed. by W. Weil. Lectures Given at the C.I.M.E. Summer School held in Martina Franca. Lecture Notes in Mathematics, vol. 1892 (Springer, Berlin, 2007), pp. 119–184
113. R. Schneider, W. Weil, *Stochastic and Integral Geometry* (Springer, Berlin, 2008)
114. R. Schneider, J.A. Wieacker, Random polytopes in a convex body. Z. Wahrsch. Verw. Gebiete **52**, 69–73 (1980)
115. J. Serra, *Image Analysis and Mathematical Morphology* (Academic, London, 1984)
116. L.A. Shepp, Covering the circle with random arcs. Isr. J. Math. **11**, 328–345 (1972)
117. A.F. Siegel, Random space filling and moments of coverage in geometrical probability. J. Appl. Probab. **15**, 340–355 (1978)
118. A.F. Siegel, L. Holst, Covering the circle with random arcs of random sizes. J. Appl. Probab. **19**, 373–381 (1982)
119. H. Solomon, Geometric Probability. CBMS-NSF Regional Conference Series in Applied Mathematics, vol. 28 (SIAM, Philadelphia, 1978)
120. J. Steiner, Über parallele Flächen. Monatsber. Preuss. Akad. Wiss., Berlin (1840), pp. 114–118

121. W.L. Stevens, Solution to a geometrical problem in probability. Ann. Eugenics **9**, 315–320 (1939)
122. D. Stoyan, Applied stochastic geometry: a survey. Biometrical J. **21**, 693–715 (1979)
123. J.J. Sylvester, Problem 1491. The Educational Times, London (April, 1864)
124. P. Valtr, Probability that n random points are in convex position. Discret. Comput. Geom. **13**, 637–643 (1995)
125. P. Valtr, The probability that n random points in a triangle are in convex position. Combinatorica **16**, 567–573 (1996)
126. V.H. Vu, Sharp concentration of random polytopes. Geom. Funct. Anal. **15**, 1284–1318 (2005)
127. W. Weil, Point processes of cylinders, particles and flats. Acta Appl. Math. **9**, 103–136 (1987)
128. V. Weiss, R. Cowan, Topological relationships in spatial tessellations. Adv. Appl. Probab. **43**, 963–984 (2011)
129. J.G. Wendel, A problem in geometric probability. Math. Scand. **11**, 109–111 (1962)
130. W.A. Whitworth, *Choice and Chance* (D. Bell, Cambridge, 1870)
131. F. Willot, D. Jeulin, Elastic behavior of composites containing Boolean random sets of inhomogeneities. Int. J. Eng. Sci. **47**, 313–324 (2009)

Chapter 2
Understanding Spatial Point Patterns Through Intensity and Conditional Intensities

Jean-François Coeurjolly and Frédéric Lavancier

Abstract This chapter deals with spatial statistics applied to point patterns. As well as in many specialized books, spatial point patterns are usually treated elaborately in books devoted to spatial statistics or stochastic geometry. Our aim is to propose a different point of view as we intend to present how we can understand and analyze a point pattern through intensity and conditional intensity functions. We present these key-ingredients theoretically and in an intuitive way. Then, we list some of the main spatial point processes models and provide their main characteristics, in particular, when available the form of their intensity and conditional intensity functions. Finally, we provide a non exhaustive list of statistical methodologies to estimate these functions and discuss the pros and cons of each method.

2.1 Introduction

Spatial point patterns arise in a broad range of applications. We are faced with such data for instance when we want to model the locations of trees in a forest, the locations of disease cases in a region, the locations of cells in a biological tissue, the locations of sunspots on the surface of the sun, the fixations of the retina (acquired with an eye-tracker) superimposed on an image or video a subject is looking at, etc. With the ability to acquire spatial data, the number of applications is increasing and the amount of data is becoming huge.

The objective of this chapter is to analyze spatial point processes through the notion of intensity. The word intensity here has to be understood in a very large sense, since it can be a simple intensity function, often interpreted as the local probability to observe a point at a fixed location, or a conditional intensity function.

J.-F. Coeurjolly (✉)
Université du Québec à Montréal, Département de Mathématiques, Montréal, QC, Canada
e-mail: coeurjolly.jean-francois@uqam.ca

F. Lavancier
Université de Nantes, Laboratoire de Mathématiques Jean Leray, Nantes, France
e-mail: frederic.lavancier@univ-nantes.fr

© Springer Nature Switzerland AG 2019
D. Coupier (ed.), *Stochastic Geometry*, Lecture Notes in Mathematics 2237,
https://doi.org/10.1007/978-3-030-13547-8_2

Again, the term conditional intensity can gather several concepts: the intensity of the process given that the process has a point at the location x (or more generally n points at $x_1, \ldots, x_n$), or the intensity of the process given that the rest of the configuration is fixed.

Section 2.2 presents the main notation of the paper, some background on spatial point processes (general definition, characterization of the distribution), and specifically the concepts of intensity functions, Palm intensities and Papangelou conditional intensities for a spatial point process. These characteristics are introduced from a theoretical and intuitive point of view. In Sect. 2.3, we review classical models of spatial point processes: Poisson point processes, Gibbs point processes, Cox processes and determinantal point processes. This list is not exhaustive but these four classes are clearly the main models used in applications to model aggregation or repulsiveness in a point pattern. For these models, we state their main properties and report, when available, the form of intensity, Palm intensity and Papangelou conditional intensity functions.

Sections 2.4 and 2.5 are devoted to statistics. In Sect. 2.4 we focus on the intensity function and present different methodologies to estimate it. We first consider the case where this function is constant (in which case the point process is assumed to be stationary). In the inhomogeneous case, we first consider the non parametric estimation of the intensity. Then, and more intensively, we review a few methods designed to estimate a parametric form of the intensity function. The presentation is not exhaustive but gathers the main methods, for which we try to give the pros and cons for their practical use. Section 2.5 focuses on the estimation of conditional intensities and especially on the Papangelou conditional intensity and, to a lesser extent, on the one-point Palm intensity. Our purpose is here again to provide the reader the key-ingredients of each method, its advantages and drawbacks. Numerical aspects concerning the methods of Sects. 2.4 and 2.5 are briefly discussed and some illustrations are provided.

Spatial point pattern analysis constitutes the topic of many specialized books. Our aim in this chapter is to propose a different way of reading. Most of the material presented in this chapter can be found in the following books or reviews (see also the references cited in Sects. 2.4 and 2.5). For theoretical and methodological aspects, we refer the reader to Daley and Vere-Jones [28], Møller and Waagepetersen [58] and to Illian et al. [45]; see also the recent reviews by Møller and Waagepetersen [59, 60]. The reader interested in numerical and computational aspects and who simply wants to analyze spatial point patterns with R is definitely invited to read the complete and excellent book by Baddeley et al. [5]. To obtain a broader perspective, the interested reader is referred to the earliest book by Ripley [67], Kingman's monograph on Poisson processes [48], Lieshout's work on Markov models [76], the accessible introduction by Diggle [35], or the work of the American school as exemplified by Cressie [25], Cressie and Wikle [26], see also the connections to random sets and random measures as presented in Chiu et al. [12].

2.2 Intensity and Conditional Intensity Functions

2.2.1 Definition and Theoretical Characterization of a Spatial Point Process

In this chapter, we consider spatial point processes in $\mathbb{R}^d$. For ease of exposition, we view a point process as a random locally finite subset $\mathbf{X}$ of a Borel set $S \subseteq \mathbb{R}^d$, $d \geq 1$. For readers interested in measure theoretical details, we refer to e.g. [58] or [28], or to Chap. 5 by David Dereudre. This setting implies the following facts. First, we consider simple point processes (two points cannot occur at the same location). Second, we exclude manifold-valued point processes (like circular or spherical point processes), spatio-temporal point processes and marked point processes, even if most of the concepts and methodologies presented hereafter exist or can be straightforwardly adapted in such contexts.

We denote by $\mathbf{X}_B = \mathbf{X} \cap B$ the restriction of $\mathbf{X}$ to a set $B \subseteq S$ and by $|B|$ the volume of any bounded $B \subset S$. Local finiteness of $\mathbf{X}$ means that $\mathbf{X}_B$ is finite almost surely (a.s.), that is the number of points $N(B)$ of $\mathbf{X}_B$ is finite a.s., whenever B is bounded. We let $\mathcal{N}$ stand for the state space consisting of the locally finite subsets (or point configurations) of S. For $\mathbf{x} \in \mathcal{N}$, we let $|\mathbf{x}|$ denote the number of elements of $\mathbf{x}$. Furthermore, $\mathcal{B}_0$ is the family of all bounded Borel subsets of S.

The counting variables $N(B) = |\mathbf{X} \cap B|$ for $B \in \mathcal{B}_0$ play a central role in the characterization, modelling and analysis of spatial point patterns. The distribution of $\mathbf{X}$ is uniquely determined by the joint distribution of $N(B_1), \ldots, N(B_m)$ for any $B_1, \ldots, B_m \in \mathcal{B}_0$ and any $m \geq 1$. Surprisingly, the distribution is also equivalently determined by its void probabilities, i.e. by the probabilities $P(\mathbf{X}_K = \emptyset) = P(N(K) = 0)$, $K \subseteq S$ compact, see again e.g. [28] for a proof of this result.

The reference model is the Poisson point process often defined as follows.

Definition 2.1 Let ρ be a locally integrable function on S. A point process $\mathbf{X}$ satisfying the following statements is called the Poisson point process on S with intensity function ρ:

- for any $m \geq 1$, and for any disjoint and bounded $B_1, \ldots, B_m \subset S$, the random variables $\mathbf{X}_{B_1}, \ldots, \mathbf{X}_{B_m}$ are independent;
- $N(B)$ follows a Poisson distribution with parameter $\int_B \rho(u)du$ for any bounded $B \subset S$.

It is easily checked that the void probabilities of a Poisson point process are given by $P(\mathbf{X}_K = \emptyset) = \exp(-\int_K \rho(u)du)$ for any compact set $K \subset S$. We also have that, for any $B \in \mathcal{B}_0$ and any non-negative measurable function h on $\{\mathbf{x} \cap B \mid \mathbf{x} \in \mathcal{N}\}$

$$\mathrm{E}h(\mathbf{X}_B) = \sum_{n=0}^{\infty} \frac{\exp(-|B|)}{n!} \int_B \cdots \int_B h(\{u_1, \ldots, u_n\})\rho(u_1)\ldots\rho(u_n)du_1 \ldots du_n,$$

$$(2.1)$$

where for $n = 0$ the term is read as $\exp(-|B|)h(\emptyset)$.

Next sections will provide additional properties of Poisson point processes and in particular will make clearer why this model generates points without any interaction.

The nice theoretical characterization of a general spatial point process $\mathbf{X}$ does not help to define new models and/or to understand interaction between points. Next sections will present more useful ways of analyzing the dependence between points. We will review the concept of intensity function and conditional intensities (actually Palm intensities and Papangelou conditional intensities). There are clearly other ways of characterizing and understanding the interaction in a point pattern. Some of them are directly based on intensities like the pair correlation function (see Sect. 2.2.2). But most of them are not: e.g. the Ripley's K function, the L-function or distance based summary statistics like the F- G- and J-functions. We refer the interested reader to such summary statistics to [45, 58–60].

2.2.2 Moment Measures Factorial Moment Measures and Intensity Functions

For $n = 1, 2, \ldots$ and $B_i \in \mathscr{B}_0$, the n-th order moment measure $\mu^{(n)}$ and the n-th order factorial moment measure $\alpha^{(n)}$ are defined by

$$\mu^{(n)}(B_1 \times B_2 \times \cdots \times B_n) = \mathrm{E} \sum_{u_1,\ldots,u_n \in \mathbf{X}} \mathbf{1}(u_1 \in B_1, \ldots, u_n \in B_n)$$

$$\alpha^{(n)}(B_1 \times B_2 \times \cdots \times B_n) = \mathrm{E} \sum_{u_1,\ldots,u_n \in \mathbf{X}}^{\neq} \mathbf{1}(u_1 \in B_1, \ldots, u_n \in B_n),$$

where $\mathbf{1}(\cdot)$ denotes the indicator function and $\neq$ over the summation sign means that $u_1, \ldots, u_n$ are pairwise distinct. We focus on $\alpha^{(n)}$ in the following which is clearly more useful. If $\alpha^{(n)}$ has a density $\rho^{(n)}$ with respect to the Lebesgue measure, $\rho^{(n)}$ is called the n-th order joint intensity function and is determined up to a Lebesgue null set. Therefore, we can assume that $\rho^{(n)}(u_1, \ldots, u_n)$ is invariant under permutations of $u_1, \ldots, u_n$, and we need only to consider the case where $u_1, \ldots, u_n \in S$ are pairwise distinct. Then $\rho^{(n)}(u_1, \ldots, u_n)\, du_1 \cdots du_n$ can be interpreted as the approximate probability for $\mathbf{X}$ having a point in each of infinitesimally small regions around $u_1, \ldots, u_n$ of volumes $du_1, \ldots du_n$, respectively. We also write $\rho(u)$ for the intensity function $\rho^{(1)}(u)$. The Campbell-Mecke theorem gives an integral representation of $\rho^{(n)}$: for any function h such that $|h|\rho^{(n)}$ is integrable with respect to the Lebesgue measure on S^n, then

$$\mathrm{E} \sum_{u_1,\ldots,u_n \in \mathbf{X}}^{\neq} h(u_1, \ldots, u_n) = \int_S \cdots \int_S h(u_1, \ldots, u_n)\rho^{(n)}(u_1, \ldots, u_n) du_1 \cdots du_n.$$

$$(2.2)$$

If $\mathbf{X}$ is a Poisson point process on S with intensity function ρ, it is straightforwardly seen that the n-th order intensity function satisfies

$$\rho^{(n)}(u_1, \ldots, u_n) = \prod_{i=1}^{n} \rho(u_i), \quad \forall u_1, \ldots, u_n \in S.$$

From this, when $n = 2$ it is natural to focus on the pair correlation function, which is the function $g : S \times S \to \mathbb{R}^+$ given for any $u, v \in S$ such that $\rho(u)\rho(v) > 0$ (otherwise we set $g(u, v) = 0$) by

$$g(u, v) = \frac{\rho^{(2)}(u, v)}{\rho(u)\rho(v)}.$$

The pair correlation function gives ideas on how a point process $\mathbf{X}$ deviates from the Poisson case since in this situation $g(u, v) = 1, \forall u, v \in S$.

At this stage, it is pertinent to recall that when $S = \mathbb{R}^d$, a point process $\mathbf{X}$ is said to be stationary (respectively isotropic) if its distribution is invariant under translations (respectively under rotations). In such a case the intensity function is necessarily constant, and the pair correlation function g depends through $\rho^{(2)}$ only on $u - v$ when $\mathbf{X}$ is stationary and on $\|u - v\|$ when $\mathbf{X}$ is isotropic. Point process models with an inhomogeneous intensity function and a pair correlation function invariant by translation are called second-order reweighted stationary point processes (see e.g. [58]). This class of processes is often considered to derive asymptotic results for parametric inference methods.

Figure 2.1 illustrates this section and reports intensity function for homogeneous and inhomogeneous models and for different classes of point processes. From Fig. 2.1a–c, we clearly see what homogeneity (i.e. constant intensity) means and that this notion is independent of the fact that the points interact with each other or not. Figure 2.1d–i show that more points are observed when the intensity function is high.

2.2.3 Palm Distributions and Palm Intensities

For any measurable $F \subseteq \mathcal{N}$, define the n-th order reduced Campbell measure $C^{(n)!}$ as the measure on $S^n \times \mathcal{N}$ given by

$$C^{(n)!}(B_1 \times B_2 \times \cdots \times B_n \times F)$$

$$= E \sum_{x_1, \ldots, x_n \in \mathbf{X}}^{\neq} 1(x_1 \in B_1, \ldots, x_n \in B_n, \mathbf{X} \setminus \{x_1, \ldots, x_n\} \in F).$$

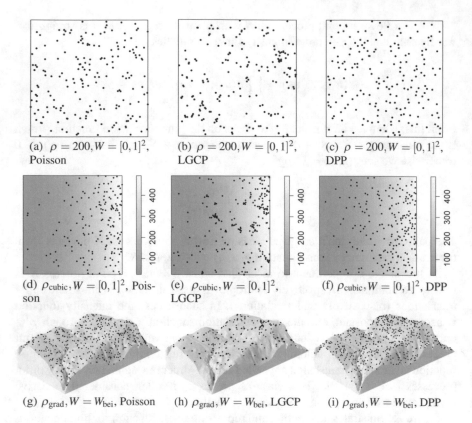

Fig. 2.1 Realization of a Poisson point process (**a**), (**d**), (**g**), a log-Gaussian Cox process (**b**), (**e**), (**h**) and a determinantal point process (**c**), (**f**), (**i**). The intensity is constant for (**a**)–(**c**) and equals 200. For (**d**)–(**f**), the intensity has the form $\log \rho_{\mathrm{cubic}}(u) = \beta + u_1 + u_1^3/2$. The latent image corresponds to the map of ρ_{cubic}. For (**g**)–(**i**), we use covariates accompanying the dataset bei (giving locations of a specific species of trees) of the R package spatstat. These covariates correspond to the elevation and slope of elevation fields of a part of the forest in Barro Colorado Island. The domain of observation is $W_{\mathrm{bei}} = [0, 1000] \times [0, 500]$. The model for the intensity is $\log \rho_{\mathrm{grad}}(\mathbf{u}) = \beta + .5z(u)$ where $z(u)$ is the slope of elevation at location u. The points are superimposed to the 3-dimensional map of the elevation field. The parameter β is adjusted to get $EN(W) = 200$ for (**d**)–(**f**) and 800 for (**g**)–(**i**); the latent Gaussian field defining the LGCP model has an exponential covariance function with variance 2 and scale parameter 0.02; the kernel defining the DPP is a Gaussian kernel with parameters 0.035 for (**c**) and 0.025 for (**f**) and (**i**). LGCPs and DPPs are presented in detail in Sects. 2.3.3.1 and 2.3.4

Note that $\mathrm{C}^{(n)!}(\cdot \times F)$, as a measure on S^n, is absolutely continuous with respect to $\alpha^{(n)}$, with a density $\mathrm{P}^!_{x_1,\ldots,x_n}(F)$ which is determined up to an $\alpha^{(n)}$ null set, and $\alpha^{(n)}(B_1 \times \cdots \times B_n) = \mathrm{C}^{(n)!}(B_1 \times B_2 \times \cdots \times B_n \times \mathcal{N})$. By the so-called Campbell-Mecke formula/theorem, we can assume that $\mathrm{P}^!_{x_1,\ldots,x_n}(\cdot)$ is a point process distribution on $\mathcal{N}$, called the n-th order reduced Palm distribution given $x_1, \ldots, x_n$ (see e.g. [29]). We denote by $\mathbf{X}^!_{x_1,\ldots,x_n}$ a point process distributed according to $\mathrm{P}^!_{x_1,\ldots,x_n}$. Again we need only to consider the case where $x_1, \ldots, x_n$ are pairwise distinct. Then $P^!_{x_1,\ldots,x_n}$ can be interpreted as the conditional distribution of $\mathbf{X} \setminus \{x_1, \ldots, x_n\}$ given that $x_1, \ldots, x_n \in \mathbf{X}$.

If $\rho^{(n)}$ exists, then by standard measure theoretical arguments we obtain the extended Campbell-Mecke formula. For any non-negative function h defined on $S^n \times \mathcal{N}$

$$
\mathrm{E} \sum_{x_1,\ldots,x_n \in \mathbf{X}}^{\neq} h(x_1, \ldots, x_n, \mathbf{X} \setminus \{x_1, \ldots, x_n\})
$$

$$
= \int_S \cdots \int_S \mathrm{E}\, h(x_1, \ldots, x_n, \mathbf{X}^!_{x_1,\ldots,x_n}) \rho^{(n)}(x_1, \ldots, x_n) \mathrm{d}x_1 \cdots \mathrm{d}x_n. \qquad (2.3)
$$

Suppose $\rho^{(m+n)}$ exists for an $m \geq 1$ and $n \geq 1$. Then, for pairwise distinct $u_1, \ldots, u_m, x_1, \ldots, x_n \in S$, it follows easily by expressing $\alpha^{(m+n)}$ as an expectation of the form (2.3) that $\mathbf{X}^!_{x_1,\ldots,x_n}$ has m-th order joint intensity function

$$
\rho^{(m)}_{x_1,\ldots,x_n}(u_1, \ldots, u_m) = \begin{cases} \dfrac{\rho^{(m+n)}(u_1,\ldots,u_m,x_1,\ldots,x_n)}{\rho^{(n)}(x_1,\ldots,x_n)} & \text{if } \rho^{(n)}(x_1, \ldots, x_n) > 0, \\ 0 & \text{otherwise.} \end{cases}
$$
$$(2.4)$$

We also write $\rho_{x_1\ldots,x_n}$ for the intensity function $\rho^{(1)}_{x_1\ldots,x_n}$. The function $\rho^{(m)}_{x_1,\ldots,x_n}$ is called the m-th order n-point Palm intensity. Obviously, when $\mathbf{X}$ is a Poisson point process, we have the equality $\rho^{(m)}_{x_1,\ldots,x_n}(u_1, \ldots, u_m) = \prod_{i=1}^{m} \rho(u_i)$; the knowledge that "$x_1, \ldots, x_n \in \mathbf{X}$" does not bring any information on the probability to observe points at $u_1, \ldots, u_m$.

When $m = n = 1$, we speak of the one-point Palm intensity function or even simpler the Palm intensity function. We can check that for any pairwise distinct $u, x \in S$, $\rho_x(u) = \rho(u)g(u, x)$. From that, we observe first, that the Palm intensity function ρ_x is independent on x for any $x \in S$ if and only if $g = 1$ (note that this is not equivalent to say that $\mathbf{X}$ is a Poisson point process). Second, if $\mathbf{X}$ is a stationary point process (resp. isotropic point process), the Palm intensity $\rho_x(\cdot)$ depends only on $x - \cdot$ (respectively on $\|x - \cdot\|$).

Equation (2.4) provides a natural interpretation of Palm intensities: $\rho^{(m)}_{x_1,\ldots,x_n}$ $(u_1, \ldots, u_m) \mathrm{d}u_1 \ldots \mathrm{d}u_m$ can be interpreted as the approximate probability for $\mathbf{X}$ having a point in each of infinitesimally small regions around $x_1, \ldots, x_n, u_1, \ldots, u_m$ of volumes $\mathrm{d}x_1, \ldots, \mathrm{d}x_n, \mathrm{d}u_1,\ldots, \mathrm{d}u_m$ respectively divided by the approximate probability for $\mathbf{X}$ having a point in each of infinitesimally small regions around $x_1, \ldots, x_n$ of volumes $\mathrm{d}x_1, \ldots, \mathrm{d}x_n$. In other words, this is also the approximate conditional probability for $\mathbf{X}$ having a point in each of infinitesimally small regions around $u_1, \ldots, u_m$ of volumes $\mathrm{d}u_1, \ldots, \mathrm{d}u_m$ conditionnally on $\mathbf{X}$ having a point in each of infinitesimally small regions around $x_1, \ldots, x_n$ of volumes $\mathrm{d}x_1, \ldots, \mathrm{d}x_n$.

Figure 2.2 illustrates this section and plots, for a homogeneous and inhomogeneous LGCP, the one-point Palm intensity function $\rho_x(u)$ for some fixed point x

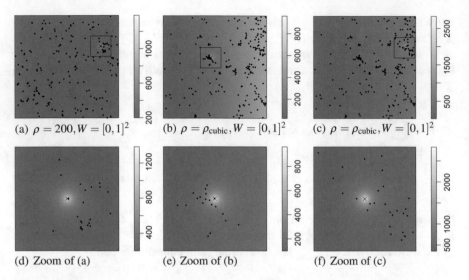

Fig. 2.2 Point patterns represented in (**a**) (respectively (**b, c**)) correspond to the LGCP from Fig. 2.1b, and e respectively (see Fig. 2.1 for details on the parameters of the LGCP). The latent image corresponds to the Palm intensity function $\rho_x(u)$ where x is the point represented by a cross ($\times$) and located at the center of the square sub-window of each plot. For (**a**), $\rho_x(u) = 200g(u, x)$ while for (**b, c**), $\rho_x(u) = \rho_{\text{cubic}}(u)g(u, x)$; see Fig. 2.1 and Sect. 2.3.3.1 for details on ρ_{cubic} and g. Figure (**d**)–(**f**) are zooms of the square sub-windows (with side-length 0.2) depicted in (**a**), (**b**) and (**c**)

represented by a cross. This class of models is presented in details in Sect. 2.3.3.1. When the LGCP is defined through an exponential (more generally a positive) covariance function, as made in Fig. 2.2, the model produces attractive patterns. The one-point Palm intensity attests this characteristic, as it is high in the neighbourhood of the fixed point x.

2.2.4 Papangelou Conditional Intensities

Let $\mathbf{Z}$ be a unit rate Poisson point process on S and assume, first, that S is bounded ($|S| < \infty$). We say that a spatial point process $\mathbf{X}$ has a density f if the distribution of $\mathbf{X}$ is absolutely continuous with respect to the one of $\mathbf{Z}$ and with density f. Thus, for any non-negative measurable function h defined on $\mathcal{N}$, $\mathrm{E}h(\mathbf{X}) = \mathrm{E}(f(\mathbf{Z})h(\mathbf{Z}))$. By (2.1)

$$\mathrm{E}h(\mathbf{X}) = \sum_{n=0}^{\infty} \frac{\exp(-|S|)}{n!} \int_B \cdots \int_B h(\{u_1, \ldots, u_n\}) f(\{u_1, \ldots, u_n\}) \mathrm{d}u_1 \ldots \mathrm{d}u_n.$$

$$(2.5)$$

Now, suppose that f is *hereditary*, i.e., for any pairwise distinct $u_0, u_1, \ldots, u_n \in S$, $f(\{u_1, \ldots, u_n\}) > 0$ whenever $f(\{u_0, u_1, \ldots, u_n\}) > 0$. We can then define the so-called n-th order *Papangelou conditional intensity* by

$$\lambda^{(n)}(u_1, \ldots, u_n, \mathbf{x}) = f(\mathbf{x} \cup \{u_1, \ldots, u_n\})/f(\mathbf{x}) \tag{2.6}$$

for pairwise distinct $u_1, \ldots, u_n \in S$ and $\mathbf{x} \in \mathcal{N} \setminus \{u_1, \ldots, u_n\}$, setting $0/0 = 0$. By the interpretation of f, $\lambda^{(n)}(u_1, \ldots, u_n, \mathbf{x}) du_1 \cdots du_n$ can be considered as the conditional probability of observing one event in each of the infinitesimally small balls B_i centered at u_i with volume du_i, conditional on that $\mathbf{X}$ outside $\cup_{i=1}^n B_i$ agrees with $\mathbf{x}$. When $n = 1$, we simply write $\lambda(u, \mathbf{x})$ and call this function the Papangelou conditional intensity of u given $\mathbf{x}$. When f is hereditary, there is a one-to-one correspondence between f and λ.

Because the notion of density for $\mathbf{Z}$ when $S = \mathbb{R}^d$ makes no sense, the Papangelou conditional intensity cannot be defined through a ratio of densities in $\mathbb{R}^d$. But it still makes sense as the n-th order Papangelou conditional intensity can actually be defined at the Radon–Nykodym derivative of $\mathrm{P}^!_{x_1, \ldots, x_n}$ with respect to P (see [28]).

A Poisson point process on S with intensity function ρ has n-th order Papangelou conditional intensity equal to

$$\lambda^{(n)}(u_1, \ldots, u_n, \mathbf{x}) = \prod_{i=1}^n \rho(u_i),$$

shedding again light on the fact that there is no interaction between points in a pattern generated by a Poisson point process.

Without further details, we mention the celebrated Georgii–Nguyen–Zessin formula (see [38, 62]), which states that for any $h : S^n \times \mathcal{N} \to \mathbb{R}$ (such that the following expectations are finite)

$$\mathrm{E} \sum_{u_1, \ldots, u_n \in \mathbf{X}}^{\neq} h(u_1, \ldots, u_n, \mathbf{X} \setminus \{u_1, \ldots, u_n\})$$

$$= \int_S \cdots \int_S \mathrm{E}\left(h(u_1, \ldots, u_n, \mathbf{X})\lambda^{(n)}(u_1, \ldots, u_n, \mathbf{X})\right) du_1 \ldots du_n. \tag{2.7}$$

By identification of (2.2) and (2.7), we see a link between the n-th order intensity and the n-th order Papangelou conditional intensity: for any pairwise distinct $u_1, \ldots, u_n \in S$

$$\rho^{(n)}(u_1, \ldots, u_n) = \mathrm{E}\left(\lambda^{(n)}(u_1, \ldots, u_n, \mathbf{X})\right).$$

We will return to the GNZ equation formula in connection to Gibbs processes in Sects. 2.3 and 2.5.

Figure 2.3 illustrates this section. For three different patterns exhibiting either repulsiveness (Fig. 2.3a,b) or attraction (Fig. 2.3c), the right plots represent the Papangelou conditional intensity $\lambda(u, \mathbf{x})$ for all $u \in W$. Specifically, Fig. 2.3a depicts a realization of a Strauss hard-core model, a Gibbs model for which points are pushed to be at a distance δ apart and repel up to a distance $R > \delta$, see Sect. 2.3.2 for details. Figure 2.3b,c show realizations of an area-interaction Gibbs model, see again Sect. 2.3.2 for details, that produces repulsive (resp. attractive) patterns when the underlying parameter θ is negative (resp. positive). To interpret the right plots of Fig. 2.3, think of where a new point is likely to appear: clearly in areas where $\lambda(u, \mathbf{x})$ is high, i.e. in the white areas. Figure 2.3a is very pedagogical. We see that outside the balls with radius R centered in $\mathbf{x}$, the Papangelou conditional intensity is constant, as expected. We also observe that a new point is forbidden within a distance δ to the point pattern. The Papangelou conditional intensity for an area-interaction point process is slightly more complex but the interpretation is similar: When $\theta < 0$, the repulsive case of Fig. 2.3b, a new point is likely to appear in the white areas, i.e. far from the points of $\mathbf{x}$. This is the converse for the attractive case of Fig. 2.3c where a new point is likely to appear close to the points of $\mathbf{x}$.

2.3 Examples on Standard Models of Spatial Point Processes

We describe in this section some widely used models of spatial point processes and we specify for each of them the intensities, Palm intensities and Papangelou conditional intensities, if these are explicitly known. Table 2.1 is a summary.

2.3.1 *Poisson Point Process*

The Poisson point process on S has already been introduced in Definition 2.1 and some realizations are shown in Fig. 2.1. This is the model for independence of the location of the events. It is characterized by its intensity function ρ, from which we deduce, see Sect. 2.2, the n-th order intensity function

$$\rho^{(n)}(u_1, \ldots, u_n) = \prod_{i=1}^{n} \rho(u_i), \quad \forall u_1, \ldots, u_n \in S,$$

the m-th order n-point Palm intensity function, given $x_1, \ldots, x_n$ in S,

$$\rho^{(m)}_{x_1, \ldots, x_n}(u_1, \ldots, u_m) = \prod_{i=1}^{m} \rho(u_i), \quad \forall u_1, \ldots, u_m \in S,$$

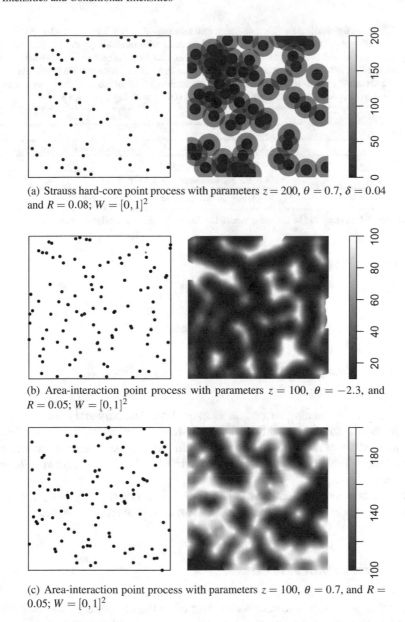

(a) Strauss hard-core point process with parameters $z = 200$, $\theta = 0.7$, $\delta = 0.04$ and $R = 0.08$; $W = [0,1]^2$

(b) Area-interaction point process with parameters $z = 100$, $\theta = -2.3$, and $R = 0.05$; $W = [0,1]^2$

(c) Area-interaction point process with parameters $z = 100$, $\theta = 0.7$, and $R = 0.05$; $W = [0,1]^2$

Fig. 2.3 Left plots: realization of point patterns in W from a Strauss hard-core model (**a**) and an area interaction point process (**b, c**). See Sect. 2.3.2 for a definition of these models. Right plots: Papangelou conditional intensity of the models, given the point pattern in the left hand side, see (2.12) and (2.13)

Table 2.1 Intensity function $\rho(u)$, one-point Palm intensity $\rho_x(u)$ and first order Papangelou conditional intensity $\lambda(u, \mathbf{x})$ for: a Poisson point process with intensity ρ; A Gibbs model with energy function H and activity parameter z; A log-Gaussian Cox process (LGCP) associated to a Gaussian random field with mean $\mu(u)$ and covariance $c(u, v)$; A Neymann Scott process (NSP) where the clusters' centres have intensity κ, the mean number of points per cluster is γ and the dispersion kernel is k_0; A determinantal point process (DPP) with kernel $C(u, v)$

Models	$\rho(u)$	$\rho_x(u)$	$\lambda(u, \mathbf{x})$
Poisson	$\rho(u)$	$\rho(u)$	$\rho(u)$
Gibbs	NA	NA	$z \exp(-(H(\mathbf{x} \cup \{u\}) - H(\mathbf{x})))$
LGCP	$\exp(\mu(u) + c(u, u)/2)$	$\exp(\mu(u) + c(u, u)/2 + c(u, x))$	NA
NSP	$\kappa\gamma$	$\kappa\gamma + \gamma \int_{\mathbb{R}^d} k_0(z)k_0(x - u + z)\mathrm{d}z$	NA
DPP	$C(u, u)$	$C(u, u) - |C(u, x)|^2/C(x, x)$	NA in general

The entries NA correspond to the cases where the formulas are not available in closed form

the one-point Palm intensity given $x \in S$

$$\rho_x(u) = \rho(u), \quad \forall u \in S,$$

and the n-th order Papangelou conditional intensity

$$\lambda^{(n)}(u_1, \ldots, u_n, \mathbf{x}) = \prod_{i=1}^{n} \rho(u_i), \quad \forall u_1, \ldots, u_n \in S, \forall \mathbf{x} \in \mathcal{N} \setminus \{u_1, \ldots, u_n\}.$$

Note that the conditioning on $x_1, \ldots, x_n$ in the Palm intensity and the conditioning on $\mathbf{x}$ in the Papangelou conditional intensity have no effect on the form of these functions. Morever all joint intensities and joint conditional intensities reduce to a product form. These observations confirm the independence property of the Poisson point process.

It is worth emphasizing that the Poisson point process is the only model for which we have an explicit and tractable expression for all characteristics above.

2.3.2 Gibbs Point Processes

Gibbs point processes have been widely treated in the literature. For a comprehensive presentation of Gibbs point processes as well as a detailed list of references, we refer to Chap. 5 by David Dereudre in this volume. These processes are characterized by an energy function H (or Hamiltonian) that maps any finite point configuration to $\mathbb{R} \cup \{\infty\}$. Specifically, if $|S| < \infty$, a Gibbs point process on S associated to H and with activity $z > 0$ admits the following density with respect to the unit rate Poisson process:

$$f(\mathbf{x}) \propto z^{|\mathbf{x}|} e^{-H(\mathbf{x})}, \tag{2.8}$$

where $\propto$ means "proportional to". This definition makes sense under some regularity conditions on H, typically non degeneracy ($H(\emptyset) < \infty$) and stability (there exists $A \in \mathbb{R}$ such that $H(\mathbf{x}) \geq A|\mathbf{x}|$ for any $\mathbf{x} \in \mathcal{N}$). Consequently, configurations $\mathbf{x}$ having a small energy $H(\mathbf{x})$ are more likely to be generated by a Gibbs point process than by a Poisson point process, and conversely for configurations having a high energy. In the extreme case where $H(\mathbf{x}) = \infty$, then $\mathbf{x}$ cannot, almost surely, be the realization of a Gibbs point process associated to H.

If $|S| = \infty$, the above definition does not make sense in general since $H(\mathbf{x})$ can be infinite or even undefined if $|\mathbf{x}| = \infty$. In this case a Gibbs point process is defined through its local specifications, which are the conditional densities on any bounded set Δ, given the outside configuration on Δ^c, with respect to the unit rate Poisson process on Δ. These conditional densities take a similar form as in (2.8), where now the Hamiltonian H becomes a family of Hamiltonian functions H_Δ that quantify the energy of $\mathbf{x}_\Delta$ given the outside configuration $\mathbf{x}_{\Delta^c}$. Some supplementary regularity assumptions on the family of H_Δ's are necessary to ensure the existence of a point process satisfying these local specifications. Again, we refer to Chap. 5 by David Dereudre for more details. The interpretation nonetheless remains similar: a (infinite) Gibbs point process associated to H_Δ tends to favor configurations $\mathbf{x}_\Delta$ on Δ having a small value $H_\Delta(\mathbf{x})$.

Gibbs point processes offer a great flexibility of modelling, depending on the choice of the energy function H that favors or penalizes some point patterns' features. This function can in particular encode inhibition between the points, or attraction, or even both inhibition and attraction depending on the distance between the points. Let us give two popular examples. They are defined hereafter for simplicity on a bounded set S, but their definition extends to $S = \mathbb{R}^d$, provided we carefully account for edge effects in the conditional specifications.

Strauss Hard-Core Model on a Bounded Set S This model is a particular case of a pairwise interaction model, for which

$$H(\mathbf{x}) = \sum_{\{u,v\} \in \mathbf{x}} \phi(u - v), \tag{2.9}$$

where ϕ is a symmetric function called the pair potential. The potential of a hardcore Strauss model depends on some hardcore parameter $\delta \geq 0$, some interaction parameter $\theta \geq 0$ and some range of interaction $R > \delta$ in the following way

$$\phi(u) = \begin{cases} \infty & \text{if} \quad |u| < \delta, \\ \theta & \text{if} \quad \delta \leq |u| < R, \\ 0 & \text{if} \quad |u| \geq R. \end{cases} \tag{2.10}$$

As encoded in the potential, this model only allows configurations where all points are δ-apart. Moreover it penalizes configurations with too many pairs of points that are R-apart. This model thus generates inhibitive point patterns. A realization is shown in Fig. 2.3.

Area-Interaction Model on a Bounded Set S The Hamiltonian H of the area-interaction process is defined for the radius $R \geq 0$ and the interacting parameter $\theta \in \mathbb{R}$ by

$$H(\mathbf{x}) = \theta \left| \bigcup_{u \in \mathbf{x}} B(u, R) \right|, \qquad (2.11)$$

where $B(u, R)$ is the ball centered at u with radius R. Depending on the sign of θ, this model yields inhibition or clustering. If $\theta > 0$, a small value of $H(\mathbf{x})$ is achieved if the union of balls in (2.11) has a small area, which happens when many balls intersect or equivalently if the points of $\mathbf{x}$ are close to each other. Conversely if $\theta < 0$, this model generates configurations with a large area in (2.11), which means that some inhibition occurs between the points to avoid intersections between the associated balls. Some realizations (in both cases) are represented in Fig. 2.3.

As illustrated in the above examples, Gibbs models provide a clear and interpretable way to introduce interactions between the points of a point pattern, these interactions being encoded in the energy function. Unfortunately, very few characteristics of a Gibbs process are explicitly known. The main reason is that the normalizing constant in (2.8) is in general intractable, making impossible the computation of the intensities and of the Palm intensities of a Gibbs process. However, from (2.6) and (2.8), this normalization cancels in the n-th order Papangelou conditional intensity to provide

$$\lambda^{(n)}(u_1, \ldots, u_n, \mathbf{x}) = z \, e^{-(H(\mathbf{x} \cup \{u_1, \ldots, u_n\}) - H(\mathbf{x}))}.$$

For instance, when $n = 1$, we obtain for a pair potential (2.9)

$$\lambda(u, \mathbf{x}) = z \, e^{-\theta \sum_{v \in \mathbf{x}} \phi(v - u)}$$

for all $u \in S$ and $\mathbf{x} \in \mathcal{N}$ such that $H(\mathbf{x}) < \infty$, while $\lambda(u, \mathbf{x}) = 0$ if $H(\mathbf{x}) = \infty$. This becomes for the Strauss hard-core model (2.10), denoting by $n_R(u, \mathbf{x})$ the number of R-closed neighbours of u in $\mathbf{x}$,

$$\lambda(u, \mathbf{x}) = z \, e^{-\theta n_R(u, \mathbf{x})} \qquad (2.12)$$

if all points in $\mathbf{x} \cup \{u\}$ are δ-apart and $\lambda(u, \mathbf{x}) = 0$ otherwise. For the area interaction model (2.11), we get

$$\lambda(u, \mathbf{x}) = z \, e^{-\theta \left(\left| \bigcup_{v \in \mathbf{x} \cup \{u\}} B(v, R) \right| - \left| \bigcup_{v \in \mathbf{x}} B(v, R) \right| \right)}. \qquad (2.13)$$

2.3.3   Cox Processes

Let $(\Lambda(u))_{u \in S}$ be a non-negative random field. A point process is a Cox process driven by Λ if conditional on Λ, it is a Poisson point process with intensity Λ. The existence is ensured if Λ is locally integrable almost surely, i.e. $\forall B \in \mathcal{B}_0$, $\int_B \Lambda(u)du < \infty$ almost surely.

If the random field Λ is not a constant random variable over S, the realization of a Cox process generates clusters of points, located around the highest values of the realization of Λ.

In general, very few mathematical properties are available for a Cox process. For instance we easily obtain that the n-th order intensity function is

$$\rho^{(n)}(u_1, \ldots, u_n) = \mathrm{E}\left(\prod_{i=1}^n \Lambda(u_i)\right) \tag{2.14}$$

but this expectation is in general not tractable. We focus below on two special cases of Cox processes, namely the log-Gaussian Cox processes (LGCP) and the shot noise Cox processes (SNCP) for which (2.14) become computable. On the other hand, the density of a Cox process on a bounded S ($|S| < \infty$) with respect to the unit rate Poisson process is

$$f(\mathbf{x}) = \mathrm{E}\left\{\exp\left(|S| - \int_S \Lambda(u)du\right) \prod_{u \in \mathbf{x}} \Lambda(u)\right\} \tag{2.15}$$

which is intractable, even for the LGCP and SNCP detailed below. Consequently the Papangelou conditional intensity is unknown for these models.

2.3.3.1   Log-Gaussian Cox Processes

These processes correspond to the particular case $\Lambda(u) = \exp(Y(u))$, $u \in S$, where Y is a Gaussian random field with mean function $\mu(u)$ and covariance function $c(u, v)$. These models are often used for modelling spatially correlated latent uncertainty. Some realizations are shown in Figs. 2.1 and 2.2.

For an LGCP, (2.14) becomes tractable, see [61]. We obtain for $n = 1$ that the intensity function is given by

$$\rho(u) = \exp(\mu(u) + c(u, u)/2), \quad \forall u \in S,$$

and combined with the case $n = 2$, we deduce the pair correlation function

$$g(u, v) = \exp(c(u, v)), \quad \forall u, v \in S.$$

We have that $g \geq 1$ if $c(u, v) \geq 0$, confirming the clustering property of an LGCP. Note that this attractive interaction is controlled through the second order moment of Y, which is a global feature, and not through the local interaction between pairs of points, as it is the case for pairwise interaction Gibbs point process with the energy function. For this reason an LGCP is not a mechanistic model in the sense that it does not describe the physical process generating the points.

More generally, the n-th order intensity function writes

$$\rho^{(n)}(u_1, \ldots, u_n) = \prod_{i=1}^{n} \rho(u_i) \prod_{1 \leq i < j \leq n} g(u_i, u_j), \quad \forall u_1, \ldots, u_n \in S.$$

From (2.4), we further obtain the m-th order n-point Palm intensity function, given $x_1, \ldots, x_n$ in S,

$$\rho_{x_1, \ldots, x_n}^{(m)}(u_1, \ldots, u_m) = \prod_{i=1}^{m} \rho_{x_1, \ldots, x_n}(u_i) \prod_{1 \leq i < j \leq m} g(u_i, u_j), \quad \forall u_1, \ldots, u_m \in S,$$

where

$$\rho_{x_1, \ldots, x_n}(u) = \rho(u) \prod_{i=1}^{n} g(u, x_i), \quad \forall u \in S,$$

corresponds to the case $m = 1$. In particular the one-point Palm intensity given $x \in S$ is

$$\rho_x(u) = \rho(u) g(u, x), \quad \forall u \in S.$$

The Palm intensities can also be derived from a stronger result proved in [24], which establishes that the n-th order reduced Palm distribution of an LGCP, given $x_1, \ldots, x_n$ in S, is still an LGCP associated to the Gaussian field $(Y(u) + \sum_{i=1}^{n} c(u, x_i))_{u \in S}$.

2.3.3.2   Shot Noise Cox Processes

They are Cox processes associated to

$$\Lambda(u) = \sum_{(c, \gamma) \in \Phi} \gamma \, k(c, u)$$

where Φ is a Poisson point process on $S \times [0, \infty)$ with intensity measure ζ and k is a kernel on S^2, i.e. $k(c, .)$ is a density function on S for all $c \in S$.

An SNCP can be viewed as a Poisson cluster process where the random set of centres (or parents) consists in the set of c's and $k(c, .)$ stands for a dispersion density around the centre c. The intensity of each cluster is encoded in γ. From this point of view, unlike LGCPs, an SNCP is a mechanistic model that provides a clear interpretation of the generating process.

It is verified in [56, Proposition 1] that the first intensity function of an SNCP is

$$\rho(u) = \int \gamma \, k(c, u) d\zeta(c, \gamma), \quad \forall u \in S,$$

and the pair correlation function writes

$$g(u, v) = 1 + \frac{\beta(u, v)}{\rho(u)\rho(v)}, \quad \forall u, v \in S$$

where

$$\beta(u, v) = \int \gamma^2 k(c, u)k(c, v) d\zeta(c, \gamma),$$

provided these integrals are finite. The fact that $g \geq 1$ confirms the clustering property of an SNCP. We deduce the one-point Palm intensity given $x \in S$

$$\rho_x(u) = \rho(u)g(u, x), \quad \forall u \in S.$$

Higher order intensities and Palm intensities are less tractable.

A widely used class of SNCP is the class of Neymann Scott processes defined on $S = \mathbb{R}^d$. They correspond to the case $k(c, u) = k_0(u - c)$, where k_0 is a density, and $\Phi = \Phi_c \times \{\gamma\}$ where $\gamma > 0$ and Φ_c is a homogeneous Poisson process with intensity $\kappa > 0$. So the point process of intensities is constant equal to γ and is independent of the locations of the centres. The resulting process is stationary. The Thomas process is the special case where k_0 is a zero-mean normal density, and the Matérn cluster process is the example where k_0 is a uniform density on a ball centered at the origin. For Neymann Scott processes the above formulae become

$$\rho(u) = \kappa\gamma, \quad \forall u \in \mathbb{R}^d,$$

and given $x \in \mathbb{R}^d$,

$$\rho_x(u) = \kappa\gamma + \gamma \int_{\mathbb{R}^d} k_0(z)k_0(x - u + z)dz, \quad \forall u \in \mathbb{R}^d.$$

2.3.4 Determinantal Point Processes

Determinantal point processes (DPPs) are models for inhibitive point patterns, introduced in their general form by Macchi [51]. We refer to [49] for their main statistical properties. They are defined through a kernel C which is a function from $S \times S$ to $\mathbb{C}$. A point process is a DPP on S with kernel C if for any n, its n-th order intensity function takes the form

$$\rho^{(n)}(u_1, \ldots, u_n) = \det[C](u_1, \ldots, u_n), \tag{2.16}$$

for every $(u_1, \ldots, u_n) \in S^n$, where $[C](u_1, \ldots, u_n)$ denotes the matrix with entries $C(u_i, u_j)$, $1 \leq i, j \leq n$. Sufficient conditions for existence are that C be a continuous covariance function on $\mathbb{R}^d \times \mathbb{R}^d$ and further satisfies a spectral condition. In the homogeneous case where $C(u, v) = C_0(v - u)$, this spectral condition is $C_0 \in L^2(\mathbb{R}^d)$ and $\mathscr{F}(C_0) \leq 1$ where $\mathscr{F}$ denotes the Fourier transform. In the inhomogeneous case, a sufficient condition is that there exists a covariance function C_0 as before such that $C_0(v - u) - C(u, v)$ remains a covariance function.

Repulsiveness of a DPP is deduced from (2.16): by continuity of C and the determinant, $\rho^{(n)}(u_1, \ldots, u_n)$ tends to 0 whenever $u_i \approx u_j$ for some $i \neq j$, showing that a DPP is unlikely to generate too close points. The intensity and the pair correlation of a DPP are easily deduced

$$\rho(u) = C(u, u), \qquad g(u, v) = 1 - \frac{|C(u, v)|^2}{C(u, u)C(v, v)},$$

and the repulsiveness property is confirmed by the fact that $g \leq 1$. These two expressions also give a clear interpretation of the kernel C: the diagonal of C encodes the intensity of the process, while the off-diagonal encodes the interaction between two points of the process. Like for LGCPs, this interaction is only controlled globally through the second order moment of the process, so DPPs are not mechanistic models. Finally, a DPP cannot be extremely inhibitive. In particular, a DPP cannot include a hardcore distance δ (this is a consequence of [69], Corollary 1.4.13), a situation where all points are at least δ-apart. The most repulsive DPP (in a certain sense) is determined in [49] and [11], where it is demonstrated that DPPs can nonetheless model a large class of point patterns exhibiting inhibition. Realizations of DPPs where C is a Gaussian covariance function are represented in Fig. 2.1c, f, and i.

An appealing feature of DPPs is that many theoretical properties are available. In particular, by the definition (2.16), the n-th order intensity function is explicitly known. Further, the m-th order n-point Palm intensity function, given $x_1, \ldots, x_n$ in S, is deduced from (2.4) and (2.16) and becomes after some algebra (see [71] Corollary 6.6)

$$\rho^{(m)}_{x_1, \ldots, x_n}(u_1, \ldots, u_m) = \det[C_{x_1, \ldots, x_n}](u_1, \ldots, u_m), \quad \forall u_1, \ldots, u_m \in S,$$

where

$$
C_{x_1,\ldots,x_n}(u, v) = \frac{1}{\det[C](x_1, \ldots, x_n)} \det
\begin{bmatrix}
C(u, v) & C(u, x_1) & \ldots & C(u, x_n) \\
C(x_1, v) & C(x_1, x_1) & \ldots & C(x_1, x_n) \\
\vdots & \vdots & \ddots & \vdots \\
C(x_n, v) & C(x_n, x_1) & \ldots & C(x_n, x_n)
\end{bmatrix}.
$$

This expression shows that the n-th order reduced Palm distribution of a DPP, given $x_1, \ldots, x_n$, is still a DPP with kernel $C_{x_1,\ldots,x_n}$. In particular, the one-point Palm intensity given $x \in S$ is

$$
\rho_x(u) = C(u, u) - \frac{|C(u, x)|^2}{C(x, x)}, \quad \forall u \in S.
$$

Concerning the n-th order Papangelou conditional intensity, its expression is

$$
\lambda^{(n)}(u_1, \ldots, u_n, \mathbf{x}) = \det[\tilde{C}](\mathbf{x} \cup \{u_1, \ldots, u_n\}) / \det[\tilde{C}](\mathbf{x}),
$$

which depends on a new kernel $\tilde{C}$ related to C by the integral equation

$$
\tilde{C}(u, v) - \int_S \tilde{C}(u, t)C(t, v)dt = C(u, v), \quad \forall u, v \in S.
$$

The solution $\tilde{C}$ is in general unknown and for this reason the Papangelou conditional intensity of a DPP does not straightforwardly follow from the knowledge of C. However, some approximations in the case where S is a rectangular region and $C(u, v) = C_0(v - u)$ are available, see [49].

2.4 Estimating the Intensity Function

In Sects. 2.4 and 2.5, we are given the realization of a spatial point process $\mathbf{X}$ defined on S. We will present very briefly asymptotic results. Asymptotics is understood as increasing-domain asymptotics. So, we assume that $\mathbf{X}$ is well-defined on $\mathbb{R}^d$ and is observed in W_n, where $(W_n)_{n \geq 1}$ is a sequence of bounded regular domains with $|W_n| \to \infty$ as $n \to \infty$. According to the problems, contexts and estimators, assumptions for the sequence $(W_n)_{n \geq 1}$ are different, but to ease the presentation think W_n as the domain $[-n, n]^d$. To lighten the text, we will also assume that the quantities used hereafter are well-defined; for instance, if an estimator has a variance depending on the pair correlation function g, it is intrinsically assumed that the second order intensity of $\mathbf{X}$ is well-defined.

In this section, we focus on estimating the intensity function. Three different settings will be considered: first, we consider the case when ρ is constant; second, the case where ρ is a d-dimensional function not specified by any parameter; third, the case where ρ is a function with a parametric form.

Because the different methodologies often start from methodologies for the reference model, i.e. the Poisson process, we mention the following result. Let $\mathbf{X}$ be a Poisson point process with intensity function ρ defined on $\mathbb{R}^d$ and observed in W_n, then $\mathbf{X}$ restricted to W_n admits a density w.r.t the unit rate Poisson process defined on W_n and the density, denoted by f, writes for $\mathbf{x} \in \mathcal{N}$

$$f(\mathbf{x}) = \exp\left(|W_n| - \int_{W_n} \rho(u)\mathrm{d}u\right) \prod_{u \in \mathbf{x}} \rho(u).$$

2.4.1   Constant Intensity

2.4.1.1   Poisson Case

When ρ is constant and $\mathbf{X}$ is a Poisson point process, the density reduces to $f(\mathbf{x}) = \exp(|W_n| - \rho|W_n|)\rho^{|\mathbf{x}|}$ and the maximum likelihood estimator (MLE) of ρ is therefore $\hat{\rho} = N(W_n)/|W_n|$. It is of course an unbiased estimator of ρ, with variance $\rho/|W_n|$ and as $n \to \infty$, $|W_n|^{1/2}(\hat{\rho} - \rho) \to N(0, \rho)$ in distribution. To build asymptotic confidence intervals, the variance stabilizing transform is sometimes used. In this setting, this leads to $|W_n|^{1/2}(\sqrt{\hat{\rho}} - \sqrt{\rho}) \to N(0, 1)$ in distribution. The MLE is, as expected, the optimal estimator, among unbiased estimators. To end this paragraph, we mention the work by Clausel et al. [14] which provides a Stein estimator of ρ. A Stein estimator is a biased estimator with mean-squared error smaller than the one of the MLE. The authors considered in particular the following estimator obtained as a small perturbation of $\hat{\rho}$: assume here that $|W_n| = B(0, 1)$ (the d-dimensional Euclidean ball)

$$\tilde{\rho} = \hat{\rho} - \gamma D_k (1 - D_k)$$

where γ is a real parameter and where D_k is the squared distance to zero of the k-th closest point to zero (set to 1 if k is larger than the number of observed points). Clausel et al. [14] showed that it is possible to adjust k and γ such that the mean-squared error of $\tilde{\rho}$ is smaller than the one of $\hat{\rho}$. Note that the bias of $\tilde{\rho}$, depending on ρ, can in principle be estimated. However, the resulting "debiased" estimator is not guaranteed to outperform the MLE in terms of mean-squared error.

2.4.1.2   General Case

In the general case, the estimator $\hat{\rho} = N(W_n)/|W_n|$ is the standard estimator: it actually corresponds to the estimator obtained from the following estimating equation

$$\sum_{u \in \mathbf{X} \cap W_n} 1 - \int_{W_n} \rho \mathrm{d}u = N(W_n) - \rho|W_n|$$

an equation which is easily shown to be unbiased using (2.2), leading to an estimator that is also unbiased. Under mixing conditions on $\mathbf{X}$ and the assumption that $\mathbf{X}$ is a second-order reweighted stationary point process [44], have shown that $|W_n|^{1/2}(\hat{\rho} - \rho) \to N(0, \sigma^2)$ where

$$\sigma^2 = \rho + \rho^2 \int_{\mathbb{R}^d} (g(u, o) - 1)du, \tag{2.17}$$

where o is the origin of $\mathbb{R}^d$. Equation (2.17) reveals an interesting fact: when $g(u, v) > 1$ (respectively $g(u, v) < 1$) for any u, v, i.e. for attractive (respectively repulsive) patterns, the asymptotic variance is larger (respectively smaller) than ρ, the asymptotic variance in the Poisson case. Heinrich and Prokešová [44] also provided the following estimator of σ^2 and established its asymptotic mean-squared error:

$$\widehat{\sigma}^2 = \widehat{\rho} + \sum_{u, v \in \mathbf{X} \cap W_n}^{\neq} \frac{\ker\left(\frac{v-u}{|W_n|^{1/d} b_n}\right)}{|(W_n - u) \cap (W_n - v)|}$$

$$- |W_n| b_n^d \hat{\rho}(\hat{\rho} - |W_n|^{-1}) \int_{W_n} \ker(u)\, du \tag{2.18}$$

where $\ker : \mathbb{R}^d \to [0, \infty)$ plays the role of a kernel and $(b_n)_{n \geq 1}$ is a sequence of real numbers playing the role of a bandwidth.

We end this section with a recent work by Coeurjolly [15] dealing with a robust estimator of ρ. Assume the domain of observation W_n can be decomposed as $W_n = \cup_{k \in \mathscr{K}_n} C_{n,k}$ where the cells $C_{n,k}$ are non-overlapping and equally sized with volume $c_n = |C_{n,k}|$ and where $\mathscr{K}_n$ is a subset of $\mathbb{Z}^d$ with finite cardinality $k_n = |\mathscr{K}_n|$. The standard estimator of ρ obviously satisfies $\hat{\rho} = k_n^{-1} \sum_{k \in \mathscr{K}_n} N(\mathbf{X} \cap C_{n,k})/c_n$ since $|W_n| = k_n c_n$, i.e. $\hat{\rho}$ is nothing else than the sample mean of intensity estimators computed in cells $C_{n,k}$. The strategy adopted by Coeurjolly [15] is to replace the sample mean by the sample median, which is known to be more robust to outliers. Quantile estimators based on count data or more generally on discrete data can cause some troubles in the asymptotic theory (see e.g. [30]). To bypass the discontinuity problem of the count variables $N(\mathbf{X} \cap C_{n,k})$, we follow a well-known technique (e.g. [52]) which introduces smoothness. Let $(U_k, k \in \mathscr{K}_n)$ be a collection of independent and identically distributed random variables, distributed as $U \sim \mathscr{U}([0, 1])$. Then, for any $k \in \mathscr{K}_n$, we define $J_{n,k} = N(\mathbf{X} \cap C_{n,k}) + U_k$ and $\mathbf{J} = (J_{n,k}, k \in \mathscr{K}_n)$. The jittering effect shows up right away: the $J_{n,k}$ admit a density at any point. The jittered median-based estimator of ρ is finally defined by

$$\hat{\rho}^{\text{med}} = \frac{\widehat{\text{Me}(\mathbf{J})}}{c_n} \tag{2.19}$$

where $\widehat{\mathrm{Me}}(\mathbf{J})$ stands for the sample median based on $\mathbf{J}$. The estimator $\hat{\rho}^{\mathrm{med}}$ is shown to be consistent and asymptotically normal and from a numerical point of view to be more robust to outliers (areas where points are unnaturally too abundant or absent). First results were mainly available for Neymann-Scott processes and LGCPs. Biscio and Coeurjolly [10] have extended them to DPPs.

2.4.2 Non Parametric Estimation of the Intensity Function

To estimate non parametrically the intensity function, the standard estimator is a kernel estimator. The analogy proposed by Baddeley et al. [5, Section 6.5.1] gives a very nice intuition of kernel intensity estimation: "imagine placing one square of chocolate on each data point. Using a hair dryer, we apply heat to the chocolate so that it melts slightly. The result is an undulating surface of chocolate; the height of the surface represents the estimated intensity function. The total mass of the chocolate is unchanged".

Let $\ker : \mathbb{R}^d \to [0, \infty)$ be a symmetric kernel function with integral one (over $\mathbb{R}^d$) and let $(b_n)_{n \geq 1}$ be a sequence of real numbers playing the role of bandwidths. Example of kernels are: the Gaussian kernel $(2\pi)^{-d/2} \exp(-\|y\|^2/2)$ or the product of Epanecnikov kernels $\prod_{i=1}^{d} \frac{3}{4}\mathbf{1}(|y_i| \leq 1)(1-|y_i|^2)$. Let $\ker_{b_n}(w) = \ker(w/b_n)b_n^{-d}$, $w \in \mathbb{R}^d$. The standard estimators are the uniformly corrected $\hat{\rho}^U$ (see [6, 34]), and the locally corrected $\hat{\rho}^L$ (suggested by Van Lieshout [77]). These estimators are respectively given for $u \in W_n$ by

$$\hat{\rho}^U(u) = \frac{1}{e(u)} \sum_{v \in \mathbf{X} \cap W_n} \ker_{b_n}(v - u) \quad \text{and} \quad \hat{\rho}^L(u) = \sum_{v \in \mathbf{X} \cap W_n} \frac{\ker_{b_n}(v - u)}{e(v)}$$

where e is an edge correction factor given by $e(w) = \int_{W_n} \ker_{b_n}(w - t)\mathrm{d}t$, for $w \in W_n$. As first outlined by Diggle [34], non parametric intensity estimators of the intensity have obvious similarities with probability density estimators introduced by Rosenblatt [68]. The distinction relies on the edge-correction factor and on the more subtle distinction that for point process data the observation domain is W_n but (most of the time) the model exists in $\mathbb{R}^d$ but is unobserved on $\mathbb{R}^d \setminus W_n$. A further difference is that the points of a sample pattern can in general no longer be seen as an i.i.d. sample from a common probability density.

We want to give hereafter the main idea why it works. Consider the non parametric estimator $\hat{\rho}^U$. By the Campbell-Mecke formula (2.2)

$$\mathrm{E}\hat{\rho}^U(u) = e(u)^{-1} \int_{W_n} \ker_{b_n}(v - u)\rho(v)\mathrm{d}v$$

$$= e(u)^{-1} \int_{(W_n - u)/b_n} \ker(w)\rho(b_n w + u)\mathrm{d}w$$

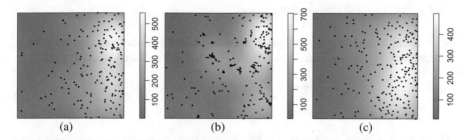

Fig. 2.4 Diggle's non parametric estimation of the intensity function $\rho = \rho_{\text{cubic}}$ described in Fig. 2.1 for the point patterns from Fig. 2.1d–f. The kernel used is the Gaussian kernel and the bandwidth is chosen using (Poisson) likelihood cross-validation (argument `bw.ppl` in the `density.ppp` function). (**a**) Poisson. (**b**) LGCP. (**c**) DPP

$$\approx e(u)^{-1} \int_{(W_n - u)/b_n} \ker(w)\rho(u)\mathrm{d}w = e(u)^{-1}\rho(u) \int_{W_n} \ker_{b_n}(v)\mathrm{d}v = \rho(u)$$

where the approximation is valid for b_n small and under regularity conditions on ρ. A similar argument can be used for $\hat{\rho}^L$. Van Lieshout [77] showed that $\hat{\rho}^L$ has better statistical performance than $\hat{\rho}^U$. As usual, the choice of bandwidth involves a tradeoff between bias and variance. Several data-driven procedures exist to select the optimal bandwidth, see e.g. [5, 6] or the recent promising work by Cronie and Van Lieshout [27] that offers a non-model based way of choosing this parameter.

Many other methods have been introduced to estimate ρ like spatially adaptive smoothing, nearest-neighbour based methods, tessellations based approaches, Bayesian estimation. We refer the interested reader to [5] and the many references therein. Following the analogy with density estimation, non parametric estimates of ρ can be obtained using the function `density` in the `spatstat` R package. This section is illustrated by Fig. 2.4.

2.4.3 Parametric Estimation of the Intensity Function

In this section, we assume that the intensity function has a parametric form, and in particular we assume that its form is log-linear:

$$\log \rho(u; \boldsymbol{\theta}) = \boldsymbol{\theta}^\top \mathbf{z}(u) \tag{2.20}$$

where $p \geq 1$, $\boldsymbol{\theta} \in \Theta$, Θ is an open bounded set of $\mathbb{R}^p$ and $\mathbf{z}(u) = (z_1(u), \ldots, z_p(u))^\top$, $u \in \mathbb{R}^d$. The functions $z_i : \mathbb{R}^d \to \mathbb{R}$ play the role of basis functions or covariates. For instance, the models for the intensity functions of Fig. 2.1c–e and h, i are respectively given by:

- $p = 3, \boldsymbol{\theta} = (\beta, 1, .5)^\top, \mathbf{z}(u) = (1, u_1, u_1^3)^\top$.
- $p = 2, \boldsymbol{\theta} = (\beta, .5)^\top, \mathbf{z}(u) = (1, \text{grad}(u))^\top$, where grad is the covariate corresponding to the slope of elevation.

2.4.3.1 Poisson Likelihood Estimator

For an inhomogeneous Poisson point process with log-linear intensity function ρ parameterized by $\boldsymbol{\theta}$, the likelihood function (up to a normalizing constant) reads

$$\ell(\boldsymbol{\theta}) = \sum_{u \in \mathbf{X} \cap W_n} \boldsymbol{\theta}^\top \mathbf{z}(u) - \int_{W_n} \exp(\boldsymbol{\theta}^\top \mathbf{z}(u)) \mathrm{d}u. \qquad (2.21)$$

[66] showed that the maximum likelihood estimator is consistent, asymptotically normal and asymptotically efficient as the sample region goes to $\mathbb{R}^d$.

Let $\boldsymbol{\theta}_0$ be the true parameter vector. By applying Campbell theorem (2.2) to the score function, i.e. the gradient vector of $\ell(\boldsymbol{\theta})$ denoted by $\ell^{(1)}(\boldsymbol{\theta})$, we have

$$\mathbb{E}\ell^{(1)}(\boldsymbol{\theta}) = \int_{W_n} \mathbf{z}(u)(\exp(\boldsymbol{\theta}_0^\top \mathbf{z}(u)) - \exp(\boldsymbol{\theta}^\top \mathbf{z}(u))) \mathrm{d}u = 0$$

when $\boldsymbol{\theta} = \boldsymbol{\theta}_0$. So, the score function of the Poisson log-likelihood appears to be an unbiased estimating equation, even though $\mathbf{X}$ is not a Poisson point process. The estimator maximizing (2.21) is referred to as the Poisson likelihood estimator. The properties of the Poisson estimator have been carefully studied. Schoenberg [70] showed that the Poisson estimator is still consistent for a class of spatio-temporal point process models. The infill asymptotic normality was obtained by Waagepetersen [78] while [41] established asymptotic normality under an increasing domain assumption and for suitable mixing point processes.

To improve the Poisson likelihood estimator when $\mathbf{X}$ is not a Poisson point process [42], proposed to consider the weighted Poisson log-likelihood function

$$\ell(w; \boldsymbol{\theta}) = \sum_{u \in \mathbf{X} \cap W_n} w(u) \log \rho(u; \boldsymbol{\theta}) - \int_{W_n} w(u) \rho(u; \boldsymbol{\theta}) \mathrm{d}u, \qquad (2.22)$$

where $w : \mathbb{R}^d \to \mathbb{R}$ is a weight surface whose objective is to reduce the asymptotic covariance matrix. Its score

$$\ell^{(1)}(w; \boldsymbol{\theta}) = \sum_{u \in \mathbf{X} \cap W_n} w(u) \mathbf{z}(u) - \int_{W_n} w(u) \mathbf{z}(u) \rho(u; \boldsymbol{\theta}) \mathrm{d}u, \qquad (2.23)$$

is easily seen to be again an unbiased estimating equation by Campbell theorem. Guan and Shen [42] proved that, under some conditions, the parameter estimates are consistent and asymptotically normal. In particular, it is shown that, in distribution

$$\boldsymbol{\Sigma}^{-1/2}(\hat{\boldsymbol{\theta}} - \boldsymbol{\theta}_0) \to N(0, \mathbf{I}_p) \text{ where } \boldsymbol{\Sigma} = \mathbf{A}^{-1}(\mathbf{B} + \mathbf{C})\mathbf{A}^{-1}$$

with

$$\mathbf{A} = \int_{W_n} w(u)\mathbf{z}(u)\mathbf{z}(u)^\top \rho(u; \boldsymbol{\theta}_0)du, \quad \mathbf{B} = \int_{W_n} w(u)^2\mathbf{z}(u)\mathbf{z}(u)^\top \rho(u; \boldsymbol{\theta}_0)du,$$

$$\mathbf{C} = \int_{W_n} \int_{W_n} w(u)w(v)\mathbf{z}(u)\mathbf{z}(v)^\top \rho(u; \boldsymbol{\theta}_0)\rho(v; \boldsymbol{\theta}_0)\,(g(u, v) - 1)\,dudv.$$

Guan and Shen [42] defined the optimal weight surface as the function w minimizing the trace of the asymptotic covariance matrix $\boldsymbol{\Sigma}$. They also came up with the following estimate of the optimal weight surface

$$\hat{w}(u) = \left(1 + \rho(u; \hat{\theta})(\hat{K}(r) - |B(0, r)|)\right)^{-1}$$

where r is the practical range of interaction the model (typically estimated by examining an empirical pair correlation function plot), $|B(0, r)|$ is the volume of the d-dimensional Euclidean ball centered at 0 with radius r and $\hat{K}(r)$ is the inhomogeneous K-function estimator (see e.g. [58]).

The problem of estimating the asymptotic covariance matrix is crucial to derive confidence intervals. The problem is not straightforward as $\boldsymbol{\Sigma}$ involves unknown second-order characteristics of $\mathbf{X}$. Guan [40], Guan and Shen [42] considered a block bootstrap procedure to achieve this problem while [17] extended (2.18) to the inhomogeneous case.

Figure 2.5 illustrates this section and depicts for the three different patterns parametric estimated intensities and relative errors.

2.4.3.2 Numerical Implementation of the Weighted Poisson Log-Likelihood

[7] developed a numerical quadrature method to approximate the Poisson likelihood (the method is actually the same for the weighted Poisson likelihood) Suppose we approximate the integral term in (2.22) by Riemann sum approximation

$$\int_{W_n} w(u)\rho(u; \boldsymbol{\theta})du \approx \sum_{i=1}^M v_i w(u_i)\rho(u_i; \boldsymbol{\theta})$$

where $u_i, i = 1, \ldots, M$ are points in W_n consisting of the $|\mathbf{x}|$ data points and $M - |\mathbf{x}|$ dummy points. The quadrature weights $v_i > 0$ sum to W_n. Examples of such weights are the volumes of the Voronoi cells obtained from the quadrature points $u_i, i = 1, \ldots, M$, see e.g. [5]. Thus, the weighted Poisson log-likelihood function (2.22) can be approximated and rewritten as

$$\ell(w; \boldsymbol{\theta}) \approx \sum_{i=1}^M v_i w_i \{y_i \log \rho_i - \rho_i\}, \tag{2.24}$$

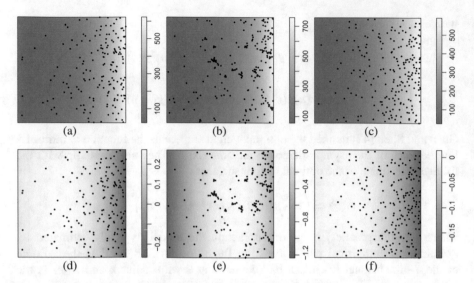

Fig. 2.5 Point patterns considered in this figure are the ones from Fig. 2.1d–f. The first row correponds to the estimated intensity function while the second one depicts the relative errors, i.e. the image with values $(\rho(u; \boldsymbol{\theta}) - \rho(u; \hat{\boldsymbol{\theta}}))/\rho(u; \boldsymbol{\theta})$. The intensity functions are parametrically estimated using the Poisson likelihood estimator based on (2.21). (**a**) Poisson. (**b**) LGCP. (**c**) DPP. (**d**) Poisson. (**e**) LGCP. (**f**) DPP

where $y_i = v_i^{-1}\mathbf{1}(u_i \in \mathbf{x})$, $w_i = w(u_i)$ and $\rho_i = \rho(u_i; \boldsymbol{\theta})$. Equation (2.24) corresponds to a quasi Poisson log-likelihood function. Maximizing (2.24) under the log-linear assumption (2.20) is equivalent to fitting a weighted Poisson generalized linear model, which can be performed using standard statistical software. This fact is in particular exploited in the ppm function in the spatstat R package [5] with option method="mpl".

2.4.3.3   Logistic Regression Likelihood

To perform well, the Berman–Turner approximation requires a large number of dummy points, actually a much larger number than the number of data points. Fitting such generalized linear models can be computationally intensive. Waagepetersen [79] proposed to consider the following contrast

$$\sum_{u \in \mathbf{X} \cap W_n} w(u) \log \left(\frac{\rho(u; \boldsymbol{\theta})}{\delta(u) + \rho(u; \boldsymbol{\theta})} \right) + \sum_{u \in \mathbf{Y} \cap W_n} w(u) \log \left(\frac{\delta(u)}{\rho(u; \boldsymbol{\theta}) + \delta(u)} \right),$$

(2.25)

where $\mathbf{Y}$ is a spatial point process with intensity function δ, independent of $\mathbf{X}$. When the intensity function δ is very large, it can be shown that (2.25) tends to (2.22). So, both criterions are actually not so far one from each other. However, (2.25) has two advantages: first, if $\rho(u; \theta)$ cannot be evaluated for $u \notin (\mathbf{X} \cup \mathbf{Y}) \cap W_n$, (2.22) cannot be computed. Second and most important, conditional on $\mathbf{X} \cup \mathbf{Y}$, (2.25) corresponds to the weighted likelihood function for Bernoulli trials with probability, $\pi(u) = \mathrm{P}(1\{u \in \mathbf{X}\} = 1)$ for $u \in \mathbf{X} \cup \mathbf{Y}$, with $\pi(u) = \rho(u; \theta)/(\delta(u) + \rho(u; \theta))$. Precisely, (2.25) is a weighted logistic regression with offset term $-\log \delta$. Thus, again, parameter estimates can be straightforwardly obtained using standard software for generalized linear models. This approach is provided in the spatstat package in R by calling the ppm function with option method="logi" [4, 5]. The estimator derived from the logistic regression likelihood, studied theoretically by Waagepetersen [79], is in general less efficient than the weighted Poisson likelihood estimator but does not suffer from the Berman–Turner approximation, which for specific situations can be very interesting. Waagepetersen [79] actually considered the case $w = 1$; an estimation of the optimal weight surface w for the logistic regression has recently been proposed by Choiruddin et al. [13] following Guan and Shen [42].

2.4.3.4 Quasi-Likelihood

Guan and Jalilian [43] extended the estimating function (2.23) by considering the following class of estimating equations

$$\sum_{u \in \mathbf{X} \cap W_n} \mathbf{h}(u; \theta) - \int_{W_n} \mathbf{h}(u; \theta)\rho(u; \theta)\mathrm{d}u, \tag{2.26}$$

where $\mathbf{h}$ is a test function to be chosen. The authors considered the Godambe information criterion, which is the inverse of the asymptotic covariance matrix, as the criterion to optimize. By denoting T the integral operator acting on $\mathbb{R}^p$ valued functions $\mathbf{h}$ defined by

$$T\mathbf{h}(u) = \int_{W_n} \mathbf{h}(v)\rho(v; \theta) \{g(v - u) - 1\} \mathrm{d}v$$

where g is the pair correlation function (assumed to be invariant under translations) [43], showed that the optimal function $\mathbf{h}$ is the solution of the Fredholm equation

$$\mathbf{h}(\cdot) + T\mathbf{h}(\cdot) = \frac{\rho^{(1)}(\cdot; \theta)}{\rho(\cdot; \theta)}.$$

The estimator derived from the estimating Eq. (2.26) with this optimal choice of $\mathbf{h}$ was shown to be consistent and asymptotically Gaussian. The procedure results in a slightly more efficient estimator than the one obtained by Guan and Shen [42] especially for very clustered patterns. The counterpart is that the derived estimator is more computationally expensive, essentially because it cannot be fitted using an analogy with generalized linear models.

2.4.3.5 Variational Approach

Previous methods require either d-dimensional integral discretization or the simulation of an extra dummy point process. This can be problematic when d is large and/or when the number of points is large. All methods share the characteristic to be implemented using optimization procedures and require $\mathbf{z}(u)$ to be observed over the whole window (or on a dense set for the logistic regression method). Coeurjolly and Møller [19] proposed an alternative to bypass all these criticisms. The authors first obtained the following equation

$$\mathrm{E} \sum_{u \in \mathbf{X} \cap W_n} h(u) \boldsymbol{\theta}^\top \operatorname{div} \mathbf{z}(u) = -\mathrm{E} \sum_{u \in \mathbf{X} \cap W_n} \operatorname{div} h(u) \tag{2.27}$$

where $\operatorname{div} \mathbf{z}(u) = (\operatorname{div} z_i(u), i = 1, \ldots, p)^\top$ and $\operatorname{div} z_i(u) = \sum_{j=1}^{d} \partial z_i(u)/\partial u_j$, for any function h compactly supported on W_n (and such that the above quantities are well-defined). Coeurjolly and Møller [19] exploited (2.27) and proposed the estimator $\hat{\boldsymbol{\theta}} = -\mathbf{A}^{-1}\mathbf{b}$ with

$$\mathbf{A} = \sum_{u \in \mathbf{X} \cap W_n} \eta(u) \operatorname{div} \mathbf{z}(u) \operatorname{div} \mathbf{z}(u)^\top \quad \text{and} \quad \mathbf{b} = \sum_{u \in \mathbf{X} \cap W_n} \operatorname{div}\left(\eta(u) \operatorname{div} \mathbf{z}(u)\right)$$

$$\tag{2.28}$$

where η is a smooth compactly supported function (which can be set to 1 in the case where the z_i are compactly supported). This estimator is called variational estimator. Such a variational approach was first introduced by Baddeley and Dereudre [2] to estimate the Papangelou conditional intensity of a Gibbs point process.

We indeed observe that the variational estimator is explicit, does not require any integral discretization and depends only on (derivatives of) the z_i where the points are observed. When the z_i correspond to covariates (like Fig. 2.1h, i) the first and second derivatives can be estimated using finite differences locally around the data points.

2.5   Higher-Order Interaction Estimation via Conditional Intensity

Conditional intensities encode the interactions of the process. In a parametric setting, they are key quantities arising in several estimating equations and contrast estimating functions to estimate the interaction parameters of the model. We describe these estimating methods in the following, whether they depend on the

Papangelou conditional intensity (Sect. 2.5.1) or the Palm likelihood (Sect. 2.5.2). The statistical setting is the same as in Sect. 2.4, i.e. we observe **X** on W_n and the asymptotic results are in the sense of an increasing domain. In view of Table 2.1, the procedures based on the Papangelou conditional intensity are mainly devoted to the estimation of Gibbs models, while the methods based on the Palm likelihood are used to estimate the parameters in some Cox and DPP models. Note that for these models, maximum likelihood estimation is generally not considered in practice: the density of a Cox model is indeed intractable, see (2.15), and the density of a Gibbs model involves an intractable normalizing constant, see (2.8). However, for some Gibbs models, this normalizing constant may be approximated by MCMC methods, in which case maximum likelihood estimation becomes a viable alternative, see [39, 55] and [32].

Beyond this parametric framework, a natural concern could be the non parametric estimation of conditional intensities, similarly as the methods described in Sect. 2.4.2 for the intensity. For Gibbs models [36], focused on the class of pairwise interaction models and proposed a non parametric estimation of the pairwise interaction function, leading to a non parametric estimation of the Papangelou conditional intensity (within the class of pairwise interaction models). For the Palm intensity, by exploiting that $\rho_x(u) = \rho(u)g(u, x)$, a non parametric estimation of ρ_x could be obtained by combining a non parametric estimation of the intensity function ρ and of the pair correlation function g, a classical problem in spatial statistics, see e.g. [45, 58]. This approach has, nevertheless, never been studied in the literature. Following the previous comments, we note that non parametric estimation of conditional intensities thus remains largely unexplored and the following subsections do not treat this aspect.

2.5.1 Parametric Estimation with the Papangelou Conditional Intensity

We assume at many places in the following that the Papangelou conditional intensity has a log-linear parametric form

$$\log \lambda(u, \mathbf{x}; \boldsymbol{\theta}) = \boldsymbol{\theta}^\top \mathbf{t}(u, \mathbf{x}), \tag{2.29}$$

for some function $\mathbf{t}(u, \mathbf{x}) = (t_1(u, \mathbf{x}), \dots, t_p(u, \mathbf{x}))^\top$, $u \in \mathbb{R}^d$, $\mathbf{x} \in \mathcal{N}$, and where $\boldsymbol{\theta} \in \Theta$, Θ is an open bounded set of $\mathbb{R}^p$. These models are sometimes called exponential models. This is a common assumption made in the literature to investigate the asymptotic properties of the estimation methods. In this setting, the inference procedures can also be much faster. Many standard models belong to the family of exponential models, see [9]. For example, the Strauss model and the area-interaction model are exponential models for the parameters $(\log z, \theta)$ but not for the range parameter R or the hard-core parameter δ.

2.5.1.1 Maximum Pseudo-Likelihood Estimator

The pseudo-likelihood function, given the observation of $\mathbf{X}$ on W_n, is

$$PL(\boldsymbol{\theta}) = e^{-\int_{W_n} \lambda(u, \mathbf{X}; \boldsymbol{\theta}) du} \prod_{u \in \mathbf{X} \cap W_n} \lambda(u, \mathbf{X}; \boldsymbol{\theta}). \qquad (2.30)$$

The concept of pseudo-likelihood was introduced by Besag [8] for Markov random fields on a lattice, in which case it is the product of the conditional densities of the field at each site of the lattice, given the neighbor sites. The extension to point processes first appears in [67], Section 4.2, where (2.30) is informally derived by a limiting argument. Specifically, given a lattice on W_n, the Markov random field consisting of the count process on each cell of the lattice is introduced. The pseudo-likelihood of this random field, in the sense of [8], tends to (2.30) as each cell of the lattice is refined to an infinitesimal point. This approach is rigorously justified in [47].

The log pseudo-likelihood (LPL) function reads

$$LPL(\boldsymbol{\theta}) = \sum_{u \in \mathbf{X} \cap W_n} \log \lambda(u, \mathbf{X}; \boldsymbol{\theta}) - \int_{W_n} \lambda(u, \mathbf{X}; \boldsymbol{\theta}) du, \qquad (2.31)$$

which becomes for exponential models

$$LPL(\boldsymbol{\theta}) = \sum_{u \in \mathbf{X} \cap W_n} \boldsymbol{\theta}^\top \mathbf{t}(u, \mathbf{X}) - \int_{W_n} \exp(\boldsymbol{\theta}^\top \mathbf{t}(u, \mathbf{X})) du. \qquad (2.32)$$

The parallel with the Poisson log-likelihood (2.21) is immediate. In virtue of the GNZ equation (2.7), the score of the LPL equation appears to be an unbiased estimating equation for $\mathbf{X}$. The estimator obtained as the maximum of $LPL(\boldsymbol{\theta})$ over Θ is the maximum pseudo-likelihood estimator.

From a theoretical point of view, the consistency of the maximum pseudo-likelihood estimator has been proved for stationary exponential models in [47], [53] and [9], and extended to general (possibly non hereditary) interactions in [31]. The asymptotic normality for this estimator is established in [46, 54] and [9] for stationary exponential models having a finite range interaction, i.e. there exists R such that $\lambda(u, \mathbf{X}; \boldsymbol{\theta}) = \lambda(u, \mathbf{X}_{B(u,R)}; \boldsymbol{\theta})$. An extension to finite range non-exponential models is carried out in [16], while the case of infinite range exponential models is treated in [18]. For finite range exponential models, a fast and consistent estimation of the asymptotic covariance matrix of the estimator is proposed in [21].

As detailed hereafter, due to the similarity between (2.31) and (2.21), the numerical approximations to get the maximum pseudo-likelihood estimator and the Poisson likelihood estimator are analogous. The ideas to improve their efficiency also follow the same lines.

However, an important difference between (2.31) and (2.21) is the presence of edge effects in (2.32). Whereas all terms in the Poisson likelihood function only depend on the observation of $\mathbf{X}$ on W_n, the LPL function involves $\lambda(u, \mathbf{X}; \theta)$ that may depend on the unobserved point pattern $\mathbf{X} \cap W_n^c$. Standard solutions to deal with this issue are: (1) to replace $\mathbf{X}$ by $\mathbf{X}_{W_n}$ in (2.31), thus accounting for the empty set configuration outside W_n; (2) to apply a periodic expansion of X_{W_n} in W_n^c, up to some distance, in order to account for a more realistic outside configuration than the empty set; (3) to apply the border correction, which consists in replacing W_n in (2.31) by $W_n \ominus R$ for some $R > 0$, that is W_n eroded by R. This border correction is particularly adapted when the interaction is finite-range with a range less than R, in which case $\lambda(u, \mathbf{X}; \theta)$, for $u \in W_n \ominus R$, is equal to $\lambda(u, \mathbf{X}_{W_n}; \theta)$. Other edge corrections are possible. The ppm function of the spatstat package in R offers six different choices, including the three options detailed above, see [3] and [5].

Beyond the edge effects, the integral in (2.31) has to be approximated. A straightforward but somewhat costly solution consists in applying a Monte-Carlo approximation of this integral. Alternatively, as pointed out and developed by Baddeley and Turner [3], the [7] device can be applied in the exact same way as described in Sect. 2.4.3.2, where $\rho(u_i; \theta)$ is replaced by $\lambda(u_i, \mathbf{X}; \theta)$. For a log-linear Papangelou conditional intensity, the maximisation of the resulting approximated LPL function reduces to fit a (weighted) Poisson generalized linear model. This is how the maximum pseudo-likelihood estimator is derived by default in the ppm function of spatstat.

The limitations of the Berman–Turner approximation are the same as for the Poisson likelihood estimation. The number of dummy points has to be large to hope for a good approximation. This is all the more important in presence of a Gibbs model generating strong interactions. If not, the Berman–Turner approximation may lead to a significant bias. This is illustrated in [4], where a logistic regression pseudo-likelihood is introduced as an alternative, in the same vein as the logistic regression likelihood of Sect. 2.4.3.3. The logistic regression pseudo-likelihood contrast function reads in the present case

$$\sum_{u \in \mathbf{X} \cap W_n} \log\left(\frac{\lambda(u, \mathbf{X}; \theta)}{\delta(u) + \lambda(u, \mathbf{X}; \theta)}\right) + \sum_{u \in \mathbf{Y} \cap W_n} \log\left(\frac{\delta(u)}{\lambda(u, \mathbf{X}; \theta) + \delta(u)}\right), \quad (2.33)$$

where $\mathbf{Y}$ is a spatial point process with intensity function δ, independent of $\mathbf{X}$. By application of the GNZ equation and the Campbell formula, it is easily seen that the associated score function is an unbiased estimating function. As for (2.31), an edge correction must be applied to be able to compute $\lambda(u, \mathbf{X}; \theta)$. The advantages of this approach are again similar to those described in Sect. 2.4.3.3 for the estimation of the intensity. In particular, for a log-linear Papangelou conditional intensity, the estimates can be simply derived using standard GLM procedures. Moreover the result does not suffer from the artificial bias of the Berman–Turner approximation. The asymptotic properties of the method, including an estimation of the variance, are studied in [4].

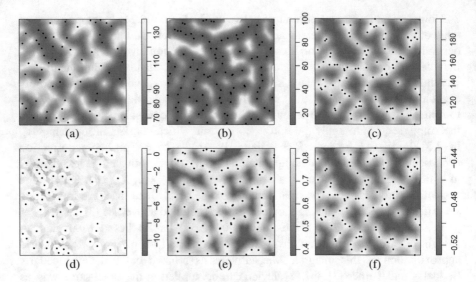

Fig. 2.6 Point patterns considered in this figure correspond to the ones from Fig. 2.3d–f. The first row corresponds to the estimated Papangelou conditional intensity function while the second one depicts the relative errors, i.e. the image with values $(\lambda(u, \mathbf{x}; \boldsymbol{\theta}) - \lambda(u, \mathbf{x}; \hat{\boldsymbol{\theta}}))/\lambda(u, \mathbf{x}; \boldsymbol{\theta})$. The Papangelou conditional intensity functions are parametrically estimated using the maximum pseudo-likelihood method. (**a**) Strauss hard-core. (**b**) Area-interaction, $\theta = -2.3$. (**c**) Area-interaction, $\theta = 0.7$. (**d**) Strauss hard-core. (**e**) Area-interaction, $\theta = -2.3$. (**f**) Area-interaction, $\theta = 0.7$

This Sect. 2.5.1 is illustrated by Fig. 2.6 which depicts for three different patterns the estimated Papangelou conditional intensities and the associated relative errors.

2.5.1.2 Takacs–Fiksel Estimator

Just like the quasi-likelihood estimator (Sect. 2.4.3.4) extends the Poisson likelihood estimator of the intensity function by fully exploiting the Campbell theorem, the Takacs–Fiksel estimator extends the pseudo-likelihood estimator of the Papangelou conditional intensity function by fully exploiting the GNZ formula (2.7). Specifically, from the GNZ formula, for any q-dimensional ($q \geq p$) function $\mathbf{h}$ (provided the following terms exist and admit an expectation), the equation

$$e_{\mathbf{h}}(\boldsymbol{\theta}) := \int_{W_n} \mathbf{h}(u, \mathbf{X}; \boldsymbol{\theta})\lambda(u, \mathbf{X}; \boldsymbol{\theta})\mathrm{d}u - \sum_{u \in \mathbf{X} \cap W_n} \mathbf{h}(u, \mathbf{X} \setminus \{u\}; \boldsymbol{\theta})$$

is an unbiased estimating equation for $\boldsymbol{\theta}$. The choice $\mathbf{h}(u, \mathbf{X}; \boldsymbol{\theta}) = \partial \log \lambda(u, \mathbf{X}; \boldsymbol{\theta})/\partial \boldsymbol{\theta}$ corresponds to the score function of the pseudo-likelihood procedure.

In its initial form [37, 73], the Takacs–Fiksel estimator is obtained by minimizing the norm of $e_{\mathbf{h}}(\boldsymbol{\theta})$. An alternative (see [5]) consists in adopting the estimating equation approach to solve $e_\mathbf{h}(\boldsymbol{\theta}) = 0$ with $q = p$. In both cases, the strength and flexibility of the method lies in the possibility to choose the function $\mathbf{h}$, as discussed now. The asymptotic properties of the Takacs–Fiksel estimator for stationary models are mainly studied by Coeurjolly et al. [22].

The first interest is to be able to find specific choices of $\mathbf{h}$ which may lead to an explicit estimator (unlike the pseudo-likelihood estimator), or may allow estimation in situations where the pseudo-likelihood estimator is unusable. We give two examples, originated from [22]. For the Strauss model, an explicit estimator of (z, θ) follows from the choice $\mathbf{h} = (h_1, h_2)$ with

$$h_1(u, \mathbf{X}; \theta) = \begin{cases} 1 & \text{if } |\mathbf{X} \cap B(u, R)| = 0 \\ 0 & \text{otherwise} \end{cases}, \quad h_2(u, \mathbf{X}; \theta) = \begin{cases} e^\theta & \text{if } |\mathbf{X} \cap B(u, R)| = 1 \\ 0 & \text{otherwise} \end{cases}.$$

It is not difficult to verify that $e_\mathbf{h}(\theta) = (zV_0 - N_0, zV_1 - e^\theta N_1)$ whose norm $\|e_\mathbf{h}(\theta)\|$ is minimal for

$$\hat{z} = N_0/V_0, \quad \hat{\theta} = \log(V_1/N_1) - \log(V_0/N_0),$$

where N_k, $k = 0, 1$, denotes the number of points $u \in \mathbf{X}_{W_n}$ such that $|B(u, R) \cap \mathbf{X}| = k + 1$, and V_k denotes the volume of the set $\{u \in W_n, |B(u, R) \cap \mathbf{X}| = k\}$. For the second example, consider the area interaction process where only the union of balls $\Gamma(\mathbf{X}_{W_n}) := \bigcup_{u \in \mathbf{X}_{W_n}} B(u, R)$ is observed, but not all individual points in $\mathbf{X}_{W_n}$. This setting comes from stochastic geometry, where the spots in a binary images can be modeled by a union of balls, see [33, 57]. In this case a pseudo-likelihood estimator cannot be applied because this estimator needs the locations of all points of $\mathbf{X}_{W_n}$ to be computed. In contrast, it is possible to find specific choices of $\mathbf{h}$ such that $e_\mathbf{h}(\theta)$ becomes computable, even if each quantity in the sum term of $e_\mathbf{h}(\theta)$ is unavailable. Define

$$h_1(u, \mathbf{X}; \theta) = \mathscr{P}\left(B(u, R) \cap \{\Gamma(\mathbf{X})\}^c\right),$$

where $\mathscr{P}$ denotes the perimeter. Neglecting edge effects, we obtain that

$$\sum_{u \in \mathbf{X} \cap W_n} h_1(u, \mathbf{X} \setminus \{u\}; \theta) = \mathscr{P}(\Gamma(\mathbf{X}_{W_n}))$$

which is a quantity computable from the observation of $\Gamma(\mathbf{X}_{W_n})$. Another choice is possible by considering isolated balls:

$$h_2(u, \mathbf{X}; \theta) = \begin{cases} 1 & \text{if } B(u, R) \cap \Gamma(\mathbf{X}) = \emptyset, \\ 0 & \text{otherwise.} \end{cases}$$

Then

$$\sum_{u \in \mathbf{X} \cap W_n} h_2(u, \mathbf{X} \setminus \{u\}; \theta) = N_{\text{iso}}(\varGamma(\mathbf{X}_{W_n})),$$

where $N_{\text{iso}}(\varGamma(\mathbf{X}_{W_n}))$ denotes the number of isolated balls in $\varGamma(\mathbf{X}_{W_n})$. Besides, the integrals terms in $e_{h_1}(\theta)$ and $e_{h_2}(\theta)$ do not suffer from the same unobservability issue as the sum terms, and they can be computed numerically. Therefore the estimation of (z, θ) in the area interaction model can be carried out using the Takacs–Fiksel estimator associated to $\mathbf{h} = (h_1, h_2)$, with the observation of $\varGamma(\mathbf{X}_{W_n})$ only.

Another interest of the Takacs–Fiksel approach is the possibility to optimize the estimating equation in $\mathbf{h}$, specifically by minimizing the associated Godambe information. This problem has been studied in [23], where the authors notice that the minimization of the Godambe information is a too difficult problem in practice. Instead they suggest a "semi-optimal" procedure, where the Godambe information is replaced by a close approximation. Let $T_{\mathbf{x}}$ be the integral operator acting on $\mathbb{R}^p$ valued functions $\mathbf{g}$

$$T_{\mathbf{x}} \mathbf{g}(u) = \int_{W_n} \mathbf{g}(v)(\lambda(v, \mathbf{x}; \theta) - \lambda(v, \mathbf{x} \cup \{u\}; \theta)) dv.$$

Coeurjolly and Guan [23] show that their semi-optimal test function $\mathbf{h}$ is the solution of the Fredholm equation

$$\mathbf{h}(., \mathbf{x}; \theta) + T_{\mathbf{x}} \mathbf{h}(., \mathbf{x}; \theta) = \frac{\lambda^{(1)}(., \mathbf{x}; \theta)}{\lambda(., \mathbf{x}; \theta)}$$

where $\lambda^{(1)}(., \mathbf{x}; \theta) = \partial \lambda(., \mathbf{x}; \theta)/\partial \theta$. In practice, this solution is obtained numerically. As a result [23], show that the Takacs–Fiksel estimator based on this semi-optimal test function outperforms the pseudo-likelihood estimator in most cases. However the gain might appear not sufficiently significant in view of the computational cost to implement the method.

2.5.1.3 Variational Estimator

An alternative inference method based on the Papangelou conditional intensity consists in a variational approach, in the same spirit as in Sect. 2.4.3.5 for the estimation of the intensity. This idea originates from [2]. We assume in this section that the Papangelou conditional intensity admits a log-linear form as in (2.29). To stress the peculiar role of the activity parameter z in this expression, we rather write

$$\log \lambda(u, \mathbf{x}; \theta) = \log z + \theta^\top \mathbf{t}(u, \mathbf{x}). \tag{2.34}$$

where $\boldsymbol{\theta} \in \mathbb{R}^p$ gathers all parameters but the activity parameter. We further assume that $u \mapsto \mathbf{t}(u, \mathbf{x})$ is differentiable. Then, an integration by part in the GNZ formula shows that for any real function h on $\mathbb{R}^d \times \mathcal{N}$, which is differentiable and compactly supported with respect to the first variable and such that the following expectations are well-defined,

$$\mathrm{E} \sum_{u \in \mathbf{X}} h(u, \mathbf{X} \setminus \{u\}) \boldsymbol{\theta}^\top \mathrm{div}_u \, \mathbf{t}(u, \mathbf{X}) = -\mathrm{E} \sum_{u \in \mathbf{X}} \mathrm{div}_u \, h(u, \mathbf{X} \setminus \{u\})$$

where $\mathrm{div}_u \, \mathbf{t}(u, \mathbf{X}) = \left(\mathrm{div}_u \, t_1(u, \mathbf{X}), \ldots, \mathrm{div}_u \, t_p(u, \mathbf{X}) \right)^\top$. Considering the empirical version of this equation for p different test functions $h_1, \ldots, h_p$, we obtain a system of linear equations that yields the estimator

$$\hat{\boldsymbol{\theta}} = -\mathbf{A}^{-1} \mathbf{b} \qquad (2.35)$$

where $\mathbf{A}$ is the matrix with entries $\sum_{u \in \mathbf{X}_{W_n}} h_i(u, \mathbf{X} \setminus \{u\}) \mathrm{div}_u \, t_j(u, \mathbf{X})$, $i, j = 1, \ldots, p$ and $\mathbf{b}$ is the vector with entries $\sum_{u \in \mathbf{X}_{W_n}} \mathrm{div}_u \, h_i(u, \mathbf{X} \setminus \{u\})$, $i = 1, \ldots, p$. Here we have assumed that $\mathbf{A}$ is invertible. Note that if z (or $\log z$) was part of $\boldsymbol{\theta}$, that is (2.29) holds instead of (2.34), then its associated $t_0(u, \mathbf{X})$ function would be constant equal to 1, so $\mathrm{div}_u \, t_0(u, \mathbf{X}) = 0$ and $\mathbf{A}$ would not be invertible. In fact, this procedure does not make the estimation of the activity parameter possible.

The asymptotic properties of (2.35) for stationary models are studied in [2]. The main interest of this variational estimator is its simple and explicit form. As to the choice of the test functions h_i, some recommendations are suggested in [2]. For pairwise exponential models, they consider $h_i(u, \mathbf{X}) = \mathrm{div}_u \, t_i(u, \mathbf{X})$. A simulation study carried out with this choice for the Lennard-Jones model [50] demonstrates the efficiency of the procedure, especially in presence of strong interactions. In this situation the variational estimator outperforms the pseudo-likelihood estimator, both in terms of accuracy and computational cost, see [2].

The assumption that $u \mapsto t_i(u, \mathbf{x})$ is differentiable is the main restriction of this procedure. For instance, for the Strauss model (2.12), $t_1(u, \mathbf{x}) = -n_R(u, \mathbf{x})$ is not differentiable, even not continuous in u. The other restriction, as noted before, is the impossibility to estimate z. However, note that given $\hat{\boldsymbol{\theta}}$ in (2.35) for the exponential model (2.34), a natural procedure to get an estimation of z is to apply the Takacs–Fiksel estimator associated to the test function $h = e^{-\hat{\boldsymbol{\theta}}^\top \mathbf{t}(u, \mathbf{x})}$. This two-step procedure provides

$$\hat{z} = \frac{1}{|W_n|} \sum_{u \in \mathbf{X}_{W_n}} e^{-\hat{\boldsymbol{\theta}}^\top \mathbf{t}(u, \mathbf{x})}.$$

For finite-range Gibbs models, [20], also proposed the estimator $\hat{z} = N_0 / V_0$ (see Sect. 2.5.1.2 for the definitions of N_0, V_0).

2.5.2 Palm Likelihood Estimation

Palm likelihood refers to the estimation procedures based on the first order Palm intensity $\rho_x(u)$. For this reason Palm likelihood only applies to models for which we have a closed form expression of the Palm intensity. This is the case for the LGCP, SNCP and DPP models, see Table 2.1.

The first contribution exploiting the properties of $\rho_x(u)$, namely in [63], was not intended to estimate the parameters in one of these models, but to estimate the fractal dimension of the point pattern. For isotropic models, implying that $\rho_x(u) = \rho_0(u)$ only depends on $|u|$ and is independent of x, this estimation boils down to estimate the parameter $H \in [0, 2]$ in the (assumed) asymptotic behavior $\rho_0(u) - \rho \sim \kappa |u|^{-H}$ as $|u| \to \infty$, where $\kappa > 0$ and ρ denotes the intensity parameter. To get an estimation of H [63], suggested a non parametric and a parametric method. In the non parametric approach, they estimate $\rho_0(u)$ for $|u| \in [r_1, r_2]$ by the number of pairs at a distance belonging to the annular region $\mathscr{A}(r_1, r_2)$ divided by the area of $\mathscr{A}(r_1, r_2)$. Then the slope of the (approximated) line in the log-log plot of $\hat{\rho}_0(u) - \hat{\rho}$ with respect to $|u|$ provides an estimate of H. For their parametric approach, they assume that $\rho_0(u) = \rho_0(u; \theta) = \rho + \kappa |u|^{-H}$ where $\theta = (\rho, \kappa, H)$ and they estimate θ by maximizing the Palm likelihood

$$\sum_{u,v \in \mathbf{X}_{W_n}, \|u-v\| < R}^{\neq} \log(\rho_0(u - v; \theta)) - |\mathbf{X}_{W_n}| \int_{B(0,R)} \rho_0(u; \theta) du, \qquad (2.36)$$

where R is a tuning parameter that is typically small relative to the size of W_n. The Palm likelihood is informally motivated in [63] by the fact that the point process consisting of pairs of points R-apart in $\mathbf{X}_{W_n}$ is approximately an inhomogeneous Poisson process with intensity $|\mathbf{X}_{W_n}| \rho_0(u)$. The Palm likelihood is thus deduced from the associated Poisson likelihood (2.22).

The Palm likelihood (2.36) has been used in [74] to estimate the parameters of stationary Neymann Scott processes. The method can also be applied to other models like LGCPs and DPPs. The theoretical aspects (consistency and asymptotic normality) have been studied in the stationary case by Prokešová and Jensen [64], where the authors used the fact that the score associated to (2.36) is nearly an unbiased estimating equation. Specifically, it follows from the Campbell theorem that the score associated to the modified Palm likelihood

$$\sum_{u \in \mathbf{X}_{W_n \ominus R}, v \in W_n, \|u-v\| < R}^{\neq} \log(\rho_0(u - v; \theta)) - |\mathbf{X}_{W_n \ominus R}| \int_{B(0,R)} \rho_0(u; \theta) du,$$

that accounts for edge effects, is an unbiased estimating equation. The difference between this modified Palm likelihood and (2.36) is proved to be asymptotically negligible in [64].

Several generalizations of (2.36) in the setting of inhomogeneous models are possible. In [65], a two step procedure is proposed for the estimation of second-order intensity-reweighted stationary models. For these models, the pair correlation is invariant by translation, $g(u, v) = g_0(v - u)$, while the intensity $\rho(x)$ is not constant, whence $\rho_x(u) = \rho(u)g_0(x - u)$. In [65], $\rho(u)$ is estimated by the Poisson likelihood method to yield $\hat{\rho}(u)$, and the remaining parameters $\boldsymbol{\theta}$ are then estimated by the following Palm likelihood

$$\sum_{u,v \in \mathbf{X}_{W_n}, \|u-v\|<R}^{\neq} \log(\hat{\rho}(v)g_0(u - v; \boldsymbol{\theta})) - \sum_{u \in \mathbf{X}_{W_n}} \int_{B(u,R)} \hat{\rho}(v)g_0(u - v; \boldsymbol{\theta})\mathrm{d}v.$$

Note that if $\rho(v) = \rho$ is constant, replacing $\hat{\rho}(v)$ by ρ in the above formula gives (2.36). As an alternative generalization of (2.36) [65] also considers the Palm likelihood

$$\sum_{u \in \mathbf{X}_{W_n \ominus R}, v \in W_n, \|u-v\|<R}^{\neq} \log(\hat{\rho}(u)\hat{\rho}(v)g_0(u - v; \boldsymbol{\theta}))$$

$$- \int_{W_n \ominus R} \int_{B(u,R)} \hat{\rho}(u)\hat{\rho}(v)g_0(u - v; \boldsymbol{\theta})\mathrm{d}u\mathrm{d}v.$$

The asymptotic properties of this two step method, with either generalization of the Palm likelihood, are established in [65], with a focus on SNCPs. For general inhomogeneous models, a natural proposition would be to use the Palm likelihood

$$\sum_{u \in \mathbf{X}_{W_n \ominus R}, v \in W_n, \|u-v\|<R}^{\neq} \log(\rho_v(u; \boldsymbol{\theta})) - \sum_{v \in \mathbf{X}_{W_n \ominus R}} \int_{B(v,R)} \rho_v(u; \boldsymbol{\theta})\mathrm{d}u,$$

whose score function is an unbiased estimating function. The estimation of all parameters could be handled in one step. Such a methodology has not been investigated so far.

2.6 Conclusion

In this chapter, we have proposed an analysis of spatial point patterns through intensity functions and conditional intensity functions. As demonstrated in the last two sections, many methods are available to estimate these quantities. However, we think there are still challenging questions that deserve to be studied.

Most inference procedures presented in this chapter are devoted to the parametric estimation of spatial point process models. We think that efficient non parametric procedures to estimate the Papangelou conditional intensity or the Palm intensity are missing. This of course can be viewed as a first step to the choice of a parametric model, but it constitutes an informative characteristic to understand the interactions in a point pattern. As briefly exploited in [63], a non parametric estimation can also serve as a basis for the estimation of specific characteristic, as the fractal index in [63], or the long-range interaction parameter, a (still open) issue raised in [72, pp. 202–210].

Modern spatial statistics face complex data sets, with a possibly huge amount of information (millions of points, thousands of covariables). Inference methods able to handle these aspects must be developed. This demands efficient numerical methods and requires the development of regularization methods [13, 75]. Similarly, the stationarity assumption is rarely realistic for complex data. There is a need to develop inference methods for inhomogeneous (conditional) intensity functions, as local Papangelou and local Palm likelihood [1].

Acknowledgements A part of the material presented here is the fruit of several collaborations. We take the opportunity to thank our main collaborators David Dereudre, Jesper Møller, Ege Rubak and Rasmus Waagepetersen. We are also grateful to Christophe Biscio, Achmad Choiruddin, Rémy Drouilhet, Yongtao Guan and Frédérique Letué. This contribution has been partly supported by the program ANR-11-LABX-0020-01.

References

1. A. Baddeley, Local composite likelihood for spatial point processes. Spat. Stat. **22**, 261–295 (2017)
2. A. Baddeley, D. Dereudre, Variational estimators for the parameters of Gibbs point process models. Bernoulli **19**(3), 905–930 (2013)
3. A. Baddeley, R. Turner, Practical maximum pseudolikelihood for spatial point patterns. Aust. N. Z. J. Stat. **42**(3), 283–322 (2000)
4. A. Baddeley, J.-F. Coeurjolly, E. Rubak, R. Waagepetersen, Logistic regression for spatial Gibbs point processes. Biometrika **101**(2), 377–392 (2014)
5. A. Baddeley, E. Rubak, R. Turner. *Spatial Point Patterns: Methodology and Applications with R* (CRC Press, Boca Raton, 2015)
6. M. Berman, P. Diggle, Estimating weighted integrals of the second-order intensity of a spatial point process. J. R. Stat. Soc. Ser. B **51**(1), 81–92 (1989)
7. M. Berman, R. Turner, Approximating point process likelihoods with GLIM. Appl. Stat. **41**, 31–38 (1992)
8. J. Besag, Spatial interaction and the statistical analysis of lattice systems. J. R. Stat. Soc. Ser. B **36**(2) 192–236 (1974).
9. J.-M. Billiot, J.-F. Coeurjolly, R. Drouilhet, Maximum pseudolikelihood estimator for exponential family models of marked Gibbs point processes. Electron. J. Stat. **2**, 234–264 (2008)
10. C.A.N. Biscio, J.-F. Coeurjolly, Standard and robust intensity parameter estimation for stationary determinantal point processes. Spat. Stat. **18**, 24–39 (2016)
11. C.A.N. Biscio, F. Lavancier, Quantifying repulsiveness of determinantal point processes. Bernoulli **22**(4), 2001–2028 (2016)
12. S.N. Chiu, D. Stoyan, W. S. Kendall, J. Mecke, *Stochastic Geometry and Its Applications*, 3rd edn. (Wiley, Chichester, 2013)
13. A. Choiruddin, J.-F. Coeurjolly, F. Letué, Convex and non-convex regularization methods for spatial point processes intensity estimation. Electron. J. Stat. **12**(1), 1210–1255 (2018)

14. M. Clausel, J.-F. Coeurjolly, J. Lelong, Stein estimation of the intensity of a spatial homogeneous Poisson point process. Ann. Appl. Probab. **26**(3), 1495–1534 (2016)
15. J.-F. Coeurjolly, Median-based estimation of the intensity of a spatial point process. Ann. Inst. Stat. Math. **69**, 303–313 (2017)
16. J.-F. Coeurjolly, R. Drouilhet, Asymptotic properties of the maximum pseudo-likelihood estimator for stationary Gibbs point processes including the Lennard-Jones model. Electron. J. Stat. **4**, 677–706 (2010)
17. J.-F. Coeurjolly, Y. Guan, Covariance of empirical functionals for inhomogeneous spatial point processes when the intensity has a parametric form. Journal of Statistical Planning and Inference **155**, 79–92 (2014)
18. J.-F. Coeurjolly, F. Lavancier, Parametric estimation of pairwise Gibbs point processes with infinite range interaction. Bernoulli **23**(2), 1299–1334 (2017)
19. J.-F. Coeurjolly, J. Møller, Variational approach to estimate the intensity of spatial point processes. Bernoulli **20**(3), 1097–1125 (2014)
20. J.-F. Coeurjolly, N. Morsli, Poisson intensity parameter estimation for stationary Gibbs point processes of finite interaction range.Spat. Stat. **4**, 45–56 (2013)
21. J.-F. Coeurjolly, E. Rubak, Fast covariance estimation for innovations computed from a spatial Gibbs point process. Scand. J. Stat. **40**(4), 669–684 (2013)
22. J.-F. Coeurjolly, D. Dereudre, R. Drouilhet, F. Lavancier, Takacs–Fiksel method for stationary marked Gibbs point processes. Scand. J. Stat. **49**(3), 416–443 (2012)
23. J.-F. Coeurjolly, Y. Guan, M. Khanmohammadi, R. Waagepetersen, Towards optimal takacs–fiksel estimation. Spat. Stat. **18**, 396–411 (2016)
24. J.-F. Coeurjolly, J. Møller, R. Waagepetersen, Palm distributions for log Gaussian Cox processes. Scand. J. Stat. **44**(1), 192–203 (2017)
25. N.A.C. Cressie, *Statistics for Spatial Data*, 2nd edn. (Wiley, New York, 1993)
26. N. Cressie, C.K. Wikle, *Statistics for Spatio-Temporal Data* (Wiley, Hoboken, 2015)
27. O. Cronie, M.N.M. Van Lieshout, A non-model-based approach to bandwidth selection for kernel estimators of spatial intensity functions. Biometrika **105**(2), 455–462 (2018)
28. D.J. Daley, D. Vere-Jones, *An Introduction to the Theory of Point Processes: Elementary Theory and Methods*, vol. I, 2nd edn. (Springer, New York, 2003).
29. D.J. Daley, D. Vere-Jones, *An Introduction to the Theory of Point Processes: General Theory and Structure*, vol. II, 2nd edn. (Springer, New York, 2008)
30. H.A. David, H.N. Nagaraja, *Order Statistics*, 3rd edn. (Wiley, Hoboken, 2003)
31. D. Dereudre, F. Lavancier, Campbell equilibrium equation and pseudo-likelihood estimation for non-hereditary Gibbs point processes. Bernoulli **15**(4), 1368–1396 (2009)
32. D. Dereudre, F. Lavancier, Consistency of likelihood estimation for Gibbs point processes. Ann. Stat. **45**(2), 744–770 (2017)
33. D. Dereudre, F. Lavancier, K. S. Helisová, Estimation of the intensity parameter of the germ-grain Quermass-interaction model when the number of germs is not observed. Scand. J. Stat. **41**(3), 809–929 (2014)
34. P. Diggle, A kernel method for smoothing point process data. J. R. Stat. Soc. Ser. C **34**(2), 138–147 (1985)
35. P. Diggle, *Statistical Analysis of Spatial and Spatio-Temporal Point Patterns* (CRC Press, Boca Raton, 2013)
36. P. Diggle, D. Gates, A. Stibbard, A nonparametric estimator for pairwise-interaction point processes. Biometrika **74**(4), 763–770 (1987)
37. T. Fiksel, Estimation of parameterized pair potentials of marked and non-marked Gibbsian point processes. Elektronische Informationsverarbeitung und Kybernetik **20**, 270–278 (1984)
38. H.-O. Georgii, Canonical and grand canonical Gibbs states for continuum systems. Commun. Math. Phys. **48**, 31–51 (1976)
39. C.J. Geyer, J. Møller, Simulation procedures and likelihood inference for spatial point processes. Scand. J. Stat. **21**(4), 359–373 (1994)
40. Y. Guan, Fast block variance estimation procedures for inhomogeneous spatial point processes. Biometrika **96**(1), 213–220 (2009)

41. Y. Guan, J.M. Loh, A thinned block bootstrap procedure for modeling inhomogeneous spatial point patterns. J. Am. Stat. Assoc. **102**, 1377–1386 (2007)
42. Y. Guan, Y. Shen, A weighted estimating equation approach for inhomogeneous spatial point processes. Biometrika **97**, 867–880 (2010)
43. Y. Guan, A. Jalilian, R. Waagepetersen, Quasi-likelihood for spatial point processes. J. R. Stat. Soc. Ser. B **77**(3), 677–697 (2015)
44. L. Heinrich, M. Prokešová, On estimating the asymptotic variance of stationary point processes. Methodol. Comput. Appl. Probab. **12**(3), 451–471 (2010)
45. J. Illian, A. Penttinen, H. Stoyan, D. Stoyan, *Statistical Analysis and Modelling of Spatial Point Patterns*. Statistics in Practice (Wiley, Chichester, 2008)
46. J.L. Jensen, H.R. Künsch, On asymptotic normality of pseudolikelihood estimates of pairwise interaction processes. Ann. Inst. Stat. Math. **46**, 475–486 (1994)
47. J.L. Jensen, J. Møller, Pseudolikelihood for exponential family models of spatial point processes. Ann. Appl. Probab. **1**, 445–461 (1991)
48. J.F.C. Kingman, *Poisson Processes*, vol. 3 (Clarendon Press, Oxford, 1992)
49. F. Lavancier, J. Møller, E. Rubak, Determinantal point process models and statistical inference. J. R. Stat. Soc. Ser. B **77**(4), 853–877 (2015)
50. J.E. Lennard-Jones, On the determination of molecular fields. Proc. R. Soc. Lond. A. **106**(738), 463–477 (1924)
51. O. Macchi, The coincidence approach to stochastic point processes. Adv. Appl. Probab. **7**, 83–122 (1975)
52. J.A.F. Machado, J.M.C. Santos Silva, Quantiles for counts. J. Am. Stat. Assoc. **100**(472), 1226–1237 (2005)
53. S. Mase, Consistency of the maximum pseudo-likelihood estimator of continuous state space Gibbs processes. Ann. Appl. Probab. **5**, 603–612 (1995)
54. S. Mase, Marked Gibbs processes and asymptotic normality of maximum pseudo-likelihood estimators. Math. Nachr. **209**, 151–169 (2000)
55. J. Mateu, F. Montes, Likelihood inference for Gibbs processes in the analysis of spatial point patterns. Int. Stat. Rev. **69**(1), 81–104 (2001)
56. J. Møller, Shot noise Cox processes. Adv. Appl. Probab. **35**, 614–640 (2003)
57. J. Møller, K. Helisová, Likelihood inference for unions of interacting discs. Scand. J. Stat. **37**(3), 365–381 (2010)
58. J. Møller, R.P. Waagepetersen, *Statistical Inference and Simulation for Spatial Point Processes* (Chapman and Hall/CRC, Boca Raton, 2004).
59. J. Møller, R.P. Waagepetersen, Modern spatial point process modelling and inference (with discussion). Scand. J. Stat. **34**, 643–711 (2007)
60. J. Møller, R. Waagepetersen, Some recent developments in statistics for spatial point patterns. Ann. Rev. Stat. Appl. **4**(1), 317–342 (2017)
61. J. Møller, A.R. Syversveen, R.P. Waagepetersen, Log Gaussian Cox processes. Scand. J. Stat. **25**, 451–482 (1998)
62. X. Nguyen, H. Zessin, Integral and differential characterizations of Gibbs processes. Math. Nachr. **88**, 105–115 (1979)
63. Y. Ogata, K. Katsura, Maximum likelihood estimates of the fractal dimension for random spatial patterns. Biometrika **78**(3), 463–474 (1991)
64. M. Prokešová, E.B.V. Jensen, Asymptotic Palm likelihood theory for stationary point processes. Ann. Inst. Stat. Math. **65**(2), 387–412 (2013)
65. M. Prokešová, J. Dvořák, E. Jensen, Two-step estimation procedures for inhomogeneous shot-noise Cox processes. Ann. Inst. Stat. Math. **69**(3), 513–542 (2017)
66. S.L. Rathbun, N. Cressie, Asymptotic properties of estimators for the parameters of spatial inhomogeneous Poisson point processes. Adv. Appl. Probab. **26**, 122–154 (1994)
67. B. Ripley, *Statistical Inference for Spatial Processes* (Cambridge University Press, Cambridge, 1988)
68. M. Rosenblatt, Remarks on some nonparametric estimates of a density function. Ann. Math. Stat. **27**(3), 832–837 (1956)

69. Z. Sasvári, *Multivariate Characteristic and Correlation Functions*, vol. 50 (Walter de Gruyter, Berlin, 2013)
70. F.P. Schoenberg, Consistent parametric estimation of the intensity of a spatial-temporal point process. J. Stat. Plann. Inference **128**, 79–93 (2005)
71. T. Shirai, Y. Takahashi, Random point fields associated with certain Fredholm determinants. I. Fermion, Poisson and Boson point processes. J. Funct. Anal. **2**, 414–463 (2003)
72. J.-L. Starck, F. Murtagh, *Astronomical Image and Data Analysis* (Springer, Berlin, 2007)
73. R. Takacs, Estimator for the pair-potential of a Gibbsian point process. Math. Oper. Stat. Ser. Stat. **17**, 429–433 (1986)
74. U. Tanaka, Y. Ogata, D. Stoyan, Parameter estimation for Neyman-Scott processes. Biom. J. **50**, 43–57 (2008)
75. A.L. Thurman, R. Fu, Y. Guan, J. Zhu, Regularized estimating equations for model selection of clustered spatial point processes. Stat. Sin. **25**(1), 173–188 (2015)
76. M.N.M. Van Lieshout, *Markov Point Processes and Their Applications* (Imperial College Press, London, 2000)
77. M.N.M. Van Lieshout, On estimation of the intensity function of a point process. Methodol. Comput. Appl. Probab. **14**(3), 567–578 (2012)
78. R. Waagepetersen, An estimating function approach to inference for inhomogeneous Neyman-Scott processes. Biometrics **63**, 252–258 (2007)
79. R. Waagepetersen, Estimating functions for inhomogeneous spatial point processes with incomplete covariate data. Biometrika **95**(2), 351–363 (2008)

Chapter 3
Stochastic Methods for Image Analysis

Agnès Desolneux

Abstract These lectures about stochastic methods for image analysis contain three parts. The first part is about visual perception and the non-accidentalness principle. It starts with an introduction to the Gestalt theory, that is a psychophysiological theory of human visual perception. It can be translated into a mathematical framework thanks to a perception principle called the non-accidentalness principle, that roughly says that "we immediately perceive in an image what has a low probability of coming from an accidental arrangement". The second part of these lectures is about the so-called "a contrario method" for the detection of geometric structures in images. The a contrario method is a generic method, based on the non-accidentalness principle, to detect meaningful geometric structures in images. We first show in details how it works in the case of the detection of straight segments. Then, we show some other detection problems (curves, vanishing points, etc.) The third part of these lectures is about stochastic models of images for the problem of modeling and synthesizing texture images. It gives an overview of some methods of texture synthesis. We also discuss two models of texture images: stationary Gaussian random fields and shot noise random fields.

3.1 Visual Perception and the Non-accidentalness Principle

Figure 3.1 presents a pure noise image: in this image we don't perceive any visual structure. This lack of perception is called *Helmholtz principle*, also called, in its stronger form, the *non-accidentalness principle*. This principle can be stated in two different ways:

1. The first way is common sensical. It simply states that "we do not perceive any structure in a uniform random image". (In this form, the principle was first stated by Attneave [3] in 1954.)

A. Desolneux (✉)
CNRS, CMLA and ENS Paris-Saclay, Paris, France
e-mail: agnes.desolneux@cmla.ens-cachan.fr

© Springer Nature Switzerland AG 2019
D. Coupier (ed.), *Stochastic Geometry*, Lecture Notes in Mathematics 2237,
https://doi.org/10.1007/978-3-030-13547-8_3

Fig. 3.1 A pure noise image. It is of size 64×64 pixels, and the grey levels are independent samples of the uniform distribution on $\{0, 1, \ldots, 255\}$

Fig. 3.2 The same alignment of dots is present in the two images. On the left, there are 30 points, and there are 80 on the right. According to the non-accidentalness principle, we don't perceive the alignment in the right image (hide the left image with your hand, not to be influenced by it) because the probability of having 8 almost aligned points just by chance when you have 80 random points is quite high, whereas it is very small when there are 30 random points

2. In its stronger form, it states that whenever some large deviation from randomness occurs, a structure is perceived. In other words "we immediately perceive whatever has a low likelihood of resulting from accidental arrangement". (Under this form, the principle was stated in Computer Vision by Zhu [43] or Lowe [27].)

On Fig. 3.2, we illustrate the non-accidentalness principle on an example with aligned dots.

Now, not all possible structures are relevant for visual perception. The "interesting" structures are geometric patterns, and they have been extensively studied and defined by the so-called *Gestalt School* of Psychophysiology.

3.1.1 Gestalt Theory of Visual Perception

The aim of Gestalt theory (Wertheimer [42], Metzger [28] and Kanizsa [25]) is to answer questions such as: How do we perceive geometric objects in images? What are the laws and principles of visual construction? In other words, how do you go from pixels (or retina cells) to visual objects (lines, triangles, etc.)?

Before Gestalt theory, the study of visual perception was done through optic-geometric illusions. See Fig. 3.3 for an example. The goal of these illusions is to ask: "what is the reliability of our visual perception?"

But Gestalt theory does not continue on the same line. The question is not why we sometimes don't see parallel lines when they are; the question is why we do see a line at all. This perceived line is the result of a construction process. And it is the aim of Gestalt theory to establish the laws of this construction process. Gestalt theory starts with the assumption that there exists a small list of active grouping laws in visual perception: vicinity, same attribute (like color, shape, size or orientation for instance), alignment, good continuation, symmetry, parallelism, convexity, closure, constant width, amodal completion, T-junctions, X-junctions, Y-junctions.

The above listed grouping laws belong, according to Kanizsa, to the so-called primary process, opposed to a more cognitive secondary process.

These different grouping laws are illustrated on the following figures (that are replications of some of the many illustrations of the book of Kanizsa [25]): on

Fig. 3.3 An example of an optic-geometric illusion: Zoellner's illusion (1860). The diagonal lines are not perceived as being parallel, but they are! An explanation is that, in some way, they partly "inherit" the horizontal or vertical orientation of the smaller segments that are on them. Therefore, their perceived orientation is not the same as their real orientation

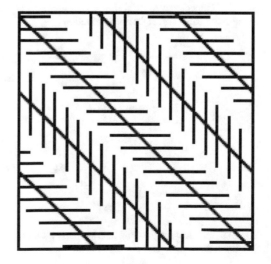

Fig. 3.4 for vicinity (we group objects that are spatially close), Fig. 3.5 for same attribute (we immediately perceive groups of objects that have the same attribute, like here color, size or orientation), Fig. 3.6 for symmetry (our visual perception is very sensitive to symmetry, we immediately perceive it when it is present), Fig. 3.7—left for good continuation (we perceive smooth curves) and Fig. 3.7—

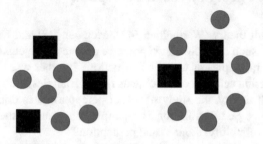

Fig. 3.4 Elementary grouping law: vicinity. We clearly perceive here two groups of "objects"

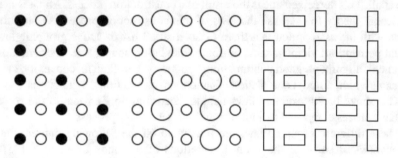

Fig. 3.5 Elementary grouping law: the objects have a common attribute and this allows us to perceive immediately vertical lines of similar objects. From left to right, the common attribute is: same color, same size and same orientation

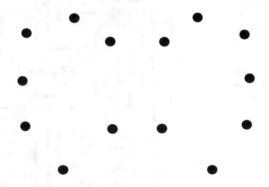

Fig. 3.6 Elementary grouping law: symmetry. Our visual perception is very sensitive to symmetry. We immediately perceive it when it is present

Fig. 3.7 Left: Elementary grouping law of Good Continuation. We perceive here two "smooth" curves, one in the horizontal direction and one in the vertical one. Right: Elementary grouping law of Closure. We perceive here two closed curves, and not an "8", as it would have been the case if we had kept the good continuation grouping of the left figure

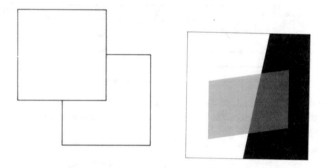

Fig. 3.8 On the left: T-junctions. They allow us to perceive occlusions. On the right: an X-junction that is typical of the transparency phenomenon

right for closure (we perceive closed curves). This last figure has to be compared with Fig. 3.7—left that illustrates good continuation. It is a typical example of *conflict* between grouping laws: here we can say that "closure wins against good continuation". But one may build examples where the converse will be true. It is one of the properties of the Gestalt grouping laws: there is no hierarchy in this small set of grouping laws, and there are sometimes conflicts between different possible interpretations of a figure.

The three types of junctions are illustrated on Figs. 3.8, 3.9 and 3.10. The T-junctions indicate occlusion (an object is partly hidden by another one). The X-junctions occur when there is some transparency. The Y-junctions indicate a 3D perspective effect (they come from the planar representation of the three principal orientations of the 3D space).

Amodal completion is one of the most applied grouping laws for our visual perception. Objects are often partly hidden by other objects (this is the occlusion phenomenon), but we are able to infer the missing contours. This is for instance illustrated on Fig. 3.11, where we "see" a black rectangle partly covered by five white shapes. Now there is no black rectangle in this figure, there are only black

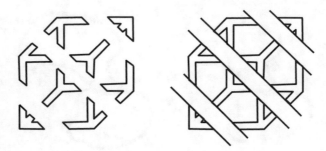

Fig. 3.9 A famous illustration of the power of T-junctions by Kanizsa. On the left, there are no T-junctions, and no occlusion is perceived. On the right: the three added bands create T-junctions, allowing contours completion across occlusions. The 3D cube is immediately perceived, whereas in the first figure one only sees strange "Y" shapes

Fig. 3.10 The famous Penrose fork: an impossible figure. In this figure, the perspective effect is created by Y-junctions. This figure also illustrates the fact that visual grouping laws can be stronger than physical sense

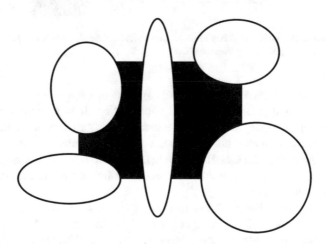

Fig. 3.11 Amodal completion: what do you see here? A black rectangle (see text)

pixels that we group according to their same color, and then thanks to alignment and T-junctions we complete the missing parts of the black rectangle.

The fact that in Fig. 3.11 we perceive a rectangle and not another shape (with round corners for instance), comes from the collaboration of several Gestalt laws,

Fig. 3.12 The recursivity of Gestalt grouping laws: black pixels are grouped (thanks to vicinity and same color) to form small rectangles. These small rectangles are grouped (thanks to parallelism and vicinity) to form "bars", that are again grouped to form larger bars, then parallel rectangles, and so on...

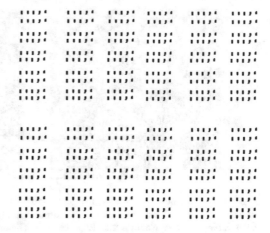

as good continuation, simplicity and "past experience". This last "law" is also part of Gestalt theory: our visual perception is very sensitive to "known shapes": digits, letters, geometric shapes, but also faces, cars, etc. This is something that is much more difficult to translate in mathematical terms.

All grouping Gestalt laws are *recursive*: they can be applied first to atomic inputs and then in the same way to partial groups already constituted. See for instance Fig. 3.12 where grouping laws are recursively applied. On this figure, the same partial Gestalt laws namely alignment, parallelism, constant width and proximity, are recursively applied not less than six times.

3.1.2 The Non-accidentalness Principle

Under its strong form, the Helmholtz or non-accidentalness principle states that we immediately perceive any regular structure that has a low probability to be there just by chance. This principle is used all the time in vision, but not only in vision, in our every day life also. For instance, if you play dice and you obtain a sequence 6,6,6,6,6,6,6—you will certainly notice it—and start wondering why!

In Computer Vision, the non-accidentalness principle is used to compute *detectability thresholds*, and this will be explained in details in the following sections.

Now to give a first simple example of computation, let us consider a black square appearing in a pure binary noise image, as illustrated on Fig. 3.13.

Let U be a random image of size $N \times N$ pixels, and such that pixels are black (with probability p) or white (with probability $1 - p$), all independently. What is the probability that U contains a $k \times k$ square of all black pixels? There is no easy exact answer to this question, since it involves counting events that may overlap. Now, let R be the random variable counting the number of $k \times k$ black squares in

Fig. 3.13 Left: a pure binary noise image of size 64×64 pixels. It contains several 3×3 black squares, and we don't pay any special attention to them. Right: the same image but containing now a 8×8 black square. We immediately perceive it. The probability of observing such a square in a pure binary noise image is less than 2×10^{-16}

U. As just explained, we cannot easily compute exactly $\mathbb{P}(R \geqslant 1)$, but thanks to *Markov Inequality*, we can bound it by the expectation of R:

$$\mathbb{P}(R \geqslant 1) = \sum_{r=1}^{+\infty} \mathbb{P}(R = r) \leqslant \sum_{r=0}^{+\infty} r\mathbb{P}(R = r) = \mathbb{E}(R).$$

And now $\mathbb{E}(R)$ is much easier to compute since

$$\mathbb{E}(R) = \mathbb{E}(\sum_{squares} \mathbb{1}_{black}) = N_{sq}\mathbb{P} \text{ (square is all black)} = N_{sq} \, p^{k^2},$$

where $N_{sq} = (N - k + 1)^2$ is the number of $k \times k$ squares in a $N \times N$ grid.

This formula for $\mathbb{E}(R)$ allows us to compute thresholds, such as the minimal value of k such that $\mathbb{E}(R)$ becomes very small, that is less than ε, with $\varepsilon = 10^{-3}$ for instance.

Let us consider the example of Fig. 3.13. Here we have $N = 64$ and $p = 0.5$. Then for $k = 3$, we can explicitly compute $\mathbb{E}(R) \simeq 7.5$, and for $k = 8$ we get $\mathbb{E}(R) \simeq 2 \times 10^{-16}$. See the comment of the figure.

Now the link between thresholds computed this way and perceptual thresholds that could be tested by psychophysicists is a difficult, but interesting question. Some studies on this have been performed for instance by Blusseau, Grompone and collaborators in [4].

3.2   A Contrario Method for the Detection of Geometric Structures in Images

The above simple computation is a first example of a general methodology to detect geometric structures in images. This methodology is called the *a contrario* method and it is closely related to statistical hypothesis testing. The aim here is not to have a statistical model of what an image is and use it as a prior in computations. Instead, we just need a simple statistical model (called the noise model, or background model, or a contrario model) of what the image **is not**, and we will perform tests to find the structures that show the image does not follow this noise model.

In this section, we will first give the general formulation of the a contrario methodology, and then we will give several instances of application of the methodology for detection purposes.

3.2.1   General Formulation of a Contrario Methods

We give here the general formulation of the a contrario methodology. Here are the different steps:

- Given n geometric objects $O_1, \ldots, O_n$, let X_i be a random variable describing an attribute of the object O_i (for instance: its position, its color, its orientation, or its size, etc. . .).
- Define an hypothesis H_0 (also called *background distribution* or *noise model* or *a contrario model*): $X_1, \ldots, X_n$ are independent identically distributed, following a "simple" distribution (the uniform one in general).
- Observe an event E on a realization $x_1, \ldots, x_n$ of $X_1, \ldots, X_n$ (for example: there is k such that $x_1, \ldots, x_k$ are very similar). Ask the question: Can this observed event happen by chance? (i.e. how likely is it under the null hypothesis H_0?)
- Perform a test. First define the *Number of False Alarm* of the event E by:

$$\text{NFA}(E) := \mathbb{E}_{H_0}[\text{nb of occurrences of } E].$$

And then test

$$\text{NFA}(E) \leqslant \varepsilon,$$

where ε is a small number (less than 1).
- If the test is positive, then the observed event E is said to be an ε-meaningful event.

This a contrario methodology is very general and can be applied in several different frameworks. However each case is not straightforward, and we need to define precisely what are the objects, the attributes, the noise model and the observed events.

3.2.2 Detection of Alignments in an Image

Detecting alignments (i.e. straight segments) in an image is often one of the first
steps for higher level object recognition. To do this task in the a contrario framework,
we need to precisely define the considered events. An alignment in a grey level
image will be defined as a group of aligned pixels such that the orientations of the
image at these pixels are also aligned (at least approximately). For more details
about this section, see [11] and [14].

3.2.2.1 Definitions and First Properties

Let us consider a discrete grey level image u of size $N \times N$ pixels. At each pixel
$\mathbf{x} = (x, y)$ we can compute the gradient of u by

$$\nabla u(\mathbf{x}) := \frac{1}{2} \begin{pmatrix} u(x+1, y+1) + u(x+1, y) - u(x, y+1) - u(x, y) \\ u(x+1, y+1) + u(x, y+1) - u(x+1, y) - u(x, y) \end{pmatrix}. \tag{3.1}$$

Then, the direction at a point $\mathbf{x}$ is defined as

$$\mathbf{d}(\mathbf{x}) = \frac{\nabla u(\mathbf{x})^{\perp}}{\|\nabla u(\mathbf{x})\|}.$$

It is a vector of unit length, and we can write it, using complex number notations, as
$\mathbf{d}(\mathbf{x}) = e^{i\theta(\mathbf{x})}$, where $\theta(\mathbf{x}) \in [0, 2\pi)$ is then called the orientation at $\mathbf{x}$. In the image,
the gradient is orthogonal to level lines ("objects" in the image), and therefore the
direction $\mathbf{d}$ will represent the tangent to the objects. Notice that here we forget
the gradient amplitude, only its orientation is considered. This implies that we will
detect straight segments independently of their contrast.

We define a discrete oriented segment $\mathbf{S}$ of length l as a sequence $\{\mathbf{x}_1, \mathbf{x}_2, \dots, \mathbf{x}_l\}$
of l aligned pixels (meaning that there is an underlying continuous line going
through them, see Fig. 3.14), that are 2-adjacent (meaning that we take them not
contiguous, but at distance 2 such that their gradient are computed on disjoint
neighborhoods). Such a segment is determined by its starting point (pixel) $\mathbf{x}_1$ and
its ending point (pixel) $\mathbf{x}_l$. As a consequence, the total number of segments in the
image is finite. If we denote it by N_S, then

$$N_S = N^2(N^2 - 9) \simeq N^4,$$

since an oriented segment is determined by its first point (N^2 choices) and its last
point ($N^2 - 9$ choices because it cannot be an immediate neighbor of the first point).

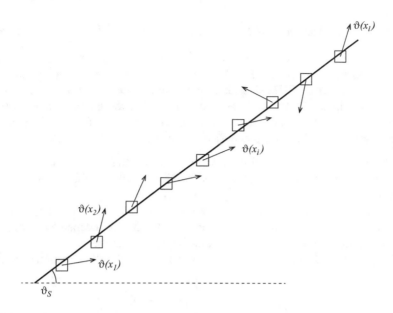

Fig. 3.14 A discrete oriented segment **S**, seen as a sequence $\{x_1, x_2, \ldots, x_l\}$ of aligned pixels. Each pixel has an orientation $\theta(\mathbf{x})$, that is then compared to the orientation θ_S of the segment **S**

Let us fix a precision p, that is a number in $(0, 1/4]$ (in general, it is taken in practice to be equal to $1/16$). We say that a point $\mathbf{x}$ is aligned with a direction $\mathbf{v}$ up to precision p if

$$|\text{Angle}(\mathbf{d}(\mathbf{x}), \mathbf{v})| \leqslant p\pi.$$

On a given segment **S** of length l, we count the number of points of **S** that are aligned with the direction of **S** up to precision p. If this number is high enough to be very unlikely under the null hypothesis H_0 (that we need to define), we will "keep" the segment, as being a large deviation from randomness.

The null hypothesis H_0 is here: The orientations at the pixels are independent and uniformly distributed on $[0, 2\pi)$. Then, as a consequence, we have

$$\mathbb{P}_{H_0}[\mathbf{d}(\mathbf{x}) \text{ is aligned with } \mathbf{v} \text{ up to precision } p] = \frac{2\pi p}{2\pi} = p. \tag{3.2}$$

Lemma 3.1 *If U is a white noise image (meaning that the grey levels are i.i.d. $\mathcal{N}(\mu, \sigma^2)$), then the orientations $\theta(\mathbf{x})$ are uniformly distributed on $[0, 2\pi)$. They are moreover independent at points taken at distance larger than 2.*

Proof Let us compute the law of ∇U at a point $\mathbf{x} = (x, y)$. Thanks to Formula (3.1), we have $\nabla U(\mathbf{x}) = AX$ where $X \in \mathbb{R}^4$ is the vector $(U(x, y), U(x+1, y), U(x, y+$

1), $U(x+1, y+1))$ and A is the 2×4 matrix given by $A = \frac{1}{2} \begin{pmatrix} -1 & 1 & -1 & 1 \\ -1 & -1 & 1 & 1 \end{pmatrix}$. When

U is a white noise image of law $\mathcal{N}(\mu, \sigma^2)$, then X is a Gaussian vector of mean (μ, μ, μ, μ) and covariance $\sigma^2 \mathbf{I}_4$ (where for $n \geq 1$, $\mathbf{I}_n$ denotes the identity matrix of size $n \times n$). Therefore $\nabla U(\mathbf{x})$ is a Gaussian vector of mean $A(\mu, \mu, \mu, \mu) = (0, 0)$ and of covariance $\sigma^2 A A^T = \sigma^2 \mathbf{I}_2$. This shows that the gradient is an isotropic Gaussian vector, and as a consequence its orientation is uniformly distributed on $[0, 2\pi)$ (and its amplitude follows a Rayleigh distribution of parameter σ). □

Let $\mathbf{S} = \{\mathbf{x}_1, \dots, \mathbf{x}_l\}$ be a discrete segment of length l (counted in independent points, i.e. at distance 2) in an image following the null hypothesis H_0 (that is for instance a Gaussian white noise image of size $N \times N$).

Let $X_i = \mathbb{1}_{|\theta_{\mathbf{S}} - \theta(\mathbf{x}_i)| \leq p\pi}$ be the random variable that has value 1 if $\mathbf{x}_i$ is aligned with $\mathbf{S}$ up to precision p, and 0 otherwise. Then the X_i are independent and thanks to Eq. (3.2), they all follow a Bernoulli distribution of parameter p.

Let $S_l = \sum_{i=1}^{l} X_i$ be the number of aligned points with the direction of $\mathbf{S}$, then:

$$\mathbb{P}_{H_0}[S_l = k] = \binom{l}{k} p^k (1-p)^{l-k}$$

and then $\mathbb{P}_{H_0}[S_l \geq k] = B(l, k, p) := \sum_{j \geq k} \binom{l}{j} p^j (1-p)^{l-j},$

where $B(l, k, p)$ denotes the tail of the binomial distribution of parameters l and p.

We can now give the main definition, that is the one of ε-meaningful segments.

Definition 3.1 Let $\mathbf{S}$ be a segment of length $l(\mathbf{S})$ containing $k(\mathbf{S})$ aligned points (for the precision p). The number of false alarms (NFA) of $\mathbf{S}$ is defined by

$$\text{NFA}(\mathbf{S}) = \text{NFA}(l(\mathbf{S}), k(\mathbf{S})) := N^4 \times B(l(\mathbf{S}), k(\mathbf{S}), p)$$

$$= N^4 \times \sum_{k=k(\mathbf{S})}^{l(\mathbf{S})} \binom{l(\mathbf{S})}{k} p^k (1-p)^{l(\mathbf{S})-k}.$$

Let $\varepsilon > 0$, be a (small) positive number. When $\text{NFA}(\mathbf{S}) \leq \varepsilon$, then the segment is said ε-meaningful. This is equivalent to have $k(\mathbf{S}) \geq k_{\min}(l(\mathbf{S}))$, where

$$k_{\min}(l) := \min \left\{ k \in \mathbb{N}, \ B(l, k, p) \leq \frac{\varepsilon}{N^4} \right\}.$$

The main property of ε-meaningful segments is that, on the average, we will not observe more than ε such segments in images following the H_0 hypothesis (for instance in a Gaussian white noise image). More precisely, we have the following proposition.

Proposition 3.1 *The expectation of the number of ε-meaningful segments in a random image of size $N \times N$ pixels following the null hypothesis H_0, is less than ε.*

Proof Let N_S be the number of segments in the $N \times N$ image. Let $e_i = 1$ if the i-th segment of the image is ε-meaningful, and 0 otherwise. Let R be the number of ε-meaningful segments in the image. Then

$$\mathbb{E}_{H_0}[R] = \sum_{i=1}^{N_S} \mathbb{E}[e_i] = \sum_{i=1}^{N_S} \mathbb{P}_{H_0}[S_{l_i} \geqslant k_{\min}(l_i)]$$

$$= \sum_{i=1}^{N_S} B(l_i, k_{\min}(l_i), p) \leqslant N_S \times \frac{\varepsilon}{N^4} \leqslant \varepsilon.$$

$\square$

The number NFA($\mathbf{S}$) measures the degree of confidence, or the "meaningfulness" of an observed alignment. The smaller NFA($\mathbf{S}$) is, the more meaningful the segment $\mathbf{S}$ is, since it has a smaller probability of having arisen just by chance.

Notice also that the NFA($\mathbf{S}$) is related to the so-called Per Family Error Rate in Statistics, when multiple tests are performed and a Bonferroni correction is applied.

Using elementary properties of the binomial distribution, we can easily deduce some elementary properties of the NFA, such that:

1. NFA$(l, 0) = N^4$. It means that a segment containing 0 aligned points will never be meaningful (this is quite natural and expected!).
2. NFA$(l, l) = N^4 p^l$. It implies that, in order to be ε-meaningful, a segment has to be of length larger than $\frac{4 \log N - \log \varepsilon}{-\log p}$. For instance for $N = 512$, $p = 1/16$ and $\varepsilon = 1$, we have $l \geqslant 9$ (that is a "true" length of at least 18 pixels).
3. NFA$(l, k + 1) <$ NFA(l, k). It means that when two segments have the same length, the most meaningful one (i.e. the one with the smallest NFA) is the one that has the largest number of aligned points. Again this is quite normal and expected.
4. NFA$(l, k) <$ NFA$(l + 1, k)$. It can be interpreted by saying that if one considers a segment, and then extends it with a non-aligned point, then one increases the NFA. Also, if one removes a non-aligned point from a segment, one decreases its NFA.
5. NFA$(l+1, k+1) <$ NFA(l, k). This is the opposite situation: here, if one extends the segment with an aligned point, then one decreases its NFA. Or, also, if one removes an aligned point from a segment, one increases its NFA.

3.2.2.2 Maximality

In the previous section, we have seen the definition of meaningful segments. Now, when a segment $\mathbf{S}$ is "very" meaningful (that is NFA($\mathbf{S}$) $\ll \varepsilon$), then many segments that contain it or that are contained in it, will also be meaningful. Therefore the detections that we obtain in an image are redundant, and some of the detected segments are just the consequence of a very meaningful segment. This is why we need to introduce a notion of "maximality", that is defined as follows (see also [13] for more details about this part).

Definition 3.2 A segment **S** is said maximal meaningful if it is meaningful and if

$$\forall \text{ segment } \mathbf{S}' \subset \mathbf{S}, \quad \text{NFA}(\mathbf{S}') \geqslant \text{NFA}(\mathbf{S}),$$

$$\forall \text{ segment } \mathbf{S}' \supset \mathbf{S}, \quad \text{NFA}(\mathbf{S}') > \text{NFA}(\mathbf{S}).$$

In other words, maximal meaningful segments are local minima of the NFA for the relationship of inclusion.

Thanks to the elementary properties of the NFA, we immediately have the two following properties of maximal meaningful segments:

- The two ending points of S are aligned with S,
- The two "adjacent" points to S (i.e. one "before" and one "after" S), are not aligned with S.

The difference between meaningful segments and maximal meaningful ones is illustrated on Fig. 3.15.

In fact, maximal meaningful segments have strong structural properties, mainly the one of being disjoint when lying on the same straight line. This property has been numerically checked for segments of length l up to 256 (see [14]), but the general result is still an open conjecture that can be stated as follows:

Conjecture If S_1 and S_2 are two distinct meaningful segments lying on the same straight line, and such that $S_1 \cap S_2 \neq \emptyset$, then

$$\min(\text{NFA}(S_1 \cup S_2), \text{NFA}(S_1 \cap S_2)) < \max(\text{NFA}(S_1), \text{NFA}(S_2)).$$

Fig. 3.15 From left to right: an image, all its ε-meaningful segments (with $\varepsilon = 10^{-3}$), its maximal meaningful segments

Fig. 3.16 Left: an original image (a painting by Uccello). Middle: the maximal meaningful segments. Right: result of the LSD algorithm (only the middle line of the meaningful regions are shown)

As a direct consequence of this conjecture, we would have that indeed two maximal meaningful segments lying on the same straight line cannot meet.

Let us also remark that the above conjecture is equivalent to say that

$$\min(B(l_1+l_2-l_\cap, k_1+k_2-k_\cap, p), B(l_\cap, k_\cap, p)) < \max(B(l_1, k_1, p), B(l_2, k_2, p)),$$

where $l_\cap$ and $k_\cap$ (resp. $l_\cup$ and $k_\cup$) denote the length and the number of aligned points of $S_1 \cap S_2$ (resp. of $S_1 \cup S_2$). Such a result may seem simple, it is however not so easy, and we haven't found a proof for it (see [14] for developments on this conjecture).

As shown on Figs. 3.15 and 3.16 (middle), the maximal meaningful segments found in an image are "satisfactory" from a perceptual viewpoint. Now, since the method is based on an almost exhaustive testing of all segments of the image (there are however some properties that can be used to have a not so complex algorithm), it has the drawbacks that it is quite slow, and we often get "bundles" of segments. An elegant solution to these two points has been proposed by Grompone et al. in [21]. It is based on the idea of considering directly "thick" segments (rectangles), and it is explained in the following subsection.

3.2.2.3 The Line Segment Detector Algorithm

The Line Segment Detector (LSD) Algorithm proposed by Grompone et al. in [21] can be summarized the following way:

1. Partition of the image (of size $N \times N$ pixels) into *Line-Support Regions* (connected sets of pixels sharing the same orientation up to precision $p\pi$).
2. Approximate these sets by rectangular regions.
3. Compute the NFA of each region: for a region (rectangle) $\mathbf{r}$ containing $l(\mathbf{r})$ points with $k(\mathbf{r})$ of them aligned with it (i.e. aligned with the principal orientation of $\mathbf{r}$ up to precision $p\pi$), define

$$\text{NFA}(\mathbf{r}) = N^5 \times B(l(\mathbf{r}), k(\mathbf{r}), p).$$

4. The rectangular regions $\mathbf{r}$ such that $\text{NFA}(\mathbf{r}) < 1$ are kept (meaningful regions).

The definition of meaningful regions in the LSD algorithm is very similar to the one of meaningful segments given in Definition 3.1. Here ε is fixed to 1, and the number of tests is now N^5, that is (approximately) the number of possible rectangles in a $N \times N$ image. Thanks to its first step, the LSD algorithm is able to directly identify the candidate meaningful regions, without the need of an exhaustive search. It makes therefore the algorithm much faster than the one of meaningful segments, and the results are much "cleaner". This is illustrated on Fig. 3.16.

The LSD algorithm can be tested online [22] on the website of the Image Processing On Line (IPOL) journal (http://www.ipol.im/), whose aim is to *emphasize the role of mathematics as a source for algorithm design and the reproducibility of the research.*

Remark The methodology and the definitions for meaningful segments (or rectangles) can be easily extended to other parametric curves, such as arcs of circle for instance. More generally, the a contrario methodology can be used to find meaningful peaks in the so-called Hough Transform (see [23] for an historical viewpoint on the Hough Transform).

3.2.3 Detection of Contrasted or Smooth Curves

3.2.3.1 Meaningful Boundaries

It is a classical problem in Image Processing to find the boundaries (contours) in an image. This problem is generally called the *edge detection* problem. In order to apply the a contrario methodology to this problem, we first need to consider "test objects", that will be curves here. Now, we don't want to restrict ourselves to parametric curves, and it is also impossible to consider all possible curves in an image. Since we are interested in curves that will have a contrast across them, some natural candidates are the level lines of the image (or pieces of them). Let us first recall how they are defined.

Definition 3.3 Let Ω be a discrete domain (rectangle of $\mathbb{Z}^2$), and let $u : \Omega \to \mathbb{R}$ be a grey level image. The upper- and lower-level sets of u are respectively the sets defined for all $\lambda \in \mathbb{R}$ by

$$\chi_\lambda(u) = \{\mathbf{x} \in \Omega \,;\, u(\mathbf{x}) \geqslant \lambda\} \quad \text{and} \quad \chi^\lambda(u) = \{\mathbf{x} \in \Omega \,;\, u(\mathbf{x}) \leqslant \lambda\}.$$

The level lines of u are then defined as being the (discrete) topological boundaries of its level sets.

Notice that the upper-level sets are decreasing: $\forall \lambda \leqslant \mu, \ \chi_\mu \subset \chi_\lambda$, whereas the lower-level sets are increasing: $\forall \lambda \leqslant \mu, \ \chi^\lambda \subset \chi^\mu$. The knowledge of all the upper (resp. lower)-level sets is enough to reconstruct u by for instance

$$u(\mathbf{x}) = \sup\{\lambda \,;\, \mathbf{x} \in \chi_\lambda\}.$$

Fig. 3.17 Left: an image (the church of Valbonne, source: INRIA). The grey levels are taking value in the range $\{0, 1, \ldots, 255\}$. Right: the level lines corresponding to levels λ multiple of 6

Another interesting property of the level sets is that they are globally invariant to contrast changes. More precisely, if g is a change of contrast (i.e. an increasing function), then u and $g(u)$ globally have the same level sets, since $\forall \lambda \, \exists \mu$ such that $\chi_\lambda(u) = \chi_\mu(g(u))$.

On Fig. 3.17, we show an example of an image and some of its level lines. We don't show all the level lines because they would cover entirely the image. A reader interested in making its own experiments on level lines can use the Image Processing On Line demo on "Image Curvature Microscope" [9].

Let u be a discrete grey level image of size $N \times N$ and let N_{ll} be the (finite) number of level lines it contains. Let L be a level line of u. It is a sequence of pixels such that each of them is neighbor of the previous one. In this sequence, we keep only points at distance 2, and then identify L with this sub-sequence of pixels, that is $L = \{\mathbf{x}_1, \ldots, \mathbf{x}_l\}$, and l is called the length of L. We define the contrast of u at a point $\mathbf{x}$ by

$$c(\mathbf{x}) = \|\nabla u(\mathbf{x})\|,$$

where ∇u is the gradient of u (computed for instance on a 2×2 neighbourhood as in the case of segments). Notice that we only consider here the amplitude of the gradient. Its orientation is almost already known, since it is orthogonal to the level lines (at least when considering an underlying continuous framework).

We can now define meaningful boundaries in an image. More details about all this can be found in [12]. Roughly speaking, a meaningful boundary is a level line that is long enough and contrasted enough not to appear just by chance. It is more precisely defined the following way.

Definition 3.4 Let u be an image of size $N \times N$, containing N_{ll} level lines. Let H be the empirical gradient amplitude distribution in u given by

$$H(\mu) = \frac{1}{N^2} \, \#\{\mathbf{x} \, ; \, \|\nabla u(\mathbf{x})\| \geqslant \mu\} \tag{3.3}$$

We define the Number of False Alarms of a level line L with discrete length l and minimal contrast $\mu = \min_{\mathbf{x} \in L} c(\mathbf{x})$ by

$$\mathrm{NFA}(L) = N_{ll} \times H(\mu)^l.$$

The level line L is said to be an ε-meaningful boundary iff $\mathrm{NFA}(L) \leqslant \varepsilon$.

Here the underlying a contrario model H_0 is given by: the contrasts at points taken at distance at least 2 are independent and distributed according to H (that is given by Eq. (3.3)). Notice that this is a noise model on the contrasts but not on the image itself. It is indeed not clear what image noise model it corresponds to (if it exists).

Meaningful boundaries have several interesting properties, that are direct consequences of their definition:

- The ε-meaningful boundaries are invariant under affine contrast changes (that is u and $g(u)$ have exactly the same ε-meaningful boundaries when g is of the form $g(t) = at + b$, with $a > 0$).
- If $l \leqslant l'$ and μ is fixed, then $H(\mu)^l \geqslant H(\mu)^{l'}$. Therefore if two level lines have the same contrast, the longest one has the smallest NFA.
- If $\mu \leqslant \mu'$ and l is fixed, then $H(\mu)^l \geqslant H(\mu')^l$. Therefore if two level lines have the same length, the most contrasted one has the smallest NFA.
- A level line with minimal contrast μ is ε-meaningful iff its length is larger than

$$l_{\min}(\mu) = \frac{\log \varepsilon - \log N_{ll}}{\log H(\mu)}.$$

- A level line with length l is ε-meaningful iff its minimal contrast μ is larger than

$$\mu_{\min}(l) = H^{-1}\left(\left(\frac{\varepsilon}{N_{ll}} \right)^{1/l} \right).$$

As in the case of segments, meaningful level lines are often organized in "bundles". When a contrasted "object" is present in the image, many parallel level lines representing the contour of this object will be detected (see for instance Fig. 3.18—middle). Therefore, we also need to define here a notion of maximality. As for segments, maximal meaningful level lines will be defined as being local

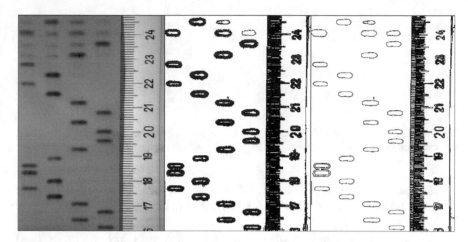

Fig. 3.18 On the left, the original image. On the middle, all meaningful boundaries with $\varepsilon = 1$. On the right, all maximal meaningful boundaries with $\varepsilon = 1$

minima of the NFA for the relationship of inclusion. Here the natural organization of level sets (and therefore of level lines) is a tree structure, and we will consider "branches" in this tree. More precisely, the definition of maximal meaningful boundaries is the following.

Definition 3.5 A monotonic branch in the tree of level lines is a branch along which the grey level is monotonic and such that each level line has a unique child. A monotonic branch is maximal if it is not contained in another monotonic branch. A level line is a maximal meaningful boundary if it is meaningful and if its NFA is minimal in its maximal monotonic branch of the tree of level lines.

An example of an image, its meaningful boundaries and the maximal meaningful ones is given on Fig. 3.18. On this figure, one can clearly see how the notion of maximality helps to describe in an "optimal" way the set of meaningful boundaries of the image.

3.2.3.2 Meaningful Good Continuations

The goal here will be to look for "smooth" curves in the image, without considering the contrast along the curve. The content of this section is taken from the work of Cao in [7].

Since all curves are discrete (they are sequences of pixels), the definition of smoothness has to be adapted. Here a discrete curvature will be defined, and a smooth curve will be a curve such that its curvature is "small". The precise value of "small" will be obtained by applying the a contrario methodology.

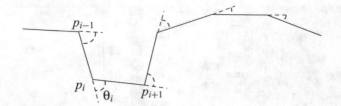

Fig. 3.19 A discrete curve Γ is a sequence of points $p_0, \ldots, p_{l+1}$, and we define its maximal discrete curvature by Eq. (3.4)

The framework is the following. Let $\Gamma = (p_0, \ldots, p_{l+1})$ be a discrete curve of length l, and let κ be its maximal discrete curvature defined by

$$\kappa_\Gamma = \max_{1 \leqslant i \leqslant l} |\text{Angle}(p_{i+1} - p_i, p_i - p_{i-1})| = \max_{1 \leqslant i \leqslant l} |\theta_i|, \qquad (3.4)$$

as illustrated on Fig. 3.19.

The a contrario noise model H_0 is here: the angles θ_i are i.i.d. with uniform law on $[-\pi, \pi)$, i.e. the curve is a discrete "random walk". Under the a contrario noise model, the probability of having l angles θ_i smaller than an observed value κ is

$$\mathbb{P}(\forall 1 \leqslant i \leqslant l, \ |\theta_i| \leqslant \kappa) = \prod_{i=1}^{l} \mathbb{P}(|\theta_i| \leqslant \kappa) = \left(\frac{\kappa}{\pi}\right)^l.$$

As in all other applications of the a contrario methodology, we have to take into account the total number of "tests" that are made. Let thus N_c be the number of considered curves in the image. In practice, it will be the number of pieces of level lines in the image. We can now state the definition of meaningful good continuations.

Definition 3.6 We say that a discrete curve Γ is an ε-meaningful good continuation if

$$\kappa_\Gamma < \frac{\pi}{2} \quad \text{and} \quad \text{NFA}(\Gamma) = N_c \left(\frac{\kappa_\Gamma}{\pi}\right)^l \leqslant \varepsilon.$$

As for segments, a meaningful good continuation with a very small NFA (i.e. much less than ε) will generate many other meaningful good continuations that are else contained or included in it. Therefore we again need here to define a notion of maximality.

Definition 3.7 Let $\mathscr{G}$ be a set of (discrete) curves. A meaningful good continuation $\Gamma \in \mathscr{G}$ is maximal meaningful if it is meaningful and if

$$\forall \Gamma' \subset \Gamma, \text{NFA}(\Gamma') \geqslant \text{NFA}(\Gamma) \quad \text{and} \quad \forall \Gamma' \supsetneq \Gamma, \text{NFA}(\Gamma') > \text{NFA}(\Gamma).$$

Fig. 3.20 Left: Image of the Church of Valbonne. Middle: Maximal meaningful good continuations. Right: Maximal meaningful contrasted boundaries

As a consequence, if $\mathscr{G}$ is the set of connected pieces of level lines of an image, then if Γ and Γ' are two maximal meaningful good continuations belonging to the same level line, we necessarily have that

$$\Gamma \cap \Gamma' = \emptyset.$$

This directly comes from the fact that if $\Gamma \cap \Gamma' \neq \emptyset$, then the curve $\Gamma \cup \Gamma'$ is an element of $\mathscr{G}$. It has a length larger than l and l' with a maximal curvature less than κ_Γ or $\kappa_{\Gamma'}$. Therefore its NFA is less than the one of Γ or of Γ', which contradicts the maximality of Γ or of Γ'.

An example of maximal meaningful good continuations on an image is given on Fig. 3.20—middle, where we also show, for comparison, the maximal meaningful boundaries. What is quite remarkable is that for good continuations we recover the boundaries of the "hand-made" objects present in the image (the church, the cars, but not the tree!), independently of their contrast. Indeed this is a direct consequence of the (in general) smoothness of hand-made objects. A not so smooth object would not be detected.

In the case of meaningful straight segments detection, we saw that the LSD algorithm was an elegant, efficient and fast solution to obtain a "clean" set of segments. Similarly here for good continuations, Grompone and Randall have proposed a smooth contour detection algorithm, that can be tested online [20].

3.2.4   Detection of Vanishing Points

In the a contrario methodology, one often has to compute the probability of appearance of such or such geometric event. These are typical questions of stochastic geometry. One of the main and characteristic example is the detection of vanishing points in an image, that we are going to develop in this section and that is inspired by the paper of Almansa et al. [2].

In a pinhole camera model, parallel straight lines in 3D are projected on the image plane as 2D lines that intersect at a single point. This principle is used, for instance by painters, to create perspective effects in paintings. The intersection points are called *vanishing points*. To detect them, we start from elementary straight segments detected in the image (by the LSD algorithm of Sect. 3.2.2.3 for instance), and then we look for regions in the image plane (inside or outside the image domain), such that "a lot of" segments converge towards these regions. As in the other examples, we will use the a contrario methodology to define "a lot of" as a mathematical threshold.

Let Ω be the image domain and let N be the number of elementary segments detected in the image. Let $D_1, \ldots D_N$ be the support lines of the N segments. We define a noise model H_0, that is here: "The N lines are i.i.d. uniform". Then, we will look for regions that are intersected by significantly more support lines than the number expected under H_0. We first need to explain what is the *uniform distribution* on the lines of the image. A line G of the plane is parametrized by its two polar coordinates: $\rho \geqslant 0$ and $\theta \in [0, 2\pi)$. That is

$$G = G(\rho, \theta) = \{(x, y) \in \mathbb{R}^2 \text{ s.t. } x \cos \theta + y \sin \theta = \rho\}.$$

Then, there exists a unique (up to a positive multiplicative constant) measure on the set of lines that is invariant under translations and rotations. It is the Poincaré measure given by (see for instance the book of Santalo [37]):

$$d\mu = d\rho \, d\theta.$$

This measure will be called the uniform measure on lines.

To detect meaningful vanishing regions, and since it is impossible to test all possible regions of the plane, we choose a partition of the plane $\mathbb{R}^2$ into a finite number M of regions (we will see later how to construct this partition):

$$\mathbb{R}^2 = \bigcup_{j=1}^{M} V_j.$$

Definition 3.8 Let V_j be a region and let $k_j = \#\{D_i \text{ s.t. } D_i \cap V_j \neq \emptyset\}$. The Number of False Alarms of V_j is defined by

$$\text{NFA}(V_j) := M \times B(N, k_j, p_j) = M \times \sum_{k=k_j}^{N} \binom{N}{k} p_j^k (1 - p_j)^{N-k},$$

where p_j is the probability that a random line (under $d\mu$) going through the image domain Ω also meets V_j. When $\text{NFA}(V_j) \leqslant \varepsilon$, then we say that the region V_j is ε-meaningful.

As usual in the a contrario methodology, we have the main property that, under H_0, the expected number of ε-meaningful regions is less than ε.

Now, two questions remain: determine a "good partition" of the plane, and compute p_j. Computing p_j is a typical question of stochastic geometry. It uses the following results (that can be found in [37]):

- Let $K \subset \mathbb{R}^2$ be a bounded closed convex set with non-empty interior. Then

$$\mu\left(\{(\rho, \theta) \text{ s.t. } G(\rho, \theta) \cap K \neq \emptyset\}\right) = \text{Per}(K), \tag{3.5}$$

where $\text{Per}(K)$ is the perimeter of K ($=$ length of the boundary).
- Let K_1 and K_2 be two bounded closed convex sets with non-empty interior. Then

$$\mu\left(\{(\rho, \theta) \text{ s. t. } G(\rho, \theta) \cap K_1 \neq \emptyset \text{ and } G \cap K_2 \neq \emptyset\}\right)$$

$$= \begin{cases} \text{Per}(K_1) & \text{if } K_1 \subset K_2, \\ L_i - L_e & \text{if } K_1 \cap K_2 = \emptyset, \\ \text{Per}(K_1) + \text{Per}(K_2) - L_e & \text{otherwise.} \end{cases} \tag{3.6}$$

where L_i and L_e respectively denote the interior perimeter and the exterior perimeter of K_1 and K_2 (see Fig. 3.21).

Fig. 3.21 The exterior perimeter is the perimeter of the convex hull of $K_1 \cup K_2$. For the interior perimeter, the definition is less simple, but one can see it as the length of an elastic band drawing an "8" around K_1 and K_2

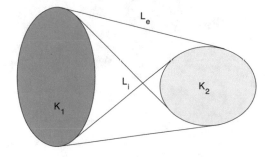

To choose the plane partition, we want to satisfy two constraints. The first one is that we would like to have all p_j equal. This will imply that all regions are equally detectable, meaning that they require the same minimal number of lines passing through them in order to become meaningful. The second constraint is about angular precision. Indeed, because of the pixelization of the image, the support line of a segment of length l (counted in pixels) is rather a cone of angle $d\theta = \arcsin\frac{1}{l} \simeq \frac{1}{l}$. Therefore, regions that are far away from the image domain Ω need to be larger than regions inside or close to Ω.

Finally, using the two above constraints, we propose (see [2]) to construct the partition of the plane the following way. We assume that the image domain is a disk of radius R, centered at the origin 0, that is $\Omega = D(0, R)$ (otherwise we include it in such a disk). We fix an angle θ. The interior regions are simply chosen as squares of side length $2R \sin\theta$. This implies by Eq. (3.5) that

$$p_j = \frac{\text{Per}(V_j)}{\text{Per}(\Omega)} = \frac{4 \sin\theta}{\pi} := p_\theta.$$

The exterior regions are then defined as portions of circular sectors with angle 2θ and determined by two distances d and d'.

Given a region $V_{d,d'}$, portion of a circular sector between the distances d and $d' > d$, then we can compute the probability that a random line intersects it knowing that it intersects Ω. Thanks to Eqs. (3.5) and (3.6) (see also Fig. 3.22) it is given by

$$pV_{d,d'} = \frac{L_i - L_e}{\text{Per}(\Omega)} = \frac{1}{\pi}\left(\tan\beta - \tan\beta' + \frac{1}{\cos\beta'} - \frac{1}{\cos\beta} + \beta' - \beta + 2\theta\right),$$

where $\beta = \arccos\left(\frac{R}{d}\cos\theta\right)$ and $\beta' = \arccos\left(\frac{R}{d'}\cos\theta\right)$.

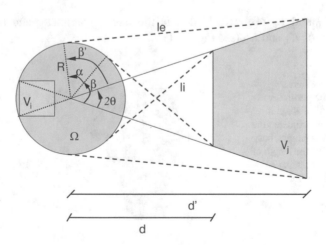

Fig. 3.22 Constructing the regions V_j of the partition of the plane

We can now exactly determine the exterior regions: we start with $d_1 = R$, then set $d_2 > d_1$ such that $pv_{d_1,d_2} = p_\theta$, then $d_3 > d_2$ such that $pv_{d_2,d_3} = p_\theta$, and so on, until being larger than d_∞, that is finite and characterized by

$$\forall d' > d_\infty, \quad pv_{d_\infty,d'} < p_\theta.$$

The last region is thus infinite and its probability is $< p_\theta$.

Now, how to choose θ? In order to have a good balance between detectability and localization, we don't fix a single value for θ, we instead use several values, having thus a multiscale approach. More precisely, we choose n angular values $\theta_s = \frac{2\pi}{2^s}$ with $s = 4, 5, \ldots, n + 3$. For each θ_s, we have a "partition" (there is some overlap between interior and exterior regions near the boundary of Ω) of the image plane into M_s regions: $\mathbb{R}^2 = \cup_{j=1}^{M_s} V_{j,s}$.

Definition 3.9 The number of false alarms of a region is then defined as

$$\mathrm{NFA}(V_{j,s}) = n \cdot M_s \cdot B(N_s, k(V_{j,s}), p_{\theta_s}),$$

where N_s is the number of segments (among the N) having a precision at least θ_s and $k(V_{j,s})$ is the number of support lines (among the N_s) that meet the region $V_{j,s}$. The region $V_{j,s}$ is said ε-meaningful iff $\mathrm{NFA}(V_{j,s}) \leqslant \varepsilon$.

As in all other applications of the a contrario methodology, we need to define a notion of maximality, related to local minima of the NFA. Indeed, when a lot of lines pass through a region $V_{j,s}$, then the neighbouring regions are also intersected by a lot of lines. We therefore need to select "the best regions".

Definition 3.10 The region $V_{j,s}$ is said maximal ε-meaningful if it is ε-meaningful and if

$$\forall s' \in [s_1, \ldots, s_n], \quad \forall j' \in [1, \ldots, M_{s'}],$$

$$\overline{V_{j',s'}} \cap \overline{V_{j,s}} \neq \emptyset \implies \mathrm{NFA}(V_{j,s}) \leqslant \mathrm{NFA}(V_{j',s'}).$$

Examples of results of maximal meaningful regions are shown on Fig. 3.23. This example also allows us to vizualise the partition of the plane that is used.

3.2.5 Detection of the Similarity of a Scalar Attribute

We have already seen several instances of application of the a contrario methodology with the definition of a Number of False Alarms (NFA) in each case: detection of segments, of contrasted curves (boundaries), of good continuations and of vanishing regions. Let us here be interested in the detection of the elementary Gestalt grouping law of similarity of a scalar attribute like grey level, or orientation, or size for instance.

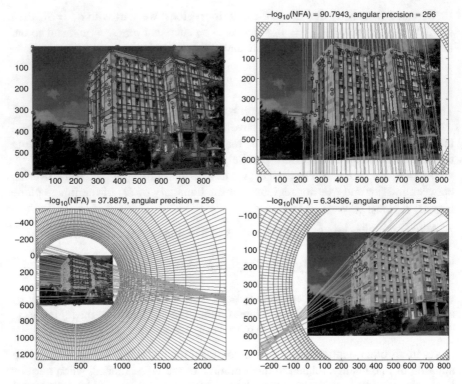

Fig. 3.23 An image (a building in Cachan), and its maximal meaningful regions. The most meaningful region is far away from the image domain, it is at "infinity" and corresponds to the set of (almost) parallel vertical lines in this image. For the two other maximal meaningful regions, the figures allow us to also see where they are among the exterior regions of the partition of the plane

The framework is the following. We assume we have M "objects", and each of them has an "attribute" $q \in \{1, 2, \ldots, L\}$. It can be its grey level, its orientation, its size. Let us consider the group $G_{a,b}$ of objects that have their scalar attribute q such that $a \leqslant q \leqslant b$. Let $k(a, b)$ denote the cardinal of $G_{a,b}$. Then, we can define its Number of False Alarm by

$$\mathrm{NFA}(G_{a,b}) = \mathrm{NFA}([a, b]) = \frac{L(L+1)}{2} \cdot B\left(M, k(a, b), \frac{b - a + 1}{L}\right),$$

where, as in Sect. 3.2.2, $B(M, k, p)$ denotes the tail of the binomial distribution of parameters M and p. The number of tests is here $\frac{L(L+1)}{2}$, it corresponds to the number of discrete intervals in $\{1, 2, \ldots, L\}$. Here again we can define a notion of maximality, saying that an interval (and therefore the corresponding group of objects) is maximal meaningful if its number of false alarms is minimal among the ones of all intervals that contain it or are contained in it.

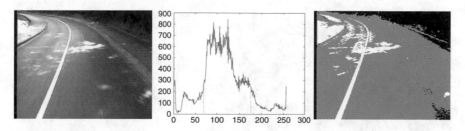

Fig. 3.24 Left: the original image. Middle: the histogram of grey levels. It contains one maximal meaningful interval, the interval [70, 175]. Right: quantized image obtained by quantifying the grey levels of the original image by the quantization function g given by $g(\lambda) = 0$ (black) if $\lambda < 70$, $g(\lambda) = 128$ (grey) if $70 \leqslant \lambda \leqslant 175$, and $g(\lambda) = 255$ (white) if $\lambda > 175$

A first example of application is the study of the grey level histogram of an image. The "objects" here are simply the pixels of the image, and their attribute is their grey level. Looking for the maximal meaningful intervals is a way to obtain an automatic histogram segmentation, and therefore a grey level quantization, as illustrated on Fig. 3.24.

A second example of application is a key point in the Gestalt theory: the recursivity of elementary grouping laws. The example is presented on Fig. 3.25. Here the considered attribute is the size (area) of the objects that are blobs detected thanks to their meaningful boundaries (Sect. 3.2.3.1).

3.2.6   Discussion

In the previous sections we have seen many examples of detection of geometric structures in images, based on the a contrario methodology. This methodology is a powerful computational tool that is the result of the combination of the non-accidentalness principle (Helmholtz principle) with the Gestalt theory of grouping laws. But beyond these detection problems, the a contrario framework has also been used for many other tasks in image processing such as: shape recognition (Musé et al. [31]), image matching (Rabin et al. [35]), epipolar geometry(Moisan and Stival [30]), motion detection and analysis (Veit et al. [41]), clustering (Cao et al. [8]), stereovision (Sabater et al. [36]), image denoising by grain filters (Coupier et al. [10]), etc.

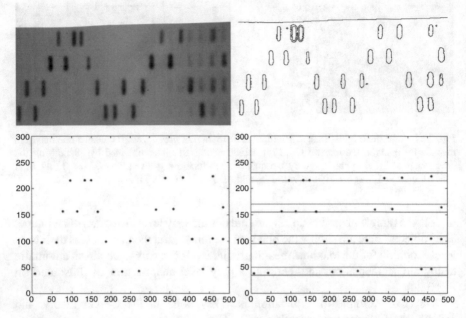

Fig. 3.25 Gestalt grouping principles at work for building an "order 3" gestalt (alignment of blobs of the same size). First row: original DNA image (left) and its maximal meaningful boundaries (right). Second row: left, barycenters of all meaningful regions whose area is inside the only maximal meaningful mode of the histogram of areas; right, meaningful alignments of these points

3.3   Stochastic Models of Images: The Problem of Modeling and Synthesizing Texture Images

In the previous sections about a contrario methodology, we needed to define a model of what the image was not (the so-called a contrario noise model). In most cases, the model was just "generalized uniform noise" (roughly saying that grey levels, orientations, lines, etc. are i.i.d. uniform). Here, at the opposite, we want to build models of what an image is. A typical example of the problem of modeling "natural" images in a stochastic framework is the case of texture images.

Many illustrations and parts of the text of this section have been prepared by Bruno Galerne[1] (MAP5, Université Paris Descartes), and are taken from the lectures that we give in the Master 2 program MVA (Mathematics, Vision, Learning) at the ENS Cachan, Université Paris-Saclay.

---

[1] Many thanks to him!

3.3.1   What is a Texture Image?

There is no clear mathematical unique definition of what a texture image is. We can just give a very general definition: a texture image is the realization of a random field, where a (more or less) random pattern is repeated in a (more or less) random way.

In fact, one of the main difficulties with texture images is that they have a huge variety of appearance. They can be roughly divided in two main classes: micro-textures (constitued of small non-discernible objects) and macro-textures (constitued of small but discernible objects). As illustrated on Fig. 3.26, a same scene, depending on the scale of observation, can generate images of various aspects, going from a micro-texture to a macro-texture, and even to an image that is not a texture anymore.

3.3.2   Texture Synthesis

One of the main question with texture images is the problem of texture synthesis. This problem can be stated as follows: given an input texture image, produce a new output texture image being both visually similar to and pixel-wise different from the input texture. The output image should ideally be perceived as another part of the same large piece of homogeneous material the input texture is taken from (see

(a)   (b)   (c)

Fig. 3.26 Depending on the viewing distance (the scale), the same scene can be perceived either as from left to right: (**a**) a micro-texture, (**b**) a macro-texture, or (**c**) a collection of individual objects (some pebbles)

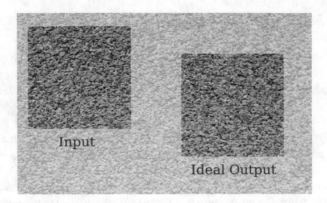

Fig. 3.27 The problem of texture synthesis: produce an image that should be perceived as another part of the same large piece of homogeneous material the input texture is taken from

Fig. 3.27). Texture synthesis is a very common issue in Computer Graphics, and it has a lot of applications for animated films or video games.

Texture synthesis algorithms can be roughly divided in two main classes: the neighborhood-based synthesis algorithms (or "copy-paste" algorithms) and the statistical constraints-based algorithms.

The general idea of neighborhood-based synthesis algorithms ("copy-paste" algorithms) is to compute sequentially an output texture such that each patch of the output corresponds to a patch of the input texture. Many variants have been proposed by: changing the scanning orders, growing pixel by pixel or patch by patch, doing a multiscale synthesis, choosing a different optimization procedure, etc. The main properties of these algorithms is that they synthesize well macro-textures, but they can have some speed and stability issues, and it is difficult to set the parameters values.

The "father" of all these algorithms is the famous Efros–Leung algorithm [16]. Its general idea relies on the Markov property assumption for the texture image I, and its aim is to provide an estimate of the probability of $I(\mathbf{x})$ knowing $I(N(\mathbf{x}))$, where $N(\mathbf{x})$ denotes a neighborhood of $\mathbf{x}$. To do this, it just searches in the input original image all similar neighborhoods, and that is then used as an estimate of the probability distribution of $I(\mathbf{x})$ knowing its neighborhood. Finally, to sample from this probability distribution, it just picks one match at random. More precisely, the algorithm works as follows:

1. Input: original texture image I_0, size of the output image I, size of the neighborhood, precision parameter ε (=0.1 in general).
2. Initialize I: take a random patch from I_0 and put it somewhere in I.
3. While the output I is not filled:

    (a) Pick an unknown pixel $\mathbf{x}$ in I with maximal known neighbor pixels, denoted $N(\mathbf{x})$.

(b) Compute the set of $\mathbf{y}$ in I_0 such that

$$d(I_0(N(\mathbf{y})), I(N(\mathbf{x}))) \leqslant (1 + \varepsilon) \min_{\mathbf{z}} d(I_0(N(\mathbf{z})), I(N(\mathbf{x}))),$$

where d is the L^2 distance.
(c) Randomly pick one of these $\mathbf{y}$ and set $I(\mathbf{x}) = I_0(\mathbf{y})$.

The mathematical analysis of this algorithm has been performed in [26], where they write it as a nonparametric algorithm for bootstrapping a stationary random field, and they show consistency results.

On Fig. 3.28, we show some examples of results obtained by this algorithm. In general, the results are visually impressive. They are however very sensitive to the

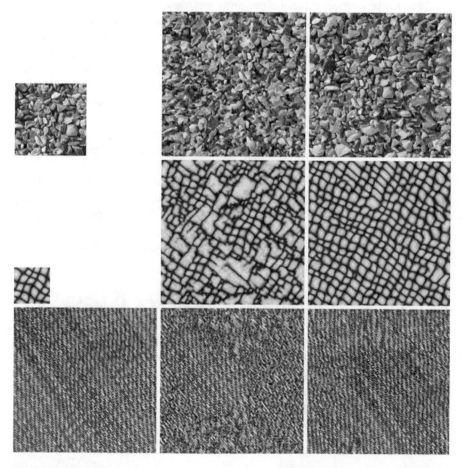

Fig. 3.28 Some results of the Efros–Leung texture synthesis algorithm, obtained from IPOL online demo [1]. Left column, the original input images: pebbles (size 128×128), cells (size 64×64) and fabric (size 256×256). Second and third column: outputs (all of size 256×256) generated by the Efros–Leung algorithm, with respective neighborhood size 5×5 and 11×11

size of the neighborhood, and one can sometimes observe two main failings of the method: it can produce "garbage" (it has "lost" the structures of the image and generates things that look like "noise") and it also often produces verbatim copy of the original image (large parts of the original image are reproduced as they are in the output image, resulting in a very limited statistical innovation).

In the other class of texture synthesis algorithms (the ones based on statistical constraints), the aim is to try to build a statistical model of the image. In a generic way, these algorithms work as follows:

1. Extract some meaningful statistics from the input image (e.g. distribution of colors, of Fourier coefficients, of wavelet coefficients, . . .).
2. Compute a random output image having the same statistics: start from a white noise and alternatively impose the statistics of the input.

The main properties of these algorithms are: they are perceptually stable (the outputs always have the same visual aspect), they allow mathematical computations, but they are generally not good enough for macro-textures (that have long-range interactions that are difficult to capture with statistics). The main representatives of these algorithms are: the Heeger–Bergen algorithm [24] (the considered statistics are the histograms of responses to multiscale and multi-orientation filters), the Portilla–Simoncelli algorithm [33] (the considered statistics are the correlations between multiscale and multi-orientation filters responses), and the Random Phase Noise [17] of Galerne et al. (that will be detailed in the next section, and where the considered statistics are the Fourier modulus). Some examples of results with these three algorithms are shown on Fig. 3.29.

To end this section, let us mention that the literature on texture synthesis methods is quite huge, and many algorithms exist that cannot be simply classified as being else "copy-paste" algorithms or "statistics-based" algorithms. They generally mix several ideas. Just to mention a few of them: [34, 39] and the recent successful algorithm of Gatys et al. [19] that uses convolutional neural networks.

3.3.3   Discrete Fourier Transform and the RPN Algorithm

We work here with digital images $u : \Omega \to \mathbb{R}$ where Ω is a rectangle of $\mathbb{Z}^2$ centered at $\mathbf{0} = (0, 0)$ and of size $M \times N$ where M and N are both assumed to be odd (this is just an assumption to make some formulas simpler, and of course things can also be written with even image sizes). We thus write: $\Omega = \{-\frac{M-1}{2}, \ldots, 0, \ldots, \frac{M-1}{2}\} \times \{-\frac{N-1}{2}, \ldots, 0, \ldots, \frac{N-1}{2}\}$. The discrete Fourier transform (DFT) of u is defined by

$$\forall \boldsymbol{\xi} \in \Omega, \quad \widehat{u}(\boldsymbol{\xi}) = \sum_{x \in \Omega} u(\mathbf{x}) e^{-2i\pi \langle \mathbf{x}, \boldsymbol{\xi} \rangle},$$

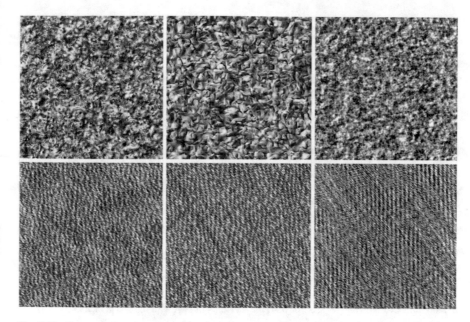

Fig. 3.29 Some results of statistics-based texture synthesis algorithm. The original input images are not shown here, they are the same as in Fig. 3.28. Left column: results of the Heeger–Bergen algorithm (obtained with the IPOL online demo [6]). Middle column: results of the Portilla–Simoncelli algorithm (obtained with the code provided by the authors). Right column: results of the RPN algorithm (obtained with the IPOL online demo [18])

where the inner product between $\mathbf{x} = (x_1, x_2)$ and $\boldsymbol{\xi} = (\xi_1, \xi_2)$ is defined by $\langle \mathbf{x}, \boldsymbol{\xi} \rangle = \frac{1}{M} x_1 \xi_1 + \frac{1}{N} x_2 \xi_2$. As usual, the image u can be recovered from its Fourier transform $\widehat{u}$ by the Inverse Discrete Fourier Transform:

$$\forall \mathbf{x} \in \Omega, \quad u(\mathbf{x}) = \frac{1}{|\Omega|} \sum_{\boldsymbol{\xi} \in \Omega} \widehat{u}(\boldsymbol{\xi}) e^{2i\pi \langle \mathbf{x}, \boldsymbol{\xi} \rangle},$$

where $|\Omega| = MN$ is the size of the domain Ω.

The modulus $|\widehat{u}|$ is called the *Fourier modulus (or Fourier amplitude)* of u and $\arg(\widehat{u})$ is the *Fourier phase* of u. Since u is real-valued, its Fourier modulus is even while its Fourier phase is odd. The Fourier modulus and the Fourier phase of an image play very different roles, in the sense that they "capture" different features of the image. Indeed, geometric contours are mostly contained in the phase [32], while (micro-)textures are mostly contained in the modulus. This fact can be experimentally checked by taking two images and by exchanging their Fourier modulus or their Fourier phases. See Figs. 3.30 and 3.31 for some illustrations of this.

We can now describe the RPN (Random Phase Noise) algorithm [17]. It is based on the above observation: for micro-textures, only the Fourier modulus is

Fig. 3.30 Exchanging Fourier modulus or phases of two images. This is an experimental check that geometric contours are mostly contained in the phase, while textures are mostly contained in the modulus. First line: Images Im1, Im2, Im3 and Im4. Second line, images obtained by taking respectively: the modulus of Im1 with the phase of Im2, the modulus of Im2 with the phase of Im1, the modulus of Im3 with the phase of Im4, the modulus of Im4 with the phase of Im3

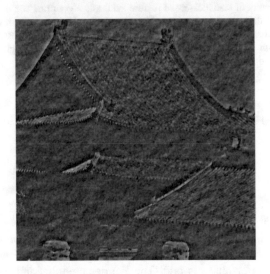

Fig. 3.31 The image that has the Fourier modulus of Im4 and the Fourier phase of Im1. This image has captured the geometry of Im1 and the texture of Im4

important, the Fourier phase can be randomly changed. The RPN algorithm works as follows:

1. Let u_0 be the original input image. To avoid artefacts in the Fourier transform due to the boundaries of Ω, replace u_0 by its *periodic component* [29].

2. Compute the Discrete Fourier Transform $\widehat{u_0}$ of the input u_0.
3. Compute a random phase Θ, with the $\Theta(\xi)$ being independent, uniformly distributed on $[0, 2\pi)$ with the constraint that $\Theta(-\xi) = -\Theta(\xi)$.
4. Set $\widehat{Z} = |\widehat{u_0}| \, e^{i\Theta}$ (or $\widehat{Z} = \widehat{u_0} e^{i\Theta}$).
5. Return Z the inverse Fourier transform of $\widehat{Z}$.

Examples of results obtained by this algorithm were shown on Fig. 3.29. As expected, this algorithm performs well on micro-textures, but fails on macro-textures that contain geometric patterns (that are lost when randomizing the phase) and long-range interactions (that are difficult to model by statistics).

As written above, the RPN algorithm generates an image Z that has the same size as the input image u_0. To obtain an output image of an arbitrary large size, one can just start by smoothly extending u_0 by a constant outside its domain (see [17] for more details) and then perform the RPN algorithm with the new extended input.

The RPN algorithm can be adapted to the case of color images, by adding the **same** random phase to the three channels. More precisely, if the input image **u** is a color image, meaning that **u** $= (u_1, u_2, u_3)$ (for the Red, Green and Blue channels) with each u_k being $\Omega \to \mathbb{R}$, then the RPN algorithm generates one random phase Θ and three output images given by $\widehat{Z_k} = \widehat{u_k} e^{i\Theta}$. Therefore we have that for all $k, j = 1, 2$ or 3, $\widehat{Z_k}\overline{\widehat{Z_j}} = \widehat{u_k}\overline{\widehat{u_j}}$, showing that it keeps the Fourier modulus of each channel but also the cross-channel Fourier correlation. This fact is related to the link between the Fourier modulus and the empirical covariance, as we will explain it in the following section.

3.3.4 Gaussian Models for Texture Images

Given two images $u : \Omega \to \mathbb{R}$ and $v : \Omega \to \mathbb{R}$, we can define the convolution of u and v as the image $u \star v : \Omega \to \mathbb{R}$ given by

$$\forall \mathbf{x} \in \Omega, \quad (u \star v)(\mathbf{x}) = \sum_{\mathbf{y} \in \Omega} u(\mathbf{y}) \, v(\mathbf{x} - \mathbf{y}),$$

where $v(\mathbf{x})$ can be defined for any $\mathbf{x} \in \mathbb{Z}^2$ by considering $\mathbf{x}$ modulo Ω (i.e. x_1 modulo M and x_2 modulo N). This amounts to extend v on $\mathbb{Z}^2$ by periodicity. The convolution is described more simply in the Fourier domain since the convolution/product exchange property states that

$$\forall \xi \in \Omega, \quad \widehat{u \star v}(\xi) = \widehat{u}(\xi)\widehat{v}(\xi).$$

Stationary Gaussian textures are widely used as models of (micro)-textures, and they are well characterized in Fourier domain. Indeed, we have the following proposition.

Proposition 3.2 *Let $(U(\mathbf{x}))_{\mathbf{x} \in \Omega}$ be a real-valued random field on Ω, and let $\Omega_+ = \{\boldsymbol{\xi} = (\xi_1, \xi_2); \xi_1 > 0 \text{ or } (\xi_1 = 0 \text{ and } \xi_2 > 0)\}$. Then $(U(\mathbf{x}))_{\mathbf{x} \in \Omega}$ is a centered Gaussian periodic stationary random field if and only if the random variables*

$$\{\widehat{U}(\mathbf{0}), \operatorname{Re} \widehat{U}(\boldsymbol{\xi}), \operatorname{Im} \widehat{U}(\boldsymbol{\xi}); \boldsymbol{\xi} \in \Omega_+\}$$

are independent centered Gaussian variables. Moreover, in this case, if Γ denotes the covariance of U defined by $\Gamma(\mathbf{x}) = \operatorname{Cov}(U(\mathbf{x}), U(\mathbf{0}))$ for all $\mathbf{x} \in \Omega$, then

$$\operatorname{Var}(\widehat{U}(\mathbf{0})) = |\Omega| \cdot \widehat{\Gamma}(\mathbf{0}) \quad \text{and} \quad \operatorname{Var}(\operatorname{Re} \widehat{U}(\boldsymbol{\xi})) = \operatorname{Var}(\operatorname{Im} \widehat{U}(\boldsymbol{\xi})) = \frac{1}{2}|\Omega| \cdot \widehat{\Gamma}(\boldsymbol{\xi})$$

for all $\boldsymbol{\xi} \in \Omega_+$.

Proof The proof is based on the fact that the Fourier transform is a linear transform. Therefore if U is a centered Gaussian periodic stationary random field, then the variables $\{\operatorname{Re} \widehat{U}(\boldsymbol{\xi}), \operatorname{Im} \widehat{U}(\boldsymbol{\xi}); \boldsymbol{\xi} \in \Omega\}$ are a centered Gaussian vector. To compute its covariance, let us for instance compute $\mathbb{E}(\operatorname{Re} \widehat{U}(\boldsymbol{\xi}) \operatorname{Re} \widehat{U}(\boldsymbol{\xi}'))$ for $\boldsymbol{\xi}$ and $\boldsymbol{\xi}'$ in Ω. Let Γ denote the covariance of U. We have

$$\mathbb{E}(\operatorname{Re} \widehat{U}(\boldsymbol{\xi}) \operatorname{Re} \widehat{U}(\boldsymbol{\xi}')) = \mathbb{E}\left(\sum_{\mathbf{x}} \sum_{\mathbf{y}} U(\mathbf{x}) \cos(2\pi \langle \mathbf{x}, \boldsymbol{\xi} \rangle) U(\mathbf{y}) \cos(2\pi \langle \mathbf{y}, \boldsymbol{\xi}' \rangle)\right)$$

$$= \sum_{\mathbf{x}} \sum_{\mathbf{z}} \mathbb{E}(U(\mathbf{x}) U(\mathbf{x} + \mathbf{z})) \cos(2\pi \langle \mathbf{x}, \boldsymbol{\xi} \rangle) \cos(2\pi \langle \mathbf{x} + \mathbf{z}, \boldsymbol{\xi}' \rangle)$$

$$= \frac{|\Omega|}{2}\left(\mathbb{1}_{\boldsymbol{\xi}=\pm\boldsymbol{\xi}'\neq\mathbf{0}} \sum_{\mathbf{z}} \Gamma(\mathbf{z}) \cos(2\pi \langle \mathbf{z}, \boldsymbol{\xi} \rangle) + 2\mathbb{1}_{\boldsymbol{\xi}=\boldsymbol{\xi}'=\mathbf{0}} \sum_{\mathbf{z}} \Gamma(\mathbf{z})\right).$$

Therefore, since $\Gamma(\mathbf{z}) = \Gamma(-\mathbf{z})$ and $\operatorname{Re} \widehat{U}(\boldsymbol{\xi}) = \operatorname{Re} \widehat{U}(-\boldsymbol{\xi})$ (because U is real), we deduce that if $\boldsymbol{\xi} \neq \boldsymbol{\xi}' \in \Omega_+$, then $\mathbb{E}(\operatorname{Re} \widehat{U}(\boldsymbol{\xi}) \operatorname{Re} \widehat{U}(\boldsymbol{\xi}')) = 0$ (these two Gaussian random variables are thus independent), $\mathbb{E}(\operatorname{Re} \widehat{U}(\mathbf{0})^2) = |\Omega|\widehat{\Gamma}(\mathbf{0})$ and $\mathbb{E}(\operatorname{Re} \widehat{U}(\boldsymbol{\xi})^2) = \frac{|\Omega|}{2}\widehat{\Gamma}(\boldsymbol{\xi})$ for $\boldsymbol{\xi} \neq \mathbf{0}$. Analogous computations hold with $(\operatorname{Re} \widehat{U}(\boldsymbol{\xi}), \operatorname{Im} \widehat{U}(\boldsymbol{\xi}'))$ and $(\operatorname{Im} \widehat{U}(\boldsymbol{\xi}), \operatorname{Im} \widehat{U}(\boldsymbol{\xi}'))$, and this ends the proof of the proposition. $\square$

The above proposition gives a characterization of periodic stationary Gaussian textures. Indeed, we have that

- The Fourier phases $(\phi(\boldsymbol{\xi}))_{\boldsymbol{\xi} \in \Omega_+}$ are independent identically distributed uniformly on $[0, 2\pi)$ and independent from the Fourier amplitudes $(|\widehat{U}(\boldsymbol{\xi})|)_{\boldsymbol{\xi} \in \Omega_+}$.
- The Fourier amplitudes $(|\widehat{U}(\boldsymbol{\xi})|)_{\boldsymbol{\xi} \in \Omega_+}$ are independent random variables following a Rayleigh distribution of parameter $\sqrt{\frac{1}{2}|\Omega|\widehat{\Gamma}(\boldsymbol{\xi})}$.
- And $\widehat{U}(\mathbf{0})$ follows a centered normal distribution with variance $|\Omega|\widehat{\Gamma}(\mathbf{0})$.

The simplest case of a Gaussian stationary random field is the white noise (of variance 1): the $U(\mathbf{x})$ are i.i.d. following the standard normal distribution $\mathcal{N}(0, 1)$.

Then, in this case $\Gamma(\mathbf{x}) = \delta_\mathbf{0}(\mathbf{x})$ (the indicator function of $\{\mathbf{0}\}$), and thus $\widehat{\Gamma}(\boldsymbol{\xi}) = 1$ for all $\boldsymbol{\xi} \in \Omega$. Therefore the Fourier transform $\widehat{U}$ (restricted to Ω_+) is also a (complex) white noise (of variance $|\Omega|$).

In the following, we will denote by GT(Γ) (GT stands for Gaussian Texture) the law of the centered Gaussian periodic stationary random field with covariance function Γ.

Starting from an original image u_0, that is assumed to have zero mean on Ω (otherwise we just add a constant to it), we can compute its empirical covariance as

$$\forall \mathbf{x} \in \Omega, \quad C_{u_0}(\mathbf{x}) = \frac{1}{|\Omega|} \sum_{\mathbf{y} \in \Omega} u_0(\mathbf{y}) u_0(\mathbf{y} + \mathbf{x}).$$

Thanks to the convolution/product property of the Fourier transform, we have that

$$\forall \boldsymbol{\xi} \in \Omega, \quad \widehat{C_{u_0}}(\boldsymbol{\xi}) = \frac{1}{|\Omega|} |\widehat{u_0}(\boldsymbol{\xi})|^2.$$

Now, let W be a white noise image of variance $\sigma^2 = 1/|\Omega|$ on Ω (i.e. all $W(\mathbf{x})$, $\mathbf{x} \in \Omega$, are i.i.d $\mathcal{N}(0, \sigma^2)$). Let us consider

$$U = u_0 \star W.$$

Then, by computing its covariance, we obtain that U follows a GT(C_{u_0}) distribution. Conversely if Γ is a covariance and U follows a GT(Γ) distribution, then by Proposition 3.2 that characterizes $\widehat{U}$, we have that U can be written as

$$U = u_\Gamma \star W,$$

where W is white noise (of variance $\sigma^2 = 1/|\Omega|$) and u_Γ is any image such that

$$\forall \boldsymbol{\xi} \in \Omega, \quad |\widehat{u_\Gamma}(\boldsymbol{\xi})| = \sqrt{|\Omega| \widehat{\Gamma}(\boldsymbol{\xi})}.$$

Notice that since Γ is a covariance, the $\widehat{\Gamma}(\boldsymbol{\xi})$ are positive (see also Proposition 3.2, where they appear as variances). There are many such u_Γ images, since one can choose any Fourier phase for it. However some choices appear to be better than others, in the sense that the image u_Γ is more "concentrated". This is in particular the choice of u_Γ being simply the inverse Fourier transform of $|\widehat{u_\Gamma}| = \sqrt{|\Omega| \widehat{\Gamma}}$ (having thus an identically null Fourier phase). This particular image, called the *texton*, is analyzed in more details in [15].

Starting from an image u_0, we can generate two samples: one with the RPN algorithm, that is given by $\widehat{U_{\mathrm{RPN}}} = \widehat{u_0} e^{i\Theta}$ where Θ is an i.i.d uniform phase field, and another one that is a sample of GT(C_{u_0}) that can be written as $U_{\mathrm{GT}} = u_0 \star W$ (where W is white noise of variance σ^2), or equivalently $\widehat{U_{\mathrm{GT}}} = \widehat{u_0} \widehat{W} = \widehat{u_0} R e^{i\Phi}$, where Φ is an i.i.d uniform phase field, independent of the $R(\boldsymbol{\xi})$ that are i.i.d following a Rayleigh distribution of parameter $\sqrt{1/2}$. The formula for

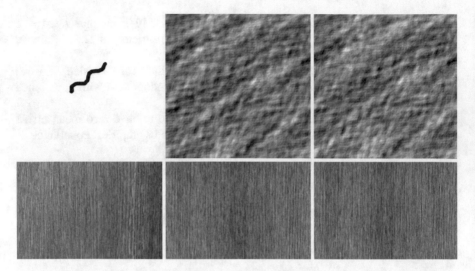

Fig. 3.32 Gaussian textures and the RPN algorithm. Left column: original images u_0 ("snake" and "wood"). Middle column: samples from the Gaussian Texture model with covariance C_{u_0}. Right column: samples from the RPN algorithm, using the same random phase as for the Gaussian samples. It shows that the random Rayleigh factor in the modulus of the Gaussian textures doesn't visually make a difference

these two samples are very similar, they have both random phases, and there is just an additional Rayleigh random variable in the Gaussian texture. However, this Rayleigh term doesn't make any visual difference, as illustrated on Fig. 3.32.

As for the RPN algorithm, Gaussian textures can be defined from an input color image $\mathbf{u} = (u_1, u_2, u_3)$ by simply considering the convolution of each channel of $\mathbf{u}$ with the same (real-valued) white noise W. This results in a random Gaussian color image such that the covariance between the channels is the same as the empirical covariance between the channels of $\mathbf{u}$.

3.3.5 Shot Noise Model and Dead Leaves Model

As it was shown by Galerne et al. in [17], as a consequence of the Central Limit Theorem, the Gaussian texture model $GT(C_{u_0})$, where u_0 is a (centered) given input image, is the limit, as n goes to infinity of the *discrete spot noise* model given by

$$U(\mathbf{x}) = \frac{1}{\sqrt{n}} \sum_{j=1}^{n} u_0(\mathbf{x} - \mathbf{y}_j),$$

where the $\mathbf{y}_j$ are independent, uniformly distributed on Ω.

The spot noise model was introduced by van Wijk in Computer Graphics [40], and its general definition in a continuous setting is called *shot noise random field*. It is (in dimension 2) a random field $X : \mathbb{R}^2 \to \mathbb{R}$ given by

$$\forall \mathbf{x} \in \mathbb{R}^2, \quad X(x) = \sum_{i \in I} g_{m_i}(\mathbf{x} - \mathbf{x}_i),$$

where the $\{\mathbf{x}_i\}_{i \in I}$ is a Poisson point process of intensity $\lambda > 0$ in $\mathbb{R}^2$, the $\{m_i\}_{i \in I}$ are independent "marks" with distribution $F(dm)$ on $\mathbb{R}^d$, and independent of $\{\mathbf{x}_i\}_{i \in I}$, and the functions g_m are real-valued functions, called *spot functions*, and such that $\int_{\mathbb{R}^d} \int_{\mathbb{R}^n} |g_m(\mathbf{y})| \, d\mathbf{y} \, F(dm) < +\infty$.

This shot noise model can be used as a model for natural images made of "objects", and it is for instance used by A. Srivastava et al. in [38] as forms for "modelling image probabilities". They write it as

$$\forall \mathbf{x} \in \mathbb{R}^2, \quad X(\mathbf{x}) = \sum_i \beta_i \, g_i \left(\frac{1}{r_i} (\mathbf{x} - \mathbf{x}_i) \right),$$

where g_i is an "object", $\mathbf{x}_i$ its position, β_i its grey level or color and r_i is a scaling factor (related to the distance at which the object is).

Such a model is also a nice model for macro-textures modelling. Now, in this model the different objects are just added, whereas in natural texture images, objects are partially hidden, in particular when some other objects are in front of them. To take into account this so-called *occlusion phenomenon*, a more "realistic" model is given by the dead leaves model introduced by G. Matheron and studied for instance in [5]. In the dead leaves model, the function g_i are indicator functions of a "grain", and an additional random variable t_i is introduced that models the time at which the grain falls. The value at a point $\mathbf{x}$ is then defined as being the value of the first grain that covers $\mathbf{x}$. Two examples of respectively a shot noise random field and a dead leaves model are given on Fig. 3.33.

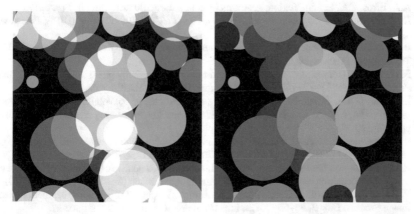

Fig. 3.33 Two examples of macro-textures: on the left, a sample of a shot noise random field, and on the right, a sample of a dead leaves model with the same color "grains"

References

1. C. Aguerrebere, Y. Gousseau, G. Tartavel, Exemplar-based texture synthesis: the Efros–Leung algorithm. Image Process. Line **3**, 223–241 (2013). https://doi.org/10.5201/ipol.2013.59
2. A. Almansa, A. Desolneux, S. Vamech, Vanishing point detection without any a priori information. IEEE Trans. Pattern Anal. Mach. Intell. **25**(4), 502–507 (2003)
3. F. Attneave, Some informational aspects of visual perception. Psychol. Rev. **61**, 183–193 (1954)
4. S. Blusseau, A. Carboni, A. Maiche, J.-M. Morel, R. Grompone von Gioi, A psychophysical evaluation of the a contrario detection theory, in *Proceedings of the 2014 IEEE International Conference on Image Processing (ICIP)* (IEEE, Piscataway, 2014), pp. 1091–1095
5. C. Bordenave, Y. Gousseau, F. Roueff, The dead leaves model: an example of a general tesselation. Adv. Appl. Probab. **38**(1), 31–46 (2006)
6. T. Briand, J. Vacher, B. Galerne, J. Rabin, The Heeger & Bergen pyramid based texture synthesis algorithm. Image Process. Line **4**, 276–299 (2014). https://doi.org/10.5201/ipol.2014.79
7. F. Cao, Good continuation in digital images, in *Proceedings of the International Conference on Computer Vision (ICCV)* (2003), pp. 440–447
8. F. Cao, J. Delon, A. Desolneux, P. Musé, F. Sur, A unified framework for detecting groups and application to shape recognition. J. Math. Imaging Vis. **27**(2), 91–119 (2007).
9. A. Ciomaga, P. Monasse, J.-M. Morel, The image curvature microscope: accurate curvature computation at subpixel resolution. Image Process. Line **7**, 197–217 (2017). https://doi.org/10.5201/ipol.2017.212
10. D. Coupier, A. Desolneux, B. Ycart, Image denoising by statistical area thresholding. J. Math. Imaging Vis. **22**(2–3), 183–197 (2005)
11. A. Desolneux, L. Moisan, J.-M. Morel, Meaningful alignments. Int. J. Comput. Vis. **40**(1), 7–23 (2000)
12. A. Desolneux, L. Moisan, J.-M. Morel, Edge detection by Helmholtz principle. J. Math. Imaging Vis. **14**(3), 271–284 (2001)
13. A. Desolneux, L. Moisan, J.-M. Morel, Maximal meaningful events and applications to image analysis. Ann. Stat. **31**(6), 1822–1851 (2003)
14. A. Desolneux, L. Moisan, J.-M. Morel, *From Gestalt Theory to Image Analysis: A Probabilistic Approach* (Springer, Berlin, 2008)
15. A. Desolneux, L. Moisan, S. Ronsin, A compact representation of random phase and Gaussian textures, in *2012 IEEE International Conference on Acoustics, Speech and Signal Processing (ICASSP)* (IEEE, Piscataway, 2012), pp. 1381–1384
16. A.A. Efros, T.K. Leung, Texture synthesis by non-parametric sampling, in *Proceedings of the IEEE International Conference on Computer Vision (ICCV 1999)*, vol. 2 (IEEE, Piscataway, 1999), pp. 1033–1038
17. B. Galerne, Y. Gousseau, J.-M. Morel, Random phase textures: theory and synthesis. IEEE Trans. Image Process. **20**(1), 257–267 (2011)
18. B. Galerne, Y. Gousseau, J.-M. Morel, Micro-texture synthesis by phase randomization. Image Process. Line **1**, 213–237 (2011). https://doi.org/10.5201/ipol.2011.ggm_rpn
19. L.A. Gatys, A.S. Ecker, M. Bethge, Texture synthesis using convolutional neural networks, in *Proceedings of the Conference on Neural Information Processing Systems, 2015* (2015), pp. 262–270
20. R. Grompone von Gioi, G. Randall, Unsupervised smooth contour detection. Image Process. Line **6**, 233–267 (2016). https://doi.org/10.5201/ipol.2016.175
21. R. Grompone von Gioi, J. Jakubowicz, J.-M. Morel, G. Randall, A fast line segment detector with a false detection control. IEEE Trans. Pattern Anal. Mach. Intell. **32**, 722–732 (2010)
22. R. Grompone von Gioi, J. Jakubowicz, J.-M. Morel, G. Randall, LSD: a line segment detector. Image Process. Line **2**, 35–55 (2012). https://doi.org/10.5201/ipol.2012.gjmr-lsd

23. P.E. Hart, How the Hough transform was invented. IEEE Signal Process. Mag. **26**(6), 18–22 (2009)
24. D.J. Heeger, J.R. Bergen, Pyramid-based texture analysis/synthesis, in *Proceedings of the Conference SIGGRAPH '95* (IEEE, Piscataway, 1995), pp. 229–238
25. G. Kanizsa, *Grammatica del Vedere/La Grammaire du Voir* (Bologna/Éditions Diderot, Arts et Sciences, IL Mulino, 1980/1997)
26. E. Levina, P.J. Bickel, Texture synthesis and nonparametric resampling of random fields. Ann. Stat. **35**(4), 1751–1773 (2006)
27. D. Lowe, *Perceptual Organization and Visual Recognition* (Kluwer Academic Publishers, Dordecht, 1985)
28. W. Metzger, *Gesetze des Sehens* (Kramer, Frankfurt, 1953).
29. L. Moisan, Periodic plus smooth image decomposition. J. Math. Imaging Vis. **39**(2), 161–179 (2011)
30. L. Moisan, B. Stival, A probabilistic criterion to detect rigid point matches between two images and estimate the fundamental matrix. Int. J. Comput. Vis. **57**(3), 201–218 (2004)
31. P. Musé, F. Sur, F. Cao, Y. Gousseau, Unsupervised thresholds for shape matching, in *Proceedings 2003 International Conference on Image Processing (ICIP 2003)*, vol. 2 (IEEE, Piscataway, 2003), pp. 647–650
32. A.V. Oppenheim, J.S. Lim, The importance of phase in signals. IEEE Proc. **69**, 529–541 (1981)
33. J. Portilla, E.P. Simoncelli, A parametric texture model based on joint statistics of complex wavelet coefficients. Int. J. Comput. Vis. **40**(1), 49–71 (2000)
34. L. Raad, A. Desolneux and J.-M. Morel, A conditional multiscale locally Gaussian texture synthesis algorithm. J. Math. Imaging Vis. **56**(2), 260–279 (2016)
35. J. Rabin, J. Delon, Y. Gousseau, A statistical approach to the matching of local features. SIAM J. Imag. Sci. **2**(3), 931–958 (2009)
36. N. Sabater, A. Almansa, J.-M. Morel, Meaningful matches in stereovision. IEEE Trans. Pattern Anal. Mach. Intell. **34**(5), 930–942 (2012)
37. L. Santalo, *Integral Geometry and Geometric Probability*, 2nd edn. (Cambridge University Press, Cambridge, 2004)
38. A. Srivastava, X. Liu, U. Grenander, Universal analytical forms for modeling image probabilities. IEEE Trans. Pattern Anal. Mach. Intell. **24**(9), 1200–1214 (2002)
39. G.Tartavel, Y. Gousseau, G. Peyré, Variational texture synthesis with sparsity and spectrum constraints. J. Math. Imaging Vis. **52**(1), 124–144 (2015)
40. J.J. van Wijk, Spot noise texture synthesis for data visualization, in *Proceedings of the 18th Annual Conference on Computer Graphics and Interactive Techniques, SIGGRAPH '91* (ACM, New York, 1991), pp. 309–318
41. T. Veit, F. Cao, P. Bouthemy, An a contrario decision framework for region-based motion detection. Int. J. Comput. Vis. **68**(2), 163–178 (2006)
42. M. Wertheimer, Unterzuchungen zur lehre der gestalt. Psychol. Forsch. **4**(1), 301–350 (1923)
43. S.C. Zhu, Embedding gestalt laws in markov random fields. IEEE Trans. Pattern Anal. Mach. Intell. **21**(11), 1170–1187 (1999)

Chapter 4
Introduction to Random Fields and Scale Invariance

Hermine Biermé

Abstract In medical imaging, several authors have proposed to characterize roughness of observed textures by their fractal dimensions. Fractal analysis of 1D signals is mainly based on the stochastic modeling using the famous fractional Brownian motion for which the fractal dimension is determined by its so-called Hurst parameter. Lots of 2D generalizations of this toy model may be defined according to the scope. This lecture intends to present some of them. After an introduction to random fields, the first part will focus on the construction of Gaussian random fields with prescribed invariance properties such as stationarity, self-similarity, or operator scaling property. Sample paths properties such as modulus of continuity and Hausdorff dimension of graphs will be settled in the second part to understand links with fractal analysis. The third part will concern some methods of simulation and estimation for these random fields in a discrete setting. Some applications in medical imaging will be presented. Finally, the last part will be devoted to geometric constructions involving Marked Poisson Point Processes and shot noise processes.

4.1 Random Fields and Scale Invariance

We recall in this section definitions and properties of random fields. Most of them can also be found in [22] but we try here to detail some important proofs. We stress on invariance properties such as stationarity, isotropy, and scale invariance and illustrate these properties with typical examples.

H. Biermé (✉)
LMA, UMR CNRS 7348, Université de Poitiers, Chasseneuil, France
e-mail: hermine.bierme@math.univ-poitiers.fr

© Springer Nature Switzerland AG 2019
D. Coupier (ed.), *Stochastic Geometry*, Lecture Notes in Mathematics 2237,
https://doi.org/10.1007/978-3-030-13547-8_4

129

4.1.1 Introduction to Random Fields

As usual when talking about randomness, we let $(\Omega, \mathscr{A}, \mathbb{P})$ be a probability space, reflecting variability.

4.1.1.1 Definitions and Distribution

Let us first recall the general definition of a stochastic process. For this purpose we have to consider a set of indices T. In this lecture we assume that $T \subset \mathbb{R}^d$ for some dimension $d \geq 1$.

Definition 4.1 A (real) stochastic process indexed by T is just a collection of real random variables meaning that for all $t \in T$, one has $X_t : (\Omega, \mathscr{A}) \to (\mathbb{R}, \mathscr{B}(\mathbb{R}))$ measurable.

Stochastic processes are very important in stochastic modeling as they can mimic numerous natural phenomena. For instance, when $d = 1$, one can choose $T \subset \mathbb{R}$ (seen as time parameters) and consider $X_t(\omega)$ as the real value of heart frequency at time $t \in T$ with noise measurement or for an individual $\omega \in \Omega$. Note that, in practice data are only available on a discrete finite subset S of T, for instance each millisecond. When $d = 2$, choosing $T = [0, 1]^2$, the value $X_t(\omega)$ may correspond to the grey level of a picture at point $t \in T$. Again, in practice, data are only available on pixels $S = \{0, 1/n, \ldots, 1\}^2 \subset T$ for an image of size $(n + 1) \times (n + 1)$. In general we talk about random fields when $d > 1$ and keep the terminology *stochastic process* only for $d = 1$. Since we have actually a map X from $\Omega \times T$ with values in $\mathbb{R}$ we can also consider it as a map from Ω to $\mathbb{R}^T$. We equip $\mathbb{R}^T$ with the smallest σ-algebra $\mathscr{C}$ such that the projections $\pi_t : (\mathbb{R}^T, \mathscr{C}) \to (\mathbb{R}, \mathscr{B}(\mathbb{R}))$, defined by $\pi_t(f) = f(t)$ are measurable. It follows that $X : (\Omega, \mathscr{A}) \to (\mathbb{R}^T, \mathscr{C})$ is measurable and its distribution is defined as the image measure of $\mathbb{P}$ by X, which is a probability measure on $(\mathbb{R}^T, \mathscr{C})$. An important consequence of Kolmogorov's consistency theorem (see [37, p. 92]) is the following equivalent definition.

Definition 4.2 The distribution of $(X_t)_{t \in T}$ is given by all its finite dimensional distribution (fdd) i.e. the distribution of all real random vectors

$$(X_{t_1}, \ldots, X_{t_k}) \text{ for } k \geq 1, t_1, \ldots, t_k \in T.$$

Note that joint distributions for random vectors of arbitrary size k are often difficult to compute. However we can infer some statistics of order one and two by considering only couples of variables.

Definition 4.3 The stochastic process $(X_t)_{t \in T}$ is a second order process if $\mathbb{E}(X_t^2) < + \infty$, for all $t \in T$. In this case we define

- its *mean function* $m_X : t \in T \to \mathbb{E}(X_t) \in \mathbb{R}$;
- its *covariance function* $K_X : (t, s) \in T \times T \to \text{Cov}(X_t, X_s) \in \mathbb{R}$.

A particular case arises when $m_X = 0$ and the process X is said *centered*. Otherwise the stochastic process $Y = X - m_X$ is also second order and now centered with the same covariance function $K_Y = K_X$. Hence we will mainly consider centered stochastic processes. The covariance function of a stochastic process must verify the following properties.

Proposition 4.1 *A function $K : T \times T \to \mathbb{R}$ is a covariance function iff*

1. *K is symmetric i.e. $K(t, s) = K(s, t)$ for all $(t, s) \in T \times T$;*
2. *K is non-negative definite: $\forall k \geq 1,\ t_1, \ldots, t_k \in T, : \lambda_1, \ldots, \lambda_k \in \mathbb{R}$,*

$$\sum_{i,j=1}^{k} \lambda_i \lambda_j K(t_i, t_j) \geq 0.$$

Proof The first implication is trivial once remarked the fact that $\mathrm{Var}\left(\sum_{i=1}^{k} \lambda_i X_{t_i}\right) = \sum_{i,j=1}^{k} \lambda_i \lambda_j K(t_i, t_j)$. For the converse, we need to introduce Gaussian processes. □

4.1.1.2   Gaussian Processes

As far as second order properties are concerned the most natural class of processes are given by Gaussian ones.

Definition 4.4 A stochastic process $(X_t)_{t \in T}$ is a Gaussian process if for all $k \geq 1$ and $t_1, \ldots, t_k \in T$

$$(X_{t_1}, \ldots, X_{t_k}) \text{ is a Gaussian vector of } \mathbb{R}^k,$$

which is equivalent to the fact that for all $\lambda_1, \ldots, \lambda_k \in \mathbb{R}$, the real random variable

$$\sum_{i=1}^{k} \lambda_i X_{t_i} \text{ is a Gaussian variable (eventually degenerate i.e. constant).}$$

Note that this definition completely characterizes the distribution of the process in view of Definition 4.2.

Proposition 4.2 *When $(X_t)_{t \in T}$ is a Gaussian process, $(X_t)_{t \in T}$ is a second order process and its distribution is determined by its mean function $m_X : t \mapsto \mathbb{E}(X_t)$ and its covariance function $K_X : (t, s) \mapsto \mathrm{Cov}(X_t, X_s)$.*

This comes from the fact that the distribution of the Gaussian vector $(X_{t_1}, \ldots, X_{t_k})$ is characterized by its mean $(\mathbb{E}(X_{t_1}), \ldots, \mathbb{E}(X_{t_k})) = (m_X(t_1), \ldots, m_X(t_k))$ and its covariance matrix $\left(\mathrm{Cov}(X_{t_i}, X_{t_j})\right)_{1 \leq i,j \leq k} = \left(K_X(t_i, t_j)\right)_{1 \leq i,j \leq k}$.

Again Kolmogorov's consistency theorem (see [37, p. 92] for instance) allows to prove the following existence result that finishes to prove Proposition 4.1.

Theorem 4.1 *Let $m : T \to \mathbb{R}$ and $K : T \times T \to \mathbb{R}$ a symmetric and non-negative definite function, then there exists a Gaussian process with mean m and covariance K.*

Let us give some insights of construction for the fundamental example of Gaussian process, namely the Brownian motion. We set here $T = \mathbb{R}^+$ and consider $(X_k)_{k \in \mathbb{N}}$ a family of independent identically distributed second order random variables with $\mathbb{E}(X_k) = 0$ and $\mathrm{Var}(X_k) = 1$. For any $n \geq 1$, we construct on T the following stochastic process

$$S_n(t) = \frac{1}{\sqrt{n}} \sum_{k=1}^{[nt]} X_k.$$

By the central limit theorem (see [28] for instance) we clearly have for $t > 0$, $\sqrt{\frac{n}{[nt]}} S_n(t) \xrightarrow[n \to +\infty]{d} \mathcal{N}(0, 1)$ so that by Slutsky's theorem (see [15] for instance) $S_n(t) \xrightarrow[n \to +\infty]{d} \mathcal{N}(0, t)$. Moreover, for $k \geq 1$, if $0 < t_1 < \ldots < t_k$, by independence of marginals,

$$(S_n(t_1), S_n(t_2) - S_n(t_1), \ldots, S_n(t_k) - S_n(t_{k-1})) \xrightarrow[n \to +\infty]{d} Z = (Z_1, \ldots, Z_k),$$

with $Z \sim \mathcal{N}(0, K_Z)$ for $K_Z = \mathrm{diag}(t_1, t_2 - t_1, \ldots, t_k - t_{k-1})$. Hence identifying the $k \times k$ matrix $P_k = \begin{pmatrix} 1 & 0 & \ldots & 0 \\ 1 & 1 & & \\ \vdots & & \ddots & \\ 1 & \ldots & \ldots & 1 \end{pmatrix}$ with the corresponding linear application on $\mathbb{R}^k$,

$$(S_n(t_1), S_n(t_2), \ldots, S_n(t_k)) = P_k(S_n(t_1), S_n(t_2) - S_n(t_1), \ldots, S_n(t_k)$$
$$- S_n(t_{k-1}))$$
$$\xrightarrow[n \to +\infty]{d} P_k Z,$$

with $P_k Z \sim \mathcal{N}(0, P_k K_Z P_k^*)$ and $P_k K_Z P_k^* = \left(\min(t_i, t_j) \right)_{1 \leq i, j \leq k}$. In particular the function

$$K(t, s) = \min(t, s) = \frac{1}{2}(t + s - |t - s|)$$

is a covariance function on the whole space $\mathbb{R}^+ \times \mathbb{R}^+$ and $(S_n)_n$ converges in finite dimensional distribution to a centered Gaussian stochastic process $X = (X_t)_{t \in \mathbb{R}^+}$

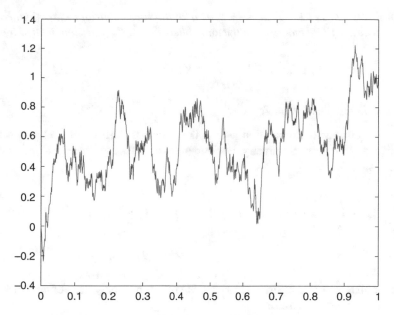

Fig. 4.1 Sample paths of a Brownian motion on [0, 1]. The realization is obtained using fast and exact synthesis presented in Sect. 4.3.1.1

with covariance K, known (up to continuity of sample paths) as the standard Brownian motion on $\mathbb{R}^+$.

Now we can extend this process on $\mathbb{R}$ by simply considering $X^{(1)}$ and $X^{(2)}$ two independent centered Gaussian processes on $\mathbb{R}^+$ with covariance function K and defining $B_t := X_t^{(1)}$ for $t \geq 0$, $B_t := X_{-t}^{(2)}$ for $t < 0$. Computing the covariance function of B yields the following definition.

Definition 4.5 A (standard) Brownian motion on $\mathbb{R}$ (Fig. 4.1) is a centered Gaussian process $(B_t)_{t \in \mathbb{R}}$ with covariance function given by

$$K_B(t, s) = \mathrm{Cov}(B_t, B_s) = \frac{1}{2}\left(|t| + |s| - |t - s|\right), \ \forall t, s \in \mathbb{R}.$$

From Gaussian stochastic processes defined on $\mathbb{R}$ we can define Gaussian random fields defined on $\mathbb{R}^d$ in several ways. We give some possibilities in the next section.

4.1.1.3   Gaussian Fields Defined from Processes

We consider on $\mathbb{R}^d$ the Euclidean norm, denoted by $\|\cdot\|$ with respect to the Euclidean scalar product $x \cdot y$ for $x, y \in \mathbb{R}^d$. The unit sphere $\{\theta \in \mathbb{R}^d; \|\theta\| = 1\}$ is denoted as S^{d-1} and we let $(e_i)_{1 \leq i \leq d}$ stand for the canonical basis of $\mathbb{R}^d$.

A first example of construction is given in the following proposition.

Proposition 4.3 *Let $K : \mathbb{R} \times \mathbb{R} \to \mathbb{R}$ be a continuous covariance function. For all μ non-negative finite measure on the unit sphere S^{d-1}, the function defined by*

$$(x, y) \in \mathbb{R}^d \times \mathbb{R}^d \mapsto \int_{S^{d-1}} K(x \cdot \theta, y \cdot \theta) d\mu(\theta) \in \mathbb{R},$$

is a covariance function on $\mathbb{R}^d \times \mathbb{R}^d$.

Proof According to Proposition 4.1, it is enough to check symmetry and non-negative definiteness. Symmetry is clear and for all $k \geq 1$, $x_1, \ldots, x_k \in \mathbb{R}^d$, $\lambda_1, \ldots, \lambda_k \in \mathbb{R}$,

$$\sum_{i,j=1}^{k} \lambda_i \lambda_j \int_{S^{d-1}} K(x_i \cdot \theta, x_j \cdot \theta) d\mu(\theta) = \int_{S^{d-1}} \left(\sum_{i,j=1}^{k} \lambda_i \lambda_j K(x_i \cdot \theta, x_j \cdot \theta) \right)$$

$$\times d\mu(\theta) \geq 0,$$

since for all $\theta \in S^{d-1}$, $x_1 \cdot \theta, \ldots, x_k \cdot \theta \in \mathbb{R}$ with K non-negative definite on $\mathbb{R} \times \mathbb{R}$ and μ non-negative measure. $\square$

As an example we can note that $\int_{S^{d-1}} |x \cdot \theta| d\theta = c_d \|x\|$, with $c_d = \int_{S^{d-1}} |e_1 \cdot \theta| d\theta$ for $e_1 = (1, 0, \ldots, 0) \in S^{d-1}$. Then, for K_B the covariance function of a standard Brownian motion on $\mathbb{R}$ we get

$$\int_{S^{d-1}} K_B(x \cdot \theta, y \cdot \theta) d\theta = \frac{c_d}{2} \left(\|x\| + \|y\| - \|x - y\| \right).$$

Definition 4.6 A (standard) Lévy Chentsov field on $\mathbb{R}^d$ (Fig. 4.2) is a centered Gaussian field $(X_x)_{x \in \mathbb{R}^d}$ with covariance function given by

$$\text{Cov}(X_x, X_y) = \frac{1}{2} \left(\|x\| + \|y\| - \|x - y\| \right), \ \forall x, y \in \mathbb{R}^d.$$

Let us note that $(X_{t\theta})_{t \in \mathbb{R}}$ is therefore a standard Brownian motion for all $\theta \in S^{d-1}$.

Another example is given using a sheet structure according to the following proposition.

Proposition 4.4 *Let $K_1, K_2, \ldots, K_d$ be covariance functions on $\mathbb{R} \times \mathbb{R}$, then the function defined by*

$$(x, y) \in \mathbb{R}^d \times \mathbb{R}^d \mapsto \prod_{i=1}^{d} K_i(x \cdot e_i, y \cdot e_i) \in \mathbb{R},$$

is a covariance function on $\mathbb{R}^d \times \mathbb{R}^d$.

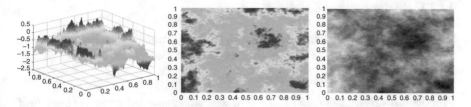

Fig. 4.2 Sample paths of a Lévy Chentsov field on $[0, 1]^2$. The realization is obtained using fast and exact synthesis presented in Sect. 4.3.1.3. On the left we draw the obtained surface, on the right the two corresponding images with colors or gray levels given according to the values on each points

Proof Since for $1 \leq i \leq d$, the function K_i is a covariance function on $\mathbb{R} \times \mathbb{R}$ we may consider independent centered Gaussian processes $X^{(i)} = (X_t^{(i)})_{t \in \mathbb{R}}$ with covariance given by K_i. For $x = (x_1, \ldots, x_d) \in \mathbb{R}^d$, we may define the random variable $X_x = \prod_{i=1}^{d} X_{x \cdot e_i}^{(i)}$ so that the random field $X = (X_x)_{x \in \mathbb{R}^d}$ is second order (but no more Gaussian!), centered, with covariance given by

$$\mathrm{Cov}(X_x, X_y) = \mathbb{E}\left(X_x X_y\right) = \prod_{i=1}^{d} \mathbb{E}\left(X_{x \cdot e_i}^{(i)} X_{y \cdot e_i}^{(i)}\right) = \prod_{i=1}^{d} K_i(x_i, y_i),$$

by independence of $X^{(1)}, \ldots, X^{(d)}$.   □

This leads to the second fundamental extension of Brownian motion on the whole space $\mathbb{R}^d$, by choosing $K_i = K_B$ for all $1 \leq i \leq d$.

Definition 4.7 A (standard) Brownian sheet on $\mathbb{R}^d$ is a centered Gaussian field $(X_x)_{x \in \mathbb{R}^d}$ with covariance function given by

$$\mathrm{Cov}(X_x, X_y) = \prod_{i=1}^{d} \frac{1}{2}(|x \cdot e_i| + |y \cdot e_i| - |x \cdot e_i - y \cdot e_i|), \quad \forall x, y \in \mathbb{R}^d.$$

Note that it implies that this field is equal to 0 on the axes $\{x \in \mathbb{R}^d; \exists i \in \{1, \ldots, d\}, x \cdot e_i = 0\}$ and corresponds to a Brownian motion (non-standard) when restricted to $\{x + te_i; t \in \mathbb{R}\}$, for $x \in \mathbb{R}^d$ with $x \cdot e_i = 0$. The following section will focus on some invariance properties.

4.1.2 Stationarity and Invariances

When considering stochastic modeling of homogeneous media it is usual to assume an invariance of distributions under translation (stationarity) or vectorial rotation (isotropy).

4.1.2.1 Stationarity and Isotropy

Definition 4.8 The random field $X = (X_x)_{x \in \mathbb{R}^d}$ is (strongly) stationary if, for all $x_0 \in \mathbb{R}^d$, the random field $(X_{x+x_0})_{x \in \mathbb{R}^d}$ has the same distribution than X.

It implies a specific structure of second order moments.

Proposition 4.5 *If $X = (X_x)_{x \in \mathbb{R}^d}$ is a stationary second order random field, then,*

- *its mean function is constant $m_X(x) = m_X$, for all $x \in \mathbb{R}^d$ and some $m_X \in \mathbb{R}$;*
- *its covariance may be written as $K_X(x, y) = c_X(x - y)$ with $c_X : \mathbb{R}^d \to \mathbb{R}$ an even function satisfying*

    (i) $c_X(0) \geq 0$;
    (ii) $|c_X(x)| \leq c_X(0) \ \forall x \in \mathbb{R}^d$;
    (iii) c_X is of non-negative type ie $\forall k \geq 1, x_1, \ldots, x_k \in \mathbb{R}^d, \lambda_1, \ldots, \lambda_k \in \mathbb{C}$,

$$\sum_{j,l=1}^{k} \lambda_j \overline{\lambda_l} c_X(x_j - x_l) \geq 0. \tag{4.1}$$

Let us remark that the two above properties characterize the weak (second-order) stationarity. Note also that they imply strong stationarity when the field is assumed to be Gaussian. This is because mean and covariance functions characterize the distribution of Gaussian fields.

Proof Since $X_x \overset{d}{=} X_0$ we get $m_X(x) = \mathbb{E}(X_x) = \mathbb{E}(X_0) := m_X$. For the covariance structure we set $c_X(z) = K_X(z, 0)$, for all $z \in \mathbb{R}^d$, and remark that for $y \in \mathbb{R}^d$, one has $(X_{z+y}, X_y) \overset{d}{=} (X_z, X_0)$ so that $\mathrm{Cov}(X_{z+y}, X_y) = c_X(z)$. Hence for $z = x - y$ we obtain $K_X(x, y) = c_X(x - y)$. Since $(X_x, X_0) \overset{d}{=} (X_0, X_{-x})$, the function c_X is even. The first point comes from the fact that $c_X(0) = \mathrm{Var}(X_0) = \mathrm{Var}(X_x) \geq 0$, the second one is obtained using Cauchy-Schwarz inequality to bound $|\mathrm{Cov}(X_x, X_0)|$. The last one is just a reformulation of the non-negative definiteness of K_X when $\lambda_1, \ldots, \lambda_k \in \mathbb{R}$. Otherwise, it follows writing $\lambda_j = a_j + ib_j$ since we have $\Re(\lambda_j \overline{\lambda_l}) = a_j a_l + b_j b_l$ and $\Im(\lambda_j \overline{\lambda_l}) = b_j a_l - b_l a_j$ with c_X even. $\square$

Remark that when a function $c : \mathbb{R}^d \to \mathbb{R}$ satisfies (4.1), then c must be even and satisfy points (i) and (ii). Actually, (i) is obtained for $k = 1, \lambda_1 = 1$ and $x_1 = 0$. Considering $k = 2, x_1 = 0$ and $x_2 = x \in \mathbb{R}^d$, we first obtain for $\lambda_1 = 1$ and $\lambda_2 = i$ that $2c(0) + ic(x) - ic(-x) \geq 0$ hence c is even, while for $\lambda_1 = \lambda_2 = 1$ it yields $-c(x) \leq c(0)$ and for $\lambda_1 = 1 = -\lambda_2$ we get $c(x) \leq c(0)$ so that (ii) is satisfied.

By Bochner's theorem (1932), a continuous function of non-negative type is a Fourier transform of a non-negative finite measure. This can be rephrased as the following theorem.

Theorem 4.2 (Bochner) *A continuous function $c : \mathbb{R}^d \to \mathbb{R}$ is of non-negative type if and only if $c(0) \geq 0$ and there exists a symmetric probability measure ν on*

$\mathbb{R}^d$ *such that*

$$\forall x \in \mathbb{R}^d, \ c(x) = c(0) \int_{\mathbb{R}^d} e^{ix \cdot \xi} d\nu(\xi).$$

In other words there exists a symmetric random vector Z on $\mathbb{R}^d$ such that

$$\forall x \in \mathbb{R}^d, \ c(x) = c(0)\mathbb{E}(e^{ix \cdot Z}).$$

When $c = c_X$ is the covariance of a random field X, the measure $\nu = \nu_X$ is called the spectral measure of X. This strong result implies in particular that we may define stationary centered Gaussian random field with a covariance function given by the characteristic function of a symmetric random vector.

Proof Note that the converse implication is straightforward so we will only prove the first one. We may assume that $c(0) > 0$, otherwise there is nothing to prove. The first step is to assume that $c \in L^1(\mathbb{R}^d)$. Note that in view of (ii), since c is bounded we also have $c \in L^2(\mathbb{R}^d)$. We will prove that its Fourier transform $\hat{c} \in L^2(\mathbb{R}^d)$ is necessarily non-negative. To this end remark that, approximating by Riemann sums for instance, we necessarily have

$$\int_{\mathbb{R}^d} \int_{\mathbb{R}^d} g(x)\overline{g(y)}c(x-y)dxdy \geq 0, \qquad (4.2)$$

for all $g \in \mathscr{S}(\mathbb{R}^d)$, the Schwartz class of infinitely differentiable function with rapid decreasing. We denote as usual $\hat{g}(\xi) = \int_{\mathbb{R}^d} e^{-ix \cdot \xi} g(x)dx$, the Fourier transform, that may be extended to any $L^2(\mathbb{R}^d)$ function. We may rewrite $\int_{\mathbb{R}^d} \int_{\mathbb{R}^d} g(x)\overline{g(y)}c(x-y)dxdy = \int_{\mathbb{R}^d} g(x)c * \overline{g}(x)dx$, where $*$ is the usual convolution product on $L^1(\mathbb{R}^d)$ and $c * \overline{g} \in L^2(\mathbb{R}^d)$ since $c \in L^1(\mathbb{R}^d)$ and $g \in \mathscr{S}(\mathbb{R}^d) \subset L^2(\mathbb{R}^d)$. Hence, by Plancherel's theorem (see [55] for instance), we have

$$\frac{1}{(2\pi)^d} \int_{\mathbb{R}^d} \hat{c}(\xi)|\hat{g}(-\xi)|^2 d\xi = \int_{\mathbb{R}^d} g(x)c * \overline{g}(x)dx, \qquad (4.3)$$

where the right hand side is non-negative in view of (4.2). Now, for $\sigma > 0$, let us denote by h_σ the density of a centered Gaussian vector of $\mathbb{R}^d$ with covariance $\sigma^2 I_d$ ie $h_\sigma(x) = \frac{1}{\sigma^d(2\pi)^{d/2}} e^{-\frac{\|x\|^2}{2\sigma^2}}$. Its characteristic function is given by the Fourier transform $\hat{h}_\sigma(\xi) = e^{-\sigma^2 \frac{\|\xi\|^2}{2}}$. In this way $(h_\sigma)_\sigma$ is an approximation of identity and $c * h_\sigma(x) \to c(x)$, as $\sigma \to 0$, since c is continuous. Moreover, $\hat{c}$ is also continuous as the Fourier transform of an $L^1(\mathbb{R}^d)$ function, so that we also have $\hat{c} * h_\sigma(\xi) \to \hat{c}(\xi)$. Now we will prove that $\hat{c} \geq 0$. Let us take $g_\sigma(x) = \frac{\sigma^{d/2}}{2^{d/4}\pi^{3d/4}} e^{-\sigma^2\|x\|^2}$ such that

$|\hat{g}_\sigma|^2 = h_\sigma$, by (4.2) and (4.3), we obtain that for all $\sigma > 0$

$$\hat{c} * h_\sigma(0) = \int_{\mathbb{R}^d} \hat{c}(\xi) h_\sigma(\xi) d\xi = (2\pi)^d \int_{\mathbb{R}^d} g_\sigma(x) c * \overline{g_\sigma}(x) dx \geq 0.$$

Letting σ tend to 0 we get $\hat{c}(0) \geq 0$. But for all $\xi \in \mathbb{R}^d$, the function $e^{-i\xi \cdot} c$ is an $L^1(\mathbb{R}^d)$ function satisfying (4.1). Hence its Fourier transform is non-negative at point 0, according to previously. But this is exactly $\hat{c}(\xi)$ and therefore $\hat{c}$ is non-negative. Using Fatou Lemma in (4.1), for $g = h_\sigma$ as σ tends to 0 we also obtain that $\int_{\mathbb{R}^d} \hat{c}(\xi) d\xi \leq c(0)$ ensuring that $\hat{c} \in L^1(\mathbb{R}^d)$. Then, by the Fourier inversion theorem, since c and $\hat{c}$ are even, we get

$$c(x) = \frac{1}{(2\pi)^d} \int_{\mathbb{R}^d} e^{ix \cdot \xi} \hat{c}(\xi) d\xi,$$

with $c(0) = \frac{1}{(2\pi)^d} \int_{\mathbb{R}^d} \hat{c}(\xi) d\xi$. Hence we can choose Z a random vector with density given by $\hat{c}/((2\pi)^d c(0))$.

For the general case we remark that $c\hat{h}_\sigma$ is also a function of non-negative type. Actually, since $\hat{h}_\sigma$ is the Fourier transform of a non-negative function, by converse of Bochner's theorem it is of non-negative type and we may consider X_σ a centered stationary Gaussian random field with covariance $\hat{h}_\sigma$. Let us also consider X a centered stationary Gaussian random field with covariance c, independent from X_σ. Then the random field XX_σ is stationary and admits $c\hat{h}_\sigma$ for covariance function. Since c is bounded by (ii), the function $c\hat{h}_\sigma$ is in $L^1(\mathbb{R}^d)$ and we may find Z_σ such that $[c\hat{h}_\sigma] = c(0)\mathbb{E}(e^{ix \cdot Z_\sigma})$. But $c\hat{h}_\sigma$ tends to c which is continuous at 0 as σ tends to 0. Hence, by Lévy's theorem (see [28] for instance), there exists a random vector Z such that $Z_\sigma \xrightarrow[\sigma \to 0]{d} Z$ and $\nu = \mathbb{P}_Z$ is convenient. Let us finally conclude that Z is symmetric since c is even. $\square$

Examples of stationary Gaussian processes are given by Ornstein Uhlenbeck processes constructed on $\mathbb{R}$ with a parameter $\theta > 0$ and B a standard Brownian motion on $\mathbb{R}^+$ (Fig. 4.3), by

$$X_t = e^{-\theta t} B_{e^{2\theta t}}, \forall t \in \mathbb{R}.$$

Then $X = (X_t)_t$ is clearly a centered Gaussian process with covariance

$$\mathrm{Cov}(X_t, X_s) = e^{-\theta|t-s|} := c_X(t-s), \forall t, s \in \mathbb{R}.$$

Hence it is weakly stationary and also strongly since it is Gaussian. Now the spectral measure is given by $\nu_X(dt) = \frac{\theta^2}{\pi(\theta^2 + t^2)} dt$, or equivalently $c_X(t) = \mathbb{E}(e^{it \cdot Z_\theta})$, with Z_θ a random variable with Cauchy distribution of parameter θ.

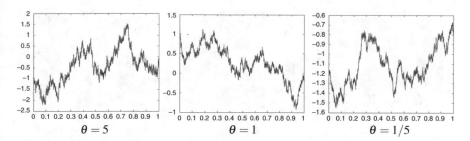

Fig. 4.3 Sample paths of Ornstein Uhlenbeck process on $[0, 1]$, using fast and exact synthesis via circulant embedding matrix method

Definition 4.9 The random field $X = (X_x)_{x \in \mathbb{R}^d}$ is isotropic if, for all R rotation of $\mathbb{R}^d$, the random field $(X_{Rx})_{x \in \mathbb{R}^d}$ has the same distribution than X.

Note that, contrarily to the stationarity, the notion of isotropy is useless in dimension 1! We already have seen one example of isotropic random field when considering the Lévy Chentsov random field. Actually, for all R rotation of $\mathbb{R}^d$,

$$\text{Cov}(X_{Rx}, X_{Ry}) = \frac{1}{2} \left(\|Rx\| + \|Ry\| - \|Rx - Ry\| \right) = \text{Cov}(X_x, X_y).$$

Since X is centered and Gaussian this implies that $(X_{Rx})_{x \in \mathbb{R}^d}$ has the same distribution than X. However X is not stationary (note that $X(0) = 0$ a.s.). An example of stationary and isotropic random field may be given by considering Gaussian covariances $\hat{h}_\sigma$, for $\sigma > 0$ (with $Z_\sigma \sim \mathcal{N}(0, \sigma I_d)$ in Bochner's theorem). Let us also remark that considering the covariance function $k_\sigma(t, s) = e^{-\sigma^2(t-s)/2}$ on $\mathbb{R} \times \mathbb{R}$ we also have

$$K_\sigma(x, y) = \hat{h}_\sigma(x - y) = \prod_{i=1}^{d} k_\sigma(x \cdot e_i, y \cdot e_i),$$

so that this field has also a sheet structure as in Proposition 4.4. Since $\hat{h}_\sigma$ is isotropic, we also have $K_\sigma(Rx, Ry) = K_\sigma(x, y)$ for $x, y \in \mathbb{R}^d$, and this allows to define a stationary isotropic centered Gaussian field with covariance K_σ (Fig. 4.4).

Another very important invariance property is the scale invariance also called self-similarity for random fields.

4.1.2.2 Self-Similarity or Scale Invariance

Definition 4.10 The random field $X = (X_x)_{x \in \mathbb{R}^d}$ is self-similar of order $H > 0$ if, for all $\lambda > 0$, the random field $(X_{\lambda x})_{x \in \mathbb{R}^d}$ has the same distribution than $\lambda^H X = (\lambda^H X_x)_{x \in \mathbb{R}^d}$.

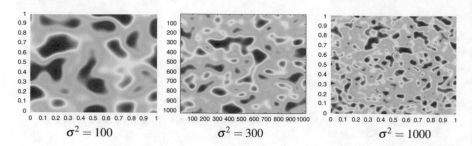

$$\sigma^2 = 100 \qquad\qquad \sigma^2 = 300 \qquad\qquad \sigma^2 = 1000$$

Fig. 4.4 Sample paths of center Gaussian random fields with Gaussian covariances on $[0, 1]^2$, obtained using circulant embedding matrix method (see [53] for details)

Note that the Lévy Chentsov field, and in particular the Brownian motion ($d = 1$), is self-similar of order $H = 1/2$ since

$$\mathrm{Cov}(X_{\lambda x}, X_{\lambda y}) = \frac{1}{2}\left(\|\lambda x\| + \|\lambda y\| - \|\lambda x - \lambda y\|\right)$$

$$= \lambda \mathrm{Cov}(X_x, X_y) = \mathrm{Cov}(\lambda^{1/2} X_x, \lambda^{1/2} X_y)$$

Recall that X is isotropic but not stationary. Actually, there does not exist a non-trivial stationary self-similar second order field since we should have $\mathrm{Var}(X_x) = \mathrm{Var}(X_{\lambda x}) = \lambda^{2H} \mathrm{Var}(X_x)$ for all $\lambda > 0$ and $x \in \mathbb{R}^d$, implying that $\mathrm{Var}(X_x) = 0$. In order to define self-similar fields for homogeneous media we must relax stationary property. This is done throughout the notion of stationary increments.

4.1.2.3 Stationary Increments

Definition 4.11 The random field $X = (X_x)_{x \in \mathbb{R}^d}$ has (strongly) stationary increments if, for all $x_0 \in \mathbb{R}^d$, the random field $(X_{x+x_0} - X_{x_0})_{x \in \mathbb{R}^d}$ has the same distribution than $(X_x - X_0)_{x \in \mathbb{R}^d}$.

Of course a stationary random field X has stationary increments but this class is larger: it also contents $X - X_0$ for instance that can not be stationary except if it is almost surely equal to 0. An example of field with stationary increments is given by the Levy Chentsov field X since we have

$$\mathrm{Cov}(X_{x+x_0} - X_{x_0}, X_{y+x_0} - X_{x_0}) = \mathrm{Cov}(X_x - X_0, X_y - X_0),$$

using the fact that $X_0 = 0$ a.s. We have an analogous of Proposition 4.5 concerning second order structure of fields with stationary increments.

Proposition 4.6 *If* $X = (X_x)_{x \in \mathbb{R}^d}$ *is a second order centered random field with stationary increments and $X_0 = 0$ a.s., then its covariance function may be written as*

$$K_X(x, y) = \frac{1}{2} \left(v_X(x) + v_X(y) - v_X(x - y) \right),$$

with the function $v_X(x) = Var(X_{x+x_0} - X_{x_0}) = Var(X_x - X_0) = Var(X_x)$ called variogram satisfying

1. $v_X(0) = 0$
2. $v_X(x) \geq 0$ *and* $v_X(-x) = v_X(x)$
3. v_X *is conditionally of negative type ie* $\forall k \geq 1, x_1, \ldots, x_k \in \mathbb{R}^d, \lambda_1, \ldots, \lambda_k \in \mathbb{C}$,

$$\sum_{j=1}^{k} \lambda_j = 0 \Rightarrow \sum_{j,l=1}^{k} \lambda_j \overline{\lambda_l} v_X(x_j - x_l) \leq 0.$$

Note that when X_0 does not vanish a.s. this proposition applies to $X - X_0$.

Proof Compute $Var(X_x - X_y) = Var(X_x) + Var(X_x) + 2K_X(x, y)$ and note that $X_x - X_y \overset{d}{=} X_{x-y} - X_0 = X_{x-y}$ to get K_X with respect to v_X. We clearly have $v_X \geq 0$ as a variance and $v_X(0) = 0$ since $X_0 = 0$ a.s. The evenness comes from $X_{-x} = X_{-x} - X_0 \overset{d}{=} X_0 - X_x = -X_x$. The last property follows from the fact that

$$Var(\sum_{j=1}^{k} \lambda_j X_{x_j}) = \frac{1}{2} \sum_{j,l=1}^{k} \lambda_j \lambda_l \left(v_X(x_j) + v_X(x_l) - v_X(x_j - x_l) \right) \geq 0,$$

for $\lambda_1, \ldots, \lambda_k \in \mathbb{R}$, using the expression of $K_X(x_j, x_l)$ with respect to v_X. The inequality is extended for $\lambda_1, \ldots, \lambda_k \in \mathbb{C}$ as in Proposition 4.5 since v_X is also even. □

In order to define centered Gaussian random fields we can use the following result.

Theorem 4.3 (Schoenberg) *Let $v : \mathbb{R}^d \to \mathbb{R}$ be a function such that $v(0) = 0$. The following are equivalent.*

i) v *is conditionally of negative type;*
ii) $K : (x, y) \in \mathbb{R}^d \times \mathbb{R}^d \mapsto \frac{1}{2} (v(x) + v(y) - v(x - y))$ *is a covariance function;*
iii) *For all $\lambda > 0$, the function $e^{-\lambda v}$ is of non-negative type.*

Proof To prove that $i) \Rightarrow ii)$, we use Proposition 4.1. Symmetry comes from the fact that v is even. Actually, taking $k = 2$, $\lambda_1 = i = -\lambda_2$ and $x_1 = x$, $x_2 = 0$ we obtain that $v(x) \leq v(-x)$ since $v(0) = 0$, such that replacing x by $-x$ we get $v(x) = v(-x)$. For the second point let $k \geq 1, x_1, \ldots, x_k \in \mathbb{R}^d, \lambda_1, \ldots, \lambda_k \in \mathbb{R}$

and set $\lambda_0 = -\sum_{i=1}^{k} \lambda_i$ and $x_0 = 0$. We compute $\sum_{i,j=1}^{k} \lambda_i \lambda_j K(x_i, x_j)$ as

$$\sum_{i,j=1}^{k} \lambda_i \lambda_j v(x_i) - \frac{1}{2} \sum_{i,j=1}^{k} \lambda_i \lambda_j v(x_i - x_j) = -\lambda_0 \sum_{i=1}^{k} \lambda_i v(x_i)$$

$$-\frac{1}{2} \sum_{j=1}^{k} \lambda_j \sum_{i=1}^{k} \lambda_i v(x_i - x_j)$$

$$= -\frac{1}{2} \lambda_0 \sum_{i=1}^{k} \lambda_i v(x_i)$$

$$-\frac{1}{2} \sum_{j=0}^{k} \lambda_j \sum_{i=1}^{k} \lambda_i v(x_i - x_j)$$

$$= -\frac{1}{2} \sum_{j=0}^{k} \lambda_j \sum_{i=0}^{k} \lambda_i v(x_i - x_j) \geq 0,$$

since v is even and conditionally of negative type.

Let us now consider $ii) \Rightarrow iii)$. Let $(X^{(n)})_n$ be a sequence of iid centered Gaussian random fields with covariance function given by K and N an independent Poisson random variable of parameter $\lambda > 0$. We may define a new random field $Y = \prod_{n=1}^{N} X^{(n)}$, with the convention that $\prod_{n=1}^{0} = 1$. Therefore, for all $x, y \in \mathbb{R}^d$, we get

$$\mathbb{E}(Y_x Y_y) = \sum_{k=0}^{+\infty} \mathbb{E}(Y_x Y_y | N = k) \mathbb{P}(N = k)$$

$$= \sum_{k=0}^{+\infty} \mathbb{E}(\prod_{n=1}^{k} X_x^{(n)} X_y^{(n)}) e^{-\lambda} \frac{\lambda^k}{k!}$$

$$= \sum_{k=0}^{+\infty} K(x, y)^k e^{-\lambda} \frac{\lambda^k}{k!} = e^{-\lambda(1 - K(x,y))},$$

by independence. Now remark that $\mathbb{E}(Y_x^2) = e^{-\lambda(1 - K(x,x))} = e^{-\lambda(1 - v(x))}$. Hence defining the random field Z by setting $Z_x = \frac{Y_x}{\sqrt{\mathbb{E}(Y_x^2)}} e^{\lambda/2}$ we get

$$\mathbb{E}(Z_x Z_y) = \frac{e^{\lambda}}{e^{-\lambda(1 - \frac{1}{2}(v(x) + v(y)))}} e^{-\lambda(1 - K(x,y))} = e^{-\frac{\lambda}{2} v(x-y)}.$$

As a consequence, for all $k \geq 1$, $x_1, \ldots, x_k \in \mathbb{R}^d$ and $\lambda_1, \ldots, \lambda_k \in \mathbb{R}$

$$\sum_{j,l}^{k} \lambda_j \lambda_l e^{-\frac{\lambda}{2}v(x_j-x_l)} = \mathbb{E}\left(\left(\sum_{j}^{k} \lambda_j Z_{x_j}\right)^2\right) \geq 0.$$

Note that v must be even since $K(x, 0) = K(0, x)$ by symmetry of a covariance function and $v(0) = 0$ so that the previous inequality extends to $\lambda_1, \ldots, \lambda_k \in \mathbb{C}$. This finishes to prove iii).

The last implication $iii) \Rightarrow i)$ comes from the fact for $k \geq 1$, $\lambda_1, \ldots, \lambda_k \in \mathbb{C}$ s.t. $\sum_{j=1}^{k} \lambda_j = 0$, $x_1, \ldots, x_k \in \mathbb{R}^d$, we may write for all $\varepsilon > 0$,

$$\sum_{j,l=1}^{k} \lambda_j \overline{\lambda_l} v(x_j - x_l) \frac{1}{\varepsilon} \int_0^{\varepsilon} e^{-\lambda v(x_j-x_l)} d\lambda = \frac{1}{\varepsilon} \sum_{j,l=1}^{k} \lambda_j \overline{\lambda_l}(1 - e^{-\varepsilon v(x_j-x_l)}) \geq 0,$$

since $\sum_{j=1}^{k} \lambda_j = 0$ and $e^{-\varepsilon v}$ is of non-negative type. Hence letting ε tend to 0 we get the result. $\qquad\square$

As an application of this result we may deduce that the function $v(x) = \|x\|^2$ is a variogram. Actually, this easily follows from bi-linearity of the Euclidean product since $v(x - y) = \|x\|^2 + \|y\|^2 - 2x \cdot y$. But we can also remark that for all $\sigma > 0$, the function $e^{-\frac{\sigma^2}{2}v}$ is the Gaussian covariance function which implies that v is conditionally of negative type. Let us remark that this variogram corresponds to a kind of trivial field since choosing $Z \sim \mathcal{N}(0, I_d)$, one can define the centered Gaussian random field $X_x = x \cdot Z$, for $x \in \mathbb{R}^d$, that admits v for variogram.

An important corollary for the construction of self-similar fields with stationary increments is the following one due to J. Istas in [35].

Corollary 4.1 *If $v : \mathbb{R}^d \to \mathbb{R}^+$ is a variogram then the function v^H is also a variogram for all $H \in (0, 1]$.*

Proof There is nothing to prove for $H = 1$ and when $H \in (0, 1)$, it is sufficient to remark that, by a change of variable, one has for $t \geq 0$,

$$t^H = c_H \int_0^{+\infty} \frac{1 - e^{-\lambda t}}{\lambda^{H+1}} d\lambda,$$

for $c_H^{-1} = \int_0^{+\infty} \frac{1-e^{-\lambda}}{\lambda^{H+1}} d\lambda \in (0, +\infty)$. Hence, for $k \geq 1$, $\lambda_1, \ldots, \lambda_k \in \mathbb{C}$ s.t. $\sum_{j=1}^{k} \lambda_j = 0$, $x_1, \ldots, x_k \in \mathbb{R}^d$, we get

$$\sum_{j,l=1}^{k} \lambda_j \overline{\lambda_l} v(x_j - x_l)^H = -c_H \int_0^{+\infty} \sum_{j,l=1}^{k} \lambda_j \overline{\lambda_l} e^{-\lambda v(x_j-x_l)} \lambda^{-H-1} d\lambda \leq 0,$$

in view of Schoenberg's theorem since v is a variogram. $\qquad\square$

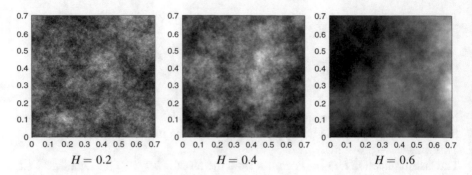

$H = 0.2$ $\qquad\qquad\qquad$ $H = 0.4$ $\qquad\qquad\qquad$ $H = 0.6$

Fig. 4.5 Sample paths realizations of fractional Brownian fields on $[0, 1]^2$ for different values of H, obtained using the fast and exact synthesis method presented in Sect. 4.3.1.3

It follows that for all $H \in (0, 1]$ the function $v_H(x) = \|x\|^{2H}$ is conditionally of negative type and leads to the next definition.

Definition 4.12 A (standard) fractional Brownian field on $\mathbb{R}^d$ (Fig. 4.5), with Hurst parameter $H \in (0, 1]$, is a centered Gaussian field $(B_H)_{x \in \mathbb{R}^d}$ with covariance function given by

$$\mathrm{Cov}(B_H(x), B_H(y)) = \frac{1}{2} \left(\|x\|^{2H} + \|y\|^{2H} - \|x - y\|^{2H} \right), \ \forall x, y \in \mathbb{R}^d.$$

Of course when $H = 1/2$, we recognize $B_{1/2}$ as the Lévy Chentsov field. The order of self-similarity is now given by H. Note also that the case $H = 1$ corresponds to a degenerate case where $B_1 = (x \cdot Z)_x$ for $Z \sim \mathcal{N}(0, I_d)$.

Proposition 4.7 *Up to a constant, the fractional Brownian field of order $H \in (0, 1]$ is the unique isotropic centered Gaussian field with stationary increments which is self-similar of order H.*

Proof This comes from the fact that the distribution of a centered Gaussian field X with stationary increments is characterized by its variogram v_X as soon as $X(0) = 0$ a.s. Self-similarity implies that $X(0) = 0$ and therefore, for all $\lambda > 0$ we have $v_X(\lambda x) = \mathrm{Var}(X_{\lambda x}) = \mathrm{Var}(\lambda^H X_x) = \lambda^{2H} \mathrm{Var}(X_x) = \lambda^{2H} v_X(x)$. Hence for all $x \neq 0$, $v_X(x) = \|x\|^{2H} v_X(\frac{x}{\|x\|})$. But isotropy also implies that $X_\theta \overset{d}{=} X_{e_1}$ for all $\theta \in S^{d-1}$ and $e_1 = (1, 0, \dots, 0) \in S^{d-1}$. Hence $v_X(\theta) = v_X(e_1)$ and $X \overset{d}{=} \sqrt{v_X(e_1)} B_H$. $\qquad\square$

Let us also remark that we cannot find a second order field with stationary increments that is self-similar for an order $H > 1$. Actually, by triangular inequality for $\| \cdot \|_2 := \sqrt{\mathrm{Var}(\cdot)}$, when X is a second order field, we have $\|X_{2x}\|_2 \leq \|X_{2x} - X_x\|_2 + \|X_x\|_2$, for all $x \in \mathbb{R}^d$. Self-similarity of order $H > 0$ implies that $X_0 = 0$ a.s. and $\|X_{2x}\|_2 = 2^H \|X_x\|_2$, while stationary increments imply that

$\|X_{2x} - X_x\|_2 = \|X_x - X_0\|_2 = \|X_x\|_2$. Therefore we must have $2^{H-1}\|X_x\|_2 \le 1$ and $H \le 1$ or $X_x = 0$ for all $x \in \mathbb{R}^d$.

In dimension $d = 1$, it is called *fractional Brownian motion*, implicitly introduced in [41] and defined in [47]. The order of self-similarity H is also called *Hurst parameter*. Hence the fractional Brownian field is an isotropic generalization of this process. Other constructions using sheet structure are known as fractional Brownian sheets but these fields loose stationary increments (see e.g. [22]).

Anisotropy may be an interesting property in applications (see [16] for instance). We also refer to [2] for many examples of anisotropic variograms. For instance, we can consider several anisotropic generalizations of fractional Brownian motions by keeping self-similarity and stationary increments properties following Proposition 4.3. Let $H \in (0, 1)$ and $v_H : t \in \mathbb{R} \mapsto |t|^{2H}$ be the variogram of a fractional Brownian motion that is conditionally of negative type. If μ is a finite positive measure on S^{d-1}, we may define on $\mathbb{R}^d$,

$$v_{H,\mu}(x) = \int_{S^{d-1}} v_H(x \cdot \theta)\mu(d\theta) = \int_{S^{d-1}} |x \cdot \theta|^{2H}\mu(d\theta) = c_{H,\mu}\left(\frac{x}{\|x\|}\right)\|x\|^{2H},$$

(4.4)

that is now a conditionally of negative type function on $\mathbb{R}^d$. Hence we may consider $X_{H,\mu} = (X_{H,\mu}(x))_{x \in \mathbb{R}^d}$ a centered Gaussian random field with stationary increments and variogram given by $v_{H,\mu}$. This new random field is still self-similar of order H but may not be isotropic according to the choice of μ. The function $c_{H,\mu}$, describing anisotropy, is called topothesy function as in [25].

For instance, when $d = 2$ we can choose $\mu(d\theta) = \mathbf{1}_{(-\alpha,\alpha)}(\theta)d\theta$ for some $\alpha \in (0, \pi/2]$. Let $\beta_H(t) = \int_0^t u^{H-1/2}(1-u)^{H-1/2}du$, $t \in [0, 1]$, be a Beta incomplete function. Then the corresponding topothesy function denoted now by $c_{H,\alpha}$ is a π periodic function defined on $(-\pi/2, \pi/2]$ by

$$c_{H,\alpha}(\theta) = 2^{2H}\begin{cases} \beta_H\left(\frac{1-\sin(\alpha-\theta)}{2}\right) + \beta_H\left(\frac{1+\sin(\alpha+\theta)}{2}\right) & \text{if } -\alpha \le \theta + \frac{\pi}{2} \le \alpha \\ \beta_H\left(\frac{1+\sin(\alpha-\theta)}{2}\right) + \beta_H\left(\frac{1-\sin(\alpha+\theta)}{2}\right) & \text{if } -\alpha \le \theta - \frac{\pi}{2} \le \alpha \\ \left|\beta_H\left(\frac{1-\sin(\alpha-\theta)}{2}\right) - \beta_H\left(\frac{1+\sin(\alpha+\theta)}{2}\right)\right| & \text{otherwise} \end{cases}$$

We refer to [14] for computations details and to Fig. 4.6 for plots of these functions.

The associated Gaussian random field denoted by $X_{H,\alpha}$ is called *elementary anisotropic fractional Brownian field*. Several realizations are presented in Fig. 4.7 for different values of parameters α and H. Note that when $\alpha = \pi/2$ the random field $X_{H,\alpha}$ is isotropic and therefore corresponds to a non-standard fractional Brownian field.

Another anisotropic generalization is obtained by considering a generalization of the self-similarity property that allows different scaling behavior according to directions.

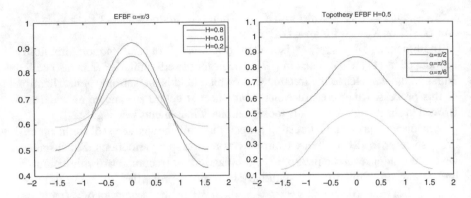

Fig. 4.6 Topothesy functions of some elementary anisotropic fractional Brownian fields

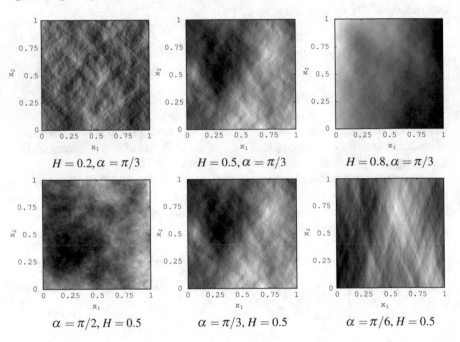

Fig. 4.7 Sample paths realizations of some elementary anisotropic fractional Brownian fields on $[0, 1]^2$ using Turning band method presented in Sect. 4.3.1.2 (see also [14])

4.1.2.4 Operator Scaling Property

Definition 4.13 Let E be a real $d \times d$ matrix with eigenvalues of positive real parts and $H > 0$. The random field $X = (X_x)_{x \in \mathbb{R}^d}$ is (E, H)-operator scaling if for all $\lambda > 0$, the random field $(X_{\lambda^E x})_{x \in \mathbb{R}^d}$ has the same distribution than $\lambda^H X = (\lambda^H X_x)_{x \in \mathbb{R}^d}$, where $\lambda^E = \exp(E \log \lambda)$ with $\exp(A) = \sum_{k=0}^{\infty} \frac{A^k}{k!}$ the matrix exponential.

Note that when E is the identity matrix we recover the definition of self-similarity of order H. Note also that (E, H)-operator scaling property is equivalent to $(E/H, 1)$ operator scaling property. We refer to [9] for general cases and focus here on a simple example where the matrix E is assumed to be diagonalizable. In particular when E is a diagonal matrix, this property may be observed for limit of aggregated discrete random fields (see [42, 52] for instance). In the general case where E is diagonalizable, we assume that its eigenvalues, denoted by $\alpha_1^{-1}, \ldots, \alpha_d^{-1}$, are all greater than 1. They are also eigenvalues of the transpose matrix E^t and we denote by $\theta_1, \ldots, \theta_d$ the corresponding eigenvectors such that $E^t \theta_i = \alpha_i^{-1} \theta_i$. We obtained in [7] the following proposition.

Proposition 4.8 *For $H \in (0, 1]$, the function defined on $\mathbb{R}^d$ by*

$$
v_{H,E}(x) = \tau_E(x)^{2H} = \left(\sum_{i=1}^d |x \cdot \theta_i|^{2\alpha_i} \right)^H = \left(\sum_{i=1}^d v_{\alpha_i} (x \cdot \theta_i) \right)^H ,
$$

is a variogram.

Proof According to Corollary 4.1, it is enough to prove that $\tau_E(x)^2$ is a variogram but this follows from the fact that $v_{\alpha_i} (x \cdot \theta_i)$ is the variogram of $\left(B_{\alpha_i}^{(i)} (x \cdot \theta_i) \right)_{x \in \mathbb{R}^d}$, where $B_{\alpha_1}^{(1)}, \ldots, B_{\alpha_d}^{(d)}$ are d independent fractional Brownian motions on $\mathbb{R}$ with Hurst parameter given by $\alpha_1, \ldots, \alpha_d \in (0, 1]$.   □

Therefore, we can consider a centered Gaussian random field $X_{H,E} = (X_{H,E}(x))_{x \in \mathbb{R}^d}$ with stationary increments and variogram given by $v_{H,E}$. Then, $X_{H,E}$ is (E, H) operator scaling since $v_{H,E}(\lambda^E x) = \lambda^{2H} v_{H,E}(x)$ for all $x \in \mathbb{R}^d$. When $\alpha_1 = \ldots = \alpha_d = \alpha \in (0, 1]$, or equivalently when $E = \alpha^{-1} I_d$, the random field $X_{H,E}$ is actually $\alpha H \in (0, 1)$ self-similar with topothesy function given by $v_{H,E}(\frac{x}{\|x\|})$. Now considering $(\theta_i)_{1 \leq i \leq d} = (e_i)_{1 \leq i \leq d}$, the canonical basis, such fields are examples of Minkowski fractional Brownian fields (see Proposition 3.3 in [49]). In particular, when $d = 2$, we get the topothesy functions as the π-periodic functions $\theta \mapsto (|\cos(\theta)|^{2\alpha} + |\sin(\theta)|^{2\alpha})^H$ (see Fig. 4.8 for some plots).

Some realizations of operator scaling self-similar random fields are presented in Fig. 4.9. Note that the case where $\alpha = 1$ corresponds to the isotropic case, whereas the case where $H = 1$ is a degenerate one obtained by adding two independent fractional Brownian processes.

The realizations of Fig. 4.10 are no more self-similar but when restricting to horizontal, respectively vertical, lines we get fractional Brownian motions of order $H\alpha_1$, respectively $H\alpha_2$. The sheet structure appearing as H increases to 1 should come from the reduction of dependency between directions. The vertical gradient comes from the fact that the self-similarity exponent $H\alpha_2$ is chosen greater than the horizontal one. This is also linked to sample paths regularity as we will see in the following section.

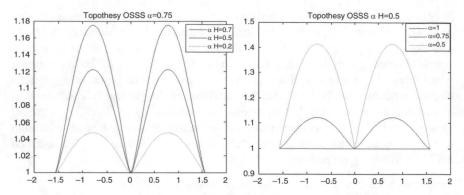

Fig. 4.8 Topothesy functions of self-similar fields $X_{H,E}$ obtained for $E = \alpha^{-1}I_d$

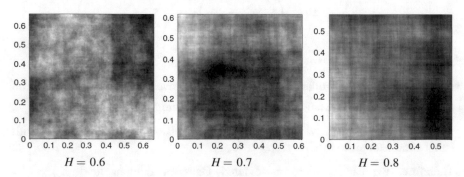

$$H = 0.6 \qquad\qquad H = 0.7 \qquad\qquad H = 0.8$$

Fig. 4.9 Sample paths realizations of self-similar fields $X_{H,E}$ obtained for $E = \alpha^{-1}I_d$ and $H\alpha = 0.5$, using the simulation method presented in Sect. 4.3.1.3

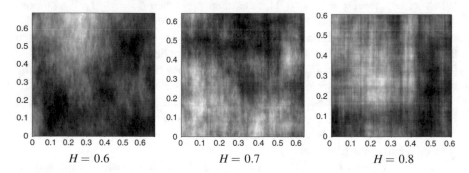

$$H = 0.6 \qquad\qquad H = 0.7 \qquad\qquad H = 0.8$$

Fig. 4.10 Sample paths realizations of operator scaling fields $X_{H,E}$ obtained for $E = \mathrm{diag}(\alpha_1^{-1}, \alpha_2^{-1})$ with $H\alpha_1 = 0.5$ and $H\alpha_2 = 0.6$, using the simulation method presented in Sect. 4.3.1.3

4.2 Sample Paths Properties

In this section we focus on Hölder sample paths properties related to fractal properties of Gaussian random fields. In particular we will see that self-similarity properties induce such fractal behaviors.

4.2.1 Sample Paths Regularity

Before considering sample paths continuity we must introduce the weak notion of stochastic continuity.

Definition 4.14 Let $X = (X_x)_{x \in \mathbb{R}^d}$ be a random field. We say that X is stochastically continuous at point $x_0 \in \mathbb{R}^d$ if

$$\forall \varepsilon > 0, \lim_{x \to x_0} \mathbb{P}(|X_x - X_{x_0}| > \varepsilon) = 0.$$

Let us emphasize that a centered Gaussian random field with stationary increments is stochastically continuous as soon as its variogram is continuous at point 0, according to Bienaymé Chebyshev's inequality. Since we have only defined random fields as a collection of real variables $(X_x)_{x \in \mathbb{R}^d}$, when studying the functions $x \mapsto X_x(\omega)$ for some typical $\omega \in \Omega$, we can in general only state results for a modification of X.

Definition 4.15 Let $X = (X_x)_{x \in \mathbb{R}^d}$ be a random field. We say that $\tilde{X} = (\tilde{X}_x)_{x \in \mathbb{R}^d}$ is a modification of X if

$$\forall x \in \mathbb{R}^d, \mathbb{P}(X_x = \tilde{X}_x) = 1.$$

Note that it follows that X and $\tilde{X}$ have the same distribution since $(X_{x_1}, \ldots, X_{x_k}) = (\tilde{X}_{x_1}, \ldots, \tilde{X}_{x_k})$ a.s. for all $k \geq 1$ and $x_1, \ldots, x_k \in \mathbb{R}^d$. This implies a weaker notion.

Definition 4.16 Let $X = (X_x)_{x \in \mathbb{R}^d}$ be a random field. We say that $\tilde{X} = (\tilde{X}_x)_{x \in \mathbb{R}^d}$ is a version of X if X and $\tilde{X}$ have the same finite dimensional distributions.

We refer to Chapter 9 of [56] for the interested reader.

4.2.1.1 Hölder Regularity

Definition 4.17 Let $K = [0, 1]^d$. Let $\gamma \in (0, 1)$. A random field $X = (X_x)_{x \in \mathbb{R}^d}$ is γ-Hölder on K if there exists a finite random variable A such that a.s.

$$|X_x - X_y| \leq A \|x - y\|^\gamma, \forall x, y \in K.$$

Note that it implies that X is a.s. continuous on K. The following theorem gives a general criterion to ensure the existence of an Hölder modification, particularly helpful for Gaussian fields, generalizing the one dimensional case (see [39, p. 53]).

Theorem 4.4 (Kolmogorov-Chentsov 1956) *If there exist $0 < \beta < \delta$ and $C > 0$ such that*

$$\mathbb{E}\left(|X_x - X_y|^\delta\right) \leq C\|x - y\|^{d+\beta}, \forall x, y \in K,$$

then there exists $\tilde{X}$ a modification of X γ-Hölder on K, for all $\gamma < \beta/\delta$.

Let us note that the assumption clearly implies stochastic continuity of X on K in view of Markov's inequality. We give the constructing proof of this result.

Proof **Step 1.** For $k \geq 1$ we introduce the dyadic points of $[0, 1]^d$

$$\mathscr{D}_k = \left\{ \frac{j}{2^k}; j = (j_1, \ldots, j_d) \in \mathbb{N}^d \text{ with } 0 \leq j_i \leq 2^k \text{ for all } 1 \leq i \leq d \right\}.$$

Note that for $x \in [0, 1]^d$, there exists $x_k \in \mathscr{D}_k$ with $\|x - x_k\|_\infty \leq 2^{-k}$ so that $\mathscr{D}_k$ is a 2^{-k} net of K for $\| \cdot \|_\infty$, where $\|x\|_\infty = \max_{1 \leq i \leq d} |x \cdot e_i|$. The sequence $(\mathscr{D}_k)_k$ is clearly increasing.

Let $\gamma \in (0, \beta/\delta)$. For $i, j \in [0, 2^k]^d \cap \mathbb{N}^d$ with $i \neq j$ define the measurable set

$$E_{i,j}^k = \{\omega \in \Omega; |X_{i/2^k}(\omega) - X_{j/2^k}(\omega)| > \|i/2^k - j/2^k\|_\infty^\gamma\}.$$

By assumption and Markov's inequality

$$\mathbb{P}(E_{i,j}^k) \leq 2^{-k(d+\beta-\gamma\delta)}\|i - j\|_\infty^{d+\beta-\gamma\delta}.$$

Set

$$E^k = \bigcup_{(i,j)\in[0,2^k];0<\|i-j\|_\infty \leq 5} E_{i,j}^k.$$

It follows that

$$\mathbb{P}(E^k) \leq 5^{d+\beta-\gamma\delta}2^{-k(d+\beta-\gamma\delta)}\#\{(i, j) \in [0, 2^k]; 0 < \|i - j\|_\infty \leq 5\}$$
$$\leq 5^{d+\beta-\gamma\delta}10^d 2^{-k(\beta-\gamma\delta)}.$$

Hence, by Borel-Cantelli Lemma we get $\mathbb{P}(\limsup_k E^k) = 0$ so that the event $\tilde{\Omega} = \cup_k \cap_{l \geq k} \Omega \setminus E^l$ satisfies $\mathbb{P}(\tilde{\Omega}) = 1$. Hence, for $\omega \in \tilde{\Omega}$, there exists $k^*(\omega)$ such that for all $l \geq k^*(\omega)$ and $x, y \in \mathscr{D}_l$ with $0 < \|x - y\|_\infty \leq 5 \times 2^{-l}$, we have

$$|X_x(\omega) - X_y(\omega)| \leq \|x - y\|_\infty^\gamma.$$

Step 2. Let us set $\mathscr{D} = \cup_k \mathscr{D}_k$. For $x, y \in \mathscr{D}$ with $0 < \|x - y\|_\infty \leq 2^{-k^*(\omega)}$, there exists a unique $l \geq k^*(\omega)$ with

$$2^{-(l+1)} < \|x - y\|_\infty \leq 2^{-l}.$$

Moreover, one can find $n \geq l+1$ such that $x, y \in \mathscr{D}_n$ and for all $k \in [l, n-1]$, there exist $x_k, y_k \in \mathscr{D}_k$ with $\|x - x_k\|_\infty \leq 2^{-k}$ and $\|y - y_k\|_\infty \leq 2^{-k}$. We set $x_n = x$ and $y_n = y$. Therefore

$$\|x_l - y_l\|_\infty \leq \|x_l - x\|_\infty + \|x - y\|_\infty + \|y - y_l\|_\infty$$
$$\leq 2 \times 2^{-l} + \|x - y\|_\infty.$$

But $2^{-l} < 2\|x - y\|_\infty$ and $\|x_l - y_l\|_\infty \leq 5\|x - y\|_\infty \leq 5 \times 2^{-l}$ and since $l \geq k^*(\omega)$

$$|X_{x_l}(\omega) - X_{y_l}(\omega)| \leq \|x_l - y_l\|_\infty^\gamma \leq 5^\gamma \|x - y\|_\infty^\gamma.$$

But for all $k \in [l, n-1]$, $\|x_k - x_{k+1}\|_\infty \leq 2^{-k} + 2^{-(k+1)} \leq 3 \times 2^{-(k+1)}$ so that

$$|X_{x_k}(\omega) - X_{x_{k+1}}(\omega)| \leq \|x_k - x_{k+1}\|_\infty^\gamma \leq (3/2)^\gamma 2^{-k\gamma}.$$

Similarly,

$$|X_{y_k}(\omega) - X_{y_{k+1}}(\omega)| \leq \|y_k - y_{k+1}\|_\infty^\gamma \leq (3/2)^\gamma 2^{-k\gamma}.$$

It follows that

$$|X_x(\omega) - X_y(\omega)| \leq \sum_{k=l}^{n-1} |X_{x_k}(\omega) - X_{x_{k+1}}(\omega)| + |X_{x_l}(\omega) - X_{y_l}(\omega)|$$

$$+ \sum_{k=l}^{n-1} |X_{y_k}(\omega) - X_{y_{k+1}}(\omega)|$$

$$\leq \frac{2 \times 3^\gamma}{2^\gamma - 1} \times 2^{-l\gamma} + 5^\gamma \|x - y\|_\infty^\gamma$$

$$\leq c_\gamma \|x - y\|_\infty^\gamma.$$

Step 3. By chaining, we obtain that for all $x, y \in \mathscr{D}$

$$|X_x(\omega) - X_y(\omega)| \leq c_\gamma 2^{k^*(\omega)} \|x - y\|_\infty^\gamma,$$

and we set $A(\omega) = c_\gamma 2^{k^*(\omega)}$. Hence we have proven that for all $\omega \in \tilde{\Omega}, x, y \in \mathscr{D}$,

$$|X_x(\omega) - X_y(\omega)| \leq A(\omega)\|x - y\|_\infty^\gamma.$$

We set $\tilde{X}_x(\omega) = 0$ if $\omega \notin \tilde{\Omega}$. For $\omega \in \tilde{\Omega}$, if $x \in \mathscr{D}$ we set $\tilde{X}_x(\omega) = X_x(\omega)$. Otherwise, there exists $(x_k)_k$ a sequence of dyadic points such that $x_k \to x$. Therefore $(X_{x_k}(\omega))$ is a Cauchy sequence and we define $\tilde{X}_x(\omega)$ as its limit. By stochastic continuity we have

$$\mathbb{P}(\tilde{X}_x = X_x) = 1,$$

ensuring that $\tilde{X}$ is a modification.   □

In order to get the best regularity we can use the notion of critical Hölder exponent, as defined in [16].

4.2.1.2   Critical Hölder Exponent

Definition 4.18 Let $\gamma \in (0, 1)$. A random field $(X_x)_{x \in \mathbb{R}^d}$ admits γ as critical Hölder exponent on $[0, 1]^d$ if there exists $\tilde{X}$ a modification of X such that:

(a) $\forall s < \gamma$, a.s. $\tilde{X}$ satisfies $H(s)$: $\exists A \geq 0$ a finite random variable such that $\forall x, y \in [0, 1]^d$,

$$\left| X_x - X_y \right| \leq A \|x - y\|^s.$$

(b) $\forall s > \gamma$, a.s. $\tilde{X}$ fails to satisfy $H(s)$.

For centered Gaussian random fields it is enough to consider second order regularity property as stated in the next proposition (see also [1]).

Proposition 4.9 *Let $(X_x)_{x \in \mathbb{R}^d}$ be a centered Gaussian random field. If for all $\varepsilon > 0$, there exist $c_1, c_2 > 0$, such that, for all $x, y \in [0, 1]^d$,*

$$c_1 \|x - y\|^{2\gamma + \varepsilon} \leq \mathbb{E}\left(X_x - X_y \right)^2 \leq c_2 \|x - y\|^{2\gamma - \varepsilon},$$

then the critical Hölder exponent of X on $[0, 1]^d$ is equal to γ.

Proof The upper bound allows to use Kolmogorov-Chentsov theorem since for all $k \in \mathbb{N}^*$ and $\varepsilon > 0$, using the fact that X is Gaussian, one can find c with

$$\mathbb{E}\left(X_x - X_y \right)^{2k} = \frac{(2k-1)!}{2^{k-1}(k-1)!} \left(\mathbb{E}\left(X_x - X_y \right)^2 \right)^k \leq c \|x - y\|^{2\gamma k - \varepsilon}.$$

Hence considering $\tilde{X}$ the modification of X constructed in the previous proof we see that $\tilde{X}$ is a.s. s-Hölder for all $s < \frac{2\gamma k - \varepsilon}{2k}$. But since $\frac{2\gamma k - \varepsilon}{2k} \to \gamma$, this is true for all $s < \gamma$. Note that according to the lower bound, for any $s > \gamma$, choosing $\varepsilon = s - \gamma$, for any $x \neq y$, the Gaussian random variable $\frac{\tilde{X}_x - \tilde{X}_y}{\|x - y\|^s}$ admits a variance greater than

$c_1\|x - y\|^{-(s-\gamma)}$, that tends to infinity as x tends to y. Therefore it is almost surely unbounded as $\|x - y\| \to 0$.   □

We have therefore a very simple condition on variogram for Gaussian random fields with stationary increments.

Corollary 4.2 *Let X be a centered Gaussian field with stationary increments. If for all $\varepsilon > 0$, there exist $c_1, c_2 > 0$, such that for all $x \in [-1, 1]^d$,*

$$c_1\|x\|^{2\gamma+\varepsilon} \le v_X(x) = \mathbb{E}((X_x - X_0)^2) \le c_2\|x\|^{2\gamma-\varepsilon},$$

then X has critical Hölder exponent on $[0, 1]^d$ equal to γ.

This allows to compute critical exponents of several examples presented above.

- Fractional Brownian fields with variogram $v_H(x) = \|x\|^{2H}$ and 2-dimensional elementary anisotropic fractional Brownian fields with variogram $v_{H,\alpha}(x) = c_{H,\alpha}(x/\|x\|)\|x\|^{2H}$ for $\alpha \in (0, \pi/2]$ have critical Hölder exponent given by H (see Figs. 4.5 and 4.7).
- Stationary Ornstein Uhlenbeck processes have variogram given by $v_X(t) = 2(c_X(0) - c_X(t)) = 2(1 - e^{-\theta|t|})$. It follows that their critical Hölder exponent is given by $1/2$ as for Brownian motion (see Fig. 4.3).
- Operator scaling random fields with variogram given by $v_{H,E}(x) = \left(\sum_{i=1}^d |x \cdot \theta_i|^{2\alpha_i}\right)^H$ admit $H \min_{1 \le i \le d} \alpha_i$ for critical Hölder exponent (see Figs. 4.9 and 4.10).

Note that this global regularity does not capture anisotropy of these last random fields. In order to enlighten it we can consider regularity along lines.

4.2.1.3 Directional Hölder Regularity

Considering a centered Gaussian random field with stationary increments X, one can extract line processes by restricting values along some lines. For $x_0 \in \mathbb{R}^d$ and $\theta \in S^{d-1}$, the line process is defined by $L_{x_0,\theta}(X) = (X(x_0 + t\theta))_{t\in\mathbb{R}}$. It is now a one-dimensional centered Gaussian process with stationary increments and variogram given by $v_\theta(t) = \mathbb{E}\left((X(x_0 + t\theta) - X(x_0))^2\right) = v_X(t\theta)$.

Definition 4.19 ([16]) Let $\theta \in S^{d-1}$. We say that X admits $\gamma(\theta) \in (0, 1)$ as directional regularity in the direction θ if, for all $\varepsilon > 0$, there exist $c_1, c_2 > 0$, such that

$$c_1|t|^{2\gamma(\theta)+\varepsilon} \le v_\theta(t) = v_X(t\theta) \le c_2|t|^{2\gamma(\theta)-\varepsilon}, \quad \forall t \in [-1, 1].$$

It follows that the process $L_{x_0,\theta}(X)$ admits $\gamma(\theta)$ as critical Hölder exponent. Actually, by stationarity of increments, the directional regularity-if exists- may not have more than d values as stated in the following proposition.

Proposition 4.10 ([16]) *If there exists* $\gamma : S^{d-1} \to (0, 1)$ *such that for all* $\theta \in S^{d-1}$, *X admits* $\gamma(\theta)$ *as directional regularity in the direction* θ, *then* γ *takes at most d values. Moreover, if* γ *takes k values* $\gamma_k < \ldots < \gamma_1$, *there exists an increasing sequence of vectorial subset* $\{0\} = V_0 \subsetneq V_1 \subsetneq \ldots \subsetneq V_k := \mathbb{R}^d$ *such that*

$$\gamma(\theta) = \gamma_i \Leftrightarrow \theta \in (V_i \smallsetminus V_{i-1}) \cap S^{d-1}.$$

In our previous examples we get the following results.

- For fractional Brownian fields with variogram $v_H(x) = \|x\|^{2H}$ and 2-dimensional elementary anisotropic fractional Brownian fields with variogram $v_H(x) = c_{H,\alpha}(x/\|x\|)\|x\|^{2H}$: for all $\theta \in S^{d-1}$, the directional Hölder regularity in direction θ is given by H and hence is constant.

- For operator scaling fields with variogram $v_{H,E}(x) = \left(\sum_{i=1}^d |x \cdot \theta_i|^{2\alpha_i}\right)^H$, for all $1 \le i \le d$, the directional Hölder regularity in direction $\tilde{\theta}_i$ is given by $H\alpha_i$ where $\tilde{\theta}_i$ is an eigenvector of E associated with the eigenvalue α_i^{-1} ie $E\tilde{\theta}_i = \alpha_i^{-1}\tilde{\theta}_i$. Moreover, assuming for instance that $\alpha_1 > \alpha_2 > \ldots > \alpha_d$, the strict subspaces defined by $V_k = \text{span}(\alpha_1, \ldots, \alpha_k)$ for $1 \le k \le d$ illustrate the previous proposition by choosing $\gamma_k = H\alpha_k$ for $1 \le k \le d$. For example, in Fig. 4.10, the greater directional regularity is given by $H\alpha_2 = 0.4$ only in vertical directions $\theta = \pm e_2$. For any other direction, the directional regularity is given by $H\alpha_1 = 0.3$. We refer to [44] for more precise results and to [24] for a general setting.

Hölder regularity properties are often linked with fractal properties as developed in the following section.

4.2.2 Hausdorff Dimension of Graphs

We will see in this section how to compute Hausdorff dimension of some Gaussian random fields graphs. The fractal nature comes from the fact that the dimension will not be an integer as usual. We shall recall basic facts on Hausdorff measures and dimensions before.

4.2.2.1 Hausdorff Measures and Dimensions

We follow [27] Chapter 2 for these definitions. Let $U \subset \mathbb{R}^d$ be a bounded Borel set and $\|\cdot\|$ be a norm on $\mathbb{R}^d$. For $\delta > 0$, a finite or countable collection of subsets $(B_i)_{i \in I}$ of $\mathbb{R}^d$ is called a δ-covering of U if $\text{diam}(B_i) \le \delta$ for all i and $U \subset \cup_{i \in I} B_i$. Then, for $s \ge 0$, we set

$$\mathcal{H}_\delta^s(U) = \inf\left\{\sum_{i \in I} \text{diam}(B_i)^s; (B_i)_{i \in I} \ \delta - \text{covering of } U\right\}.$$

Note that for all $\delta < \delta'$, since a δ covering is also a δ'-covering we get $\mathscr{H}^s_\delta(U) \geq \mathscr{H}^s_{\delta'}(U)$. The sequence $(\mathscr{H}^s_\delta(U))_\delta$ being monotonic we can give the following definition.

Definition 4.20 The s-dimensional Hausdorff measure of U is defined by

$$\mathscr{H}^s(U) = \lim_{\delta \to 0} \mathscr{H}^s_\delta(U) \in [0, +\infty].$$

Note that Hausdorff measures define measures on $(\mathbb{R}^d, \mathscr{B}(\mathbb{R}^d))$ that generalize Lebesgue measures so that $\mathscr{H}^1(U)$ gives the length of a curve U, $\mathscr{H}^2(U)$ gives the area of a surface U, etc... Let us also remark that for $s' > s$ and $(B_i)_{i \in I}$ a δ-covering of U we easily see that for all $\delta > 0$

$$\sum_{i \in I} \mathrm{diam}(B_i)^{s'} \leq \delta^{s'-s} \sum_{i \in I} \mathrm{diam}(B_i)^s,$$

so that $\mathscr{H}^{s'}_\delta(U) \leq \delta^{s'-s} \mathscr{H}^s_\delta(U) \leq \delta^{s'-s} \mathscr{H}^s(U)$. Hence, if $\mathscr{H}^s(U) < +\infty$, we get $\mathscr{H}^{s'}(U) = 0$. Conversely, if $\mathscr{H}^{s'}(U) > 0$, we obtain that $\mathscr{H}^s(U) = +\infty$. Actually, the function $s \in [0, +\infty) \mapsto \mathscr{H}^s(U)$ jumps from $+\infty$ to 0.

Definition 4.21 The Hausdorff dimension of U is defined as

$$\dim_H(U) = \inf \left\{ s \geq 0; \mathscr{H}^s(U) = 0 \right\} = \sup \left\{ s \geq 0; \mathscr{H}^s(U) = +\infty \right\}.$$

Let us emphasize that in general we do not know the value of $\mathscr{H}^{s^*}(U) \in [0, +\infty]$ at $s^* = \dim_H(U)$. But we always have that $\mathscr{H}^s(U) > 0$ implies $\dim_H(U) \geq s$, while $\mathscr{H}^s(U) < +\infty$ implies that $\dim_H(U) \leq s$, allowing to compute $\dim_H(U)$.

For instance, when $U = [0, 1]^d$, choosing $\| \cdot \|_\infty$, we can cover U by cubes $\delta i + [0, \delta]^d$ of diameter δ, for $i \in \mathbb{N}^d$ satisfying $0 \leq i_k \leq \delta^{-1} - 1$. Hence we need around δ^{-d} such cubes to cover U so that $\mathscr{H}^s_\delta(U) \leq c\delta^{-d} \times \delta^s$. It follows that for $s \geq d$, $\mathscr{H}^s(U) < +\infty$ and $\dim_H(U) \leq d$. But if $(B_i)_{i \in I}$ is a δ covering of U with $\mathrm{diam}(B_i) = r_i$ we get $1 = \mathscr{L}eb(U) \leq \sum_{i \in I} r_i^d$ and therefore $\mathscr{H}^d(U) > 0$ and $\dim_H(U) \geq d$. In conclusion we obtain that $\dim_H(U) = d$. This can be generalized as in the following Proposition (see [27] for instance).

Proposition 4.11 *If U is a non-empty open bounded set of $\mathbb{R}^d$ then $\dim_H(U) = d$.*

Following similar computations we can deduce an upper bound of Hausdorff dimension for graphs of Hölder functions.

4.2.2.2 Upper Bound of Graphs Hausdorff Dimension

Let $f : [0, 1]^d \to \mathbb{R}$ and denote its graph by

$$\mathscr{G}_f = \{(x, f(x)); x \in [0, 1]^d\} \subset \mathbb{R}^{d+1}.$$

Note that we clearly have $\dim_H \mathscr{G}_f \geq d$. An upper bound may be set according to the Hölder regularity of the function.

Proposition 4.12 *If there exist $\gamma \in (0, 1]$ and $C > 0$ such that for all $x, y \in [0, 1]^d$ one has $|f(x) - f(y)| \leq C\|x - y\|_\infty^\gamma$, then $\dim_H \mathscr{G}_f \leq d + 1 - \gamma$.*

Proof Write $[0, 1]^d \subset \bigcup\limits_{i=1}^{N_\delta} \left(x_i + [0, \delta]^d\right)$, where we can choose N_δ of the order of δ^{-d} and $x_i \in \delta\mathbb{N}^d$ for $1 \leq i \leq N_\delta$ as previously. Then

$$\mathscr{G}_f \subset \bigcup\limits_{i=1}^{N_\delta} \left(x_i + [0, \delta]^d\right) \times \left(f(x_i) + [-C\delta^\gamma, C\delta^\gamma]\right)$$

$$\subset \bigcup\limits_{i=1}^{N_\delta} \bigcup\limits_{j=1}^{N_\delta^\gamma} \left(x_i + [0, \delta]^d\right) \times \left(f(x_i) + I_j(\delta)\right),$$

choosing N_δ^γ intervals $(I_j(\delta))_j$ of size δ to cover $[-C\delta^\gamma, C\delta^\gamma]$, with N_δ^γ of the order of $\delta^{\gamma-1}$. Hence $\mathscr{H}_\delta^s(\mathscr{G}_f) \leq N_\delta N_\delta^\gamma \delta^s \leq c\delta^{-d+\gamma-1+s}$. Therefore choosing $s > d + 1 - \gamma$ implies that $\mathscr{H}^s(\mathscr{G}_f) = 0$ and $\dim_H \mathscr{G}_f \leq s$. Since this holds for all $s < d + 1 - \gamma$ we obtain that $\dim_H \mathscr{G}_f \leq d + 1 - \gamma$.   □

Lower bounds for Hausdorff dimension are usually more difficult to obtain.

4.2.2.3 Lower Bound of Graphs Hausdorff Dimension

One usual way to get lower bounds is to use Frostman criteria [27]. For second order random fields, this can be formulated as in the following theorem (see Lemma 2 of [3]).

Theorem 4.5 *Let $(X_x)_{x\in\mathbb{R}^d}$ be a second order field a.s. continuous on $[0, 1]^d$ such that there exists $s \in (d, d + 1]$,*

$$\int_{[0,1]^d \times [0,1]^d} \mathbb{E}\left(\left(|X_x - X_y|^2 + \|x - y\|^2\right)^{-s/2}\right) dxdy < +\infty,$$

then a.s. $\dim_H \mathscr{G}_X \geq s$.

For Gaussian random fields this can be used in the following way.

Corollary 4.3 *Let $(X(x))_{x\in\mathbb{R}^d}$ be a Gaussian random field. If there exists $\gamma \in (0, 1)$ such that for all $\varepsilon > 0$, there exist $c_1, c_2 > 0$,*

$$c_1\|x - y\|^{2\gamma+\varepsilon} \leq \mathbb{E}\left(X(x) - X(y)\right)^2 \leq c_2\|x - y\|^{2\gamma-\varepsilon},$$

then, for any continuous modification $\tilde{X}$ of X, one has

$$\dim_H \mathscr{G}_{\tilde{X}} = d + 1 - \gamma \quad a.s.$$

Proof The upper bound allows to construct a modification $\tilde{X}$ of X that is β-Hölder on $[0, 1]^d$ for any $\beta < \gamma$ in view of Kolmogorov-Chentsov theorem so that $\dim_H \mathcal{G}_{\tilde{X}} \leq d + 1 - \gamma$ a.s. according to the previous part. Following [3] and [4], since X is Gaussian, one can prove that for any $s > 1$ and $\beta > \gamma$, there exists $c > 0$ such that

$$\mathbb{E}\left(\left(|X_x - X_y|^2 + \|x - y\|^2\right)^{-s/2}\right) \leq c\|x - y\|^{1-\beta-s},$$

using the fact that $\mathbb{E}(X(x) - X(y))^2 \geq c_1\|x - y\|^{2\beta}$ by assumption. It follows that for $1 - \beta - s + d > 0$ the integral in Theorem 4.5 is finite. Hence a.s. $\dim_H \mathcal{G}_{\tilde{X}} \geq s$. Since this holds for all $\beta > \gamma$ and $s < d + 1 - \beta$ we get the desired lower bound and then the result. □

Note that in particular we obtain that the Hausdorff dimension of fractional Brownian fields graphs, respectively 2-dimensional elementary anisotropic fractional Brownian fields graphs, with Hurst parameter $H \in (0, 1)$, is given by $d + 1 - H \notin \mathbb{N}$, respectively $2 + 1 - H \notin \mathbb{N}$. For (E, H)-operator scaling random fields graphs, it is given by $d + 1 - H \min_{1 \leq i \leq d} \alpha_i \notin \mathbb{N}$, where $(\alpha_i^{-1})_{1 \leq i \leq d}$ are the eigenvalues of E (see [9]).

4.3 Simulation and Estimation

This section focuses on some exact methods of simulation for some previously studied Gaussian random fields. We also give one way of estimation for fractal roughness and some applications in medical imaging.

4.3.1 Simulation

In order to simulate a centered Gaussian random field X on $[0, 1]^d$ for instance, the first step is to choose the mesh of discretization, let say $1/n$ for some $n \in \mathbb{N}$. Then one want to simulate the centered Gaussian random vector $(X_{k/n})_{k \in [[0,n]]^d}$ that is of size $(n + 1)^d$. Choleski's method for diagonalizing the covariance matrix becomes quickly unpractical as n increases. Some helpful results concerning diagonalization of circulant matrices by discrete Fourier transforms may be sometimes used under an assumption of stationarity implying a Toeplitz structure of the covariance matrix. We refer to [26] for general framework and only illustrate these ideas to set fast and exact algorithms for some fractional or operator scaling 2-dimensional fields. The first step is to simulate one-dimensional fractional Brownian motion.

4.3.1.1 Fast and Exact Synthesis of Fractional Brownian Motion

Let $H \in (0, 1)$ and $B_H = (B_H(t))_{t \in \mathbb{R}}$ be a fractional Brownian motion and recall that for $n \in \mathbb{N}$ we want to simulate $(B_H(k/n))_{0 \le k \le n}$. By self-similarity,

$$(B_H(k/n))_{0 \le k \le n} \overset{d}{=} n^{-H} (B_H(k))_{0 \le k \le n}, \text{ with } B_H(k) = \sum_{j=0}^{k-1} (B_H(j+1) - B_H(j))$$

for $k \ge 1$, since $B_H(0) = 0$ a.s. Hence, let us define the fractional Gaussian noise as $Y_j = B_H(j+1) - B_H(j)$, for $j \in \mathbb{Z}$. Since B_H has stationary increments, $(Y_j)_{j \in \mathbb{Z}}$ is a centered stationary Gaussian sequence with covariance given by

$$c_k = \mathrm{Cov}(Y_{k+j}, Y_j) = \frac{1}{2} \left(|k+1|^{2H} - 2|k|^{2H} + |k-1|^{2H} \right), \forall k \in \mathbb{Z}.$$

It follows that the Gaussian vector $Y = (Y_0, \ldots, Y_n)$ has a Toeplitz covariance

matrix given by $K_Y = \begin{pmatrix} c_0 & c_1 & \cdots & c_n \\ & \ddots & & \vdots \\ & & \ddots & c_1 \\ & & & c_0 \end{pmatrix}$. The idea of the circulant embedding

matrix method [26] is to embed K_Y in the symmetric circulant matrix $S = circ(s)$ of size $2n$ with

$$s = (c_0 \, c_1 \, \ldots \, c_n \, c_{n-1} \, \ldots \, c_1) = (s_0 \, s_1 \, \ldots \, s_n \, s_{n+1} \, \ldots \, s_{2n-1}),$$

and more precisely

$$S = \begin{pmatrix} s_0 & s_{2n-1} & \cdots & s_2 & s_1 \\ s_1 & s_0 & s_{2n-1} & & s_2 \\ \vdots & s_1 & s_0 & \ddots & \vdots \\ s_{2n-2} & & & \ddots & s_{2n-1} \\ s_{2n-1} & s_{2n-2} & \cdots & s_1 & s_0 \end{pmatrix} = \begin{pmatrix} K_Y & S_1 \\ S_1^t & S_2 \end{pmatrix}.$$

Then $S = \frac{1}{2n} F_{2n}^* \mathrm{diag}(F_{2n}s) F_{2n}$ with $F_{2n} = \left(e^{\frac{2i\pi(j-1)(k-1)}{2n}} \right)_{1 \le j, k \le 2n}$ the matrix of discrete Fourier transform.

The symmetric matrix S may be used as a covariance matrix as soon as its eigenvalues are non-negative, which is equivalent to the fact that $F_{2n}s \ge 0$. This is in general difficult to establish and sometimes only checked numerically. However as far as fractional Gaussian noises are concerned we have a theoretical positive result established in [23, 51]. So we may consider a square root of S given by $R_{2n} = \frac{1}{\sqrt{2n}} F_{2n}^* \mathrm{diag}(F_{2n}s)^{1/2} \in \mathcal{M}_{2n}(\mathbb{C})$. Hence, choosing two independent centered Gaussian vectors $\varepsilon^{(1)}, \varepsilon^{(2)}$ with covariance matrix I_{2n} (hence iid marginals

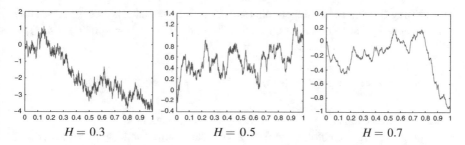

Fig. 4.11 Simulation of $(B_H(k/n))_{0 \leq k \leq n}$ for $n = 2^{12}$ using circulant embedding matrix method

of standard Gaussian variables), we get using the fact that $R_{2n} R_{2n}^* = S$,

$$R_{2n}[\varepsilon^{(1)} + i\varepsilon^{(2)}] = Z^{(1)} + iZ^{(2)},$$

with $Z^{(1)}$, $Z^{(2)}$ iid $\mathcal{N}(0, S)$. It follows that

$$Y \stackrel{d}{=} \left(Z_k^{(1)}\right)_{0 \leq k \leq n} \stackrel{d}{=} \left(Z_k^{(2)}\right)_{0 \leq k \leq n} \sim \mathcal{N}(0, K_Y).$$

Choosing $n = 2^p$ we can use fast discrete Fourier transforms to get a cost of simulation $O(n \log(n))$ to compare with $O(n^3)$ for Choleski method. On Fig. 4.11, we can now illustrate previous results on the regularity and graphs Hausdorff dimension of fractional Brownian motions given respectively by H and $2 - H$ in dimension 1.

When considering 2-dimensional Gaussian fields, several extensions are possible. For instance, if $(Y_{k_1,k_2})_{(k_1,k_2) \in \mathbb{Z}^2}$ is stationary, its covariance function may be written as $\mathrm{Cov}(Y_{k_1+l_1,k_2+l_2}, Y_{l_1,l_2}) = r_{k_1,k_2}$. Hence we may use a block Toeplitz covariance matrix with Toeplitz blocks and embed in a block circulant matrix (see [21, 26]). When only stationary increments are assumed one can still try to simulate the increments but in dimension $d > 1$ the initial conditions now correspond to values on axis and are correlated with increments [38].

We present two other possible generalizations for considered random fields based on more general ideas.

4.3.1.2   Turning Band Method for 2-Dimensional Anisotropic Self-Similar Fields

The turning band method was introduced by Matheron in [48]. It is mainly based on similar ideas developed in Proposition 4.3 when considering isotropic fields constructed from processes. Actually, when Y is a centered second order stationary process with covariance $K_Y(t, s) = c_Y(t - s)$ one can define the field

$$Z(x) = Y(x \cdot U) \text{ for } x \in \mathbb{R}^2,$$

by choosing $U \sim \mathcal{U}(S^1)$ independent from Y. It follows that Z is a centered station-ary isotropic field such that identifying $\theta \in [-\pi, \pi]$ with $u(\theta) = (\cos(\theta), \sin(\theta)) \in S^1$,

$$c_Z(x) = \mathrm{Cov}(Z(x + y), Z(y)) = \frac{1}{\pi} \int_{-\pi/2}^{\pi/2} c_Y(x \cdot u(\theta)) d\theta.$$

Let us note that even if Y is Gaussian Z is not a Gaussian field.

Assuming that we are able to simulate Y one can define for $K \geq 1$, $\theta_1, \ldots, \theta_K \in [-\pi/2, \pi/2]$ and $\lambda_1, \ldots, \lambda_K \in \mathbb{R}^+$, an approximated field

$$Z_K(x) = \sum_{i=1}^{K} \sqrt{\lambda_i} Y^{(i)}(x \cdot u(\theta_i)),$$

with $Y^{(1)}, \ldots, Y^{(K)}$ independent realizations of Y. The field Z_K is a centered stationary field with covariance

$$c_{Z_K}(x) = \sum_{i=1}^{K} \lambda_i c_Y(x \cdot u(\theta_i)),$$

such that choosing convenient weights it can be an approximation of the covariance c_Z. For Gaussian random field, Matheron proposes to use the central limit theorem and considers $\frac{1}{\sqrt{N}}(Z_K^{(1)} + \ldots + Z_K^{(N)})$, with $Z_K^{(1)}, \ldots, Z_K^{(N)}$ independent realizations of Z_K. In [14] we have exploited these ideas to propose simulations of anisotropic self-similar fields. Let $H \in (0, 1)$, μ be a finite non-negative measure on S^1, and $X_{H,\mu} = (X_{H,\mu}(x))_{x \in \mathbb{R}^2}$ be a centered Gaussian random field with stationary increments and variogram given by

$$v_{H,\mu}(x) = \int_{S^1} |x \cdot \theta|^{2H} \mu(d\theta) = C_{H,\mu}\left(\frac{x}{\|x\|}\right) \|x\|^{2H}.$$

We recall that $X_{H,\mu}$ is self-similar of order H. Note also that choosing μ the uniform measure on S^1, the corresponding field $X_{H,\mu}$ is isotropic and therefore it is a fractional Brownian field. When μ_K is a discrete measure ie $\mu_K = \sum_{i=1}^{K} \lambda_i \delta_{\theta_i}$ for some $\theta_1, \ldots, \theta_K \in S^1$ and $\lambda_1, \ldots, \lambda_K \in \mathbb{R}^+$, we get

$$v_{H,\mu_K}(x) = \sum_{i=1}^{K} \lambda_i |x \cdot \theta_i|^{2H} = \sum_{i=1}^{K} \lambda_i \mathrm{Var}(B_H(x \cdot \theta_i)).$$

Hence, considering, $(B_H^{(i)})_{1 \leq i \leq K}$ independent realizations of the one dimensional H-fractional Brownian motion,

$$X_{H,\mu_K}(x) := \sum_{i=1}^{K} \sqrt{\lambda_i} B_H^{(i)}(x \cdot \theta_i), \ x \in \mathbb{R}^2,$$

is a centered Gaussian random field with stationary increments and variogram v_{H,μ_K}.

Hence the simulation of these random fields depends on the simulation of B_H on some specific points. We can exploit further on this fact using specific choices of lines and weights. Actually, to simulate $\left(X_{H,\mu_K}\left(\frac{k_1}{n}, \frac{k_2}{n}\right)\right)_{0 \le k_1, k_2 \le n}$ one has to simulate for $1 \le i \le K$,

$$B_H^{(i)}\left(\frac{k_1}{n}\cos(\theta_i) + \frac{k_2}{n}\sin(\theta_i)\right) \text{ for } 0 \le k_1, k_2 \le n.$$

When $\cos(\theta_i) \ne 0$, by choosing θ_i with $\tan(\theta_i) = \frac{p_i}{q_i}$ for $p_i \in \mathbb{Z}$ and $q_i \in \mathbb{N}$, using self-similarity we get

$$\left(B_H^{(i)}\left(\frac{k_1}{n}\cos(\theta_i) + \frac{k_2}{n}\sin(\theta_i)\right)\right)_{k_1,k_2} \overset{fdd}{=} \left(\frac{\cos(\theta_i)}{nq_i}\right)^H \left(B_H^{(i)}(k_1 q_i + k_2 p_i)\right)_{k_1,k_2}.$$

Using the previous algorithm we are able to simulate this with a cost given by $O(n(|p_i|+q_i)\log(n(|p_i|+q_i)))$. For $\mu(d\theta) = c(\theta)d\theta$, in particular for elementary anisotropic fractional fields (see Fig. 4.7), Riemann approximation for convenient μ_K yields to error bounds between the distribution of X_{H,μ_K} and the distribution of $X_{H,\mu}$ so that X_{H,μ_K} may be used as an approximation to simulate $X_{H,\mu}$. Note that contrarily to the original Turning band method, the simulated fields are all Gaussian, with stationary increments, and self-similar of order H. We refer to Fig. 4.12 where we can see the effect of the number of chosen lines on induced realizations as well as on corresponding variograms.

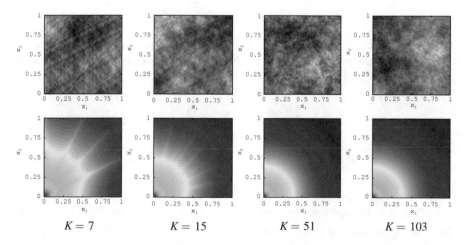

Fig. 4.12 Top: realizations of X_{H,μ_K} with $H = 0.2$ to approximate the isotropic field $X_{H,\mu}$ for $\mu(d\theta) = d\theta$ and $n = 512$; Bottom: corresponding variograms v_{H,μ_K} (see [14])

Actually, a fast and exact method of simulation has been set up in [57] for isotropic fractional Brownian fields $X_{H,\mu}$. This is based on a local stationary approximation of this field by considering a stationary field with compactly supported covariance function for which the 2-dimensional circulant embedding matrix algorithm is running. These ideas have also been exploited in [29] and may be partially used for more general operator scaling random field [7]. We briefly present this in the following section.

4.3.1.3 Stein Method for 2-Dimensional Operator Scaling fields

For the sake of simplicity we consider here the diagonal case and set $E = diag(\alpha_1^{-1}, \alpha_2^{-1})$ for some $\alpha_1, \alpha_2 \in (0, 1]$ and $\tau_E(x)^2 := |x_1|^{2\alpha_1} + |x_2|^{2\alpha_2}$. Recall that, by Proposition 4.8, for $H \in (0, 1]$ we can define $X_{H,E} = (X_{H,E}(x))_{x \in \mathbb{R}^2}$ a centered Gaussian random field with stationary increments and variogram given by

$$v_{H,E}(x) = \tau_E(x)^{2H} = \left(|x_1|^{2\alpha_1} + |x_2|^{2\alpha_2}\right)^H,$$

so that $X_{H,E}$ is (E, H)-operator scaling. Let us also note that for $\alpha_1 = \alpha_2 = \alpha \in (0, 1]$, the field $X_{H,E}$ is αH-self-similar but it is isotropic only when the common value α is equal to 1. In this case $X_{H,E}$ corresponds to a fractional Brownian field of order $\alpha H = H$. In [7], we extend the procedure developed in [57] for fast and exact synthesis of these last fields. Let us define for $c_H = 1 - H$, the real compactly supported function

$$K_{H,E}(x) = \begin{cases} c_H - \tau_E(x)^{2H} + (1 - c_H)\tau_E(x)^2 & \text{if } \tau_E(x) \leq 1 \\ 0 & \text{else} \end{cases} \quad \text{for } x \in \mathbb{R}^2.$$

Assuming that $K_{H,E}$ is a covariance function on $\mathbb{R}^2$ we can define $Y_{H,E}$ a stationary centered Gaussian random field with covariance $K_{H,E}$. Then, computing covariance functions we get that

$$\left\{X_{H,E}(x); x \in [0, M]^2\right\}$$

$$\stackrel{fdd}{=} \left\{Y_{H,E}(x) - Y_{H,E}(0) + \sqrt{1 - c_H}B_{\alpha_1}^{(1)}(x_1) + \sqrt{1 - c_H}B_{\alpha_2}^{(2)}(x_2); x \in [0, M]^2\right\},$$

for $M = \max\left\{0 \leq r \leq 1; r^{2\alpha_1} + r^{2\alpha_2} \leq 1\right\}$ and $B_{\alpha_1}^{(1)}$, $B_{\alpha_2}^{(2)}$ two standard independent 1-dimensional fractional Brownian motions, independent from $Y_{H,E}$. Since we are already able to simulate $\left(B_{\alpha_i}^{(i)}(k/n)\right)_{0 \leq k \leq n}$ in view of Sect. 4.3.1.1, it is enough to simulate $(Y_{H,E}(k_1/n, k_2/n))_{0 \leq k_1, k_2 \leq n}$ in order to simulate $(X_{H,E}(k_1/n, k_2/n))_{0 \leq k_1, k_2 \leq Mn}$ But if $K_{H,E}$ is a covariance function, since it has compact support in $[-1, 1]^2$, its periodization

$$K_{H,E}^{per}(x) = \sum_{k_1,k_2 \in \mathbb{Z}^2} K_{H,E}(x_1 + 2k_1, x_2 + 2k_2),$$

will also be a periodic covariance function on $\mathbb{R}^2$. Denoting by $Y_{H,E}^{per}$ a stationary periodic centered Gaussian random field with covariance function $K_{H,E}^{per}$ and remarking that $K_{H,E}(x) = K_{H,E}(|x_1|, |x_2|)$, the random vector $\left(Y_{H,E}^{per}\left(\frac{k_1}{n}, \frac{k_2}{n}\right)\right)_{0 \leq k_1, k_2 \leq 2n}$ has a block circulant covariance matrix diagonalized by 2D discrete Fourier transform. Following [26], a fast and exact synthesis of $\left(Y_{H,E}\left(\frac{k_1}{n}, \frac{k_2}{n}\right)\right)_{0 \leq k_1, k_2 \leq n} \overset{d}{=} \left(Y_{H,E}^{per}\left(\frac{k_1}{n}, \frac{k_2}{n}\right)\right)_{0 \leq k_1, k_2 \leq n}$ is possible with a cost of the order $O(n^2 \log(n))$. Note that according to Theorem 4.2, by the Fourier inverse theorem, the function $K_{H,E}$ is a covariance matrix if and only if its Fourier transform, defined for $\xi \in \mathbb{R}^2$ by $\hat{K}_{H,E}(\xi) = \int_{\mathbb{R}^2} e^{-ix \cdot \xi} K_{H,E}(x) dx$, is non-negative. This was proven in [57] in the isotropic case for $\alpha_1 = \alpha_2 = \alpha = 1$ and $H \in (0, 3/4)$. Note also that in this case we simply have $(B_{\alpha_i}^{(i)}(x_i)) \overset{d}{=} (x_i N^{(i)})$ with $N^{(i)} \sim \mathcal{N}(0, 1)$ for $i = 1, 2$. As long as we only want to synthesize the vector $(X_{H,E}(k_1/n, k_2/n))_{0 \leq k_1, k_2 \leq Mn}$ it is actually sufficient to check numerically the non-negativeness of eigenvalues for the covariance matrix of $\left(Y_{H,E}^{per}\left(\frac{k_1}{n}, \frac{k_2}{n}\right)\right)_{0 \leq k_1, k_2 \leq 2n}$. We refer to Figs. 4.13 and 4.14 for some realizations.

4.3.2 Estimation

We describe here one way of estimation for Hölder directional regularity from a discretized sample paths observation. Our procedure is based on the variogram method for estimation of Hurst parameter of one-dimensional fractional Brownian motions.

4.3.2.1 1D Estimation Based on Variograms

We first assume to observe $(B_H(k))_{0 \leq k \leq n}$ for some large n, where $B_H = (B_H(t))_{t \in \mathbb{R}}$ is a fractional Brownian motion with Hurst parameter $H \in (0, 1)$. We exploit the stationarity of increments by considering the increments of B_H with step $u \in \mathbb{N}$

$$\Delta_u B_H(k) = B_H(k + u) - B_H(k)$$

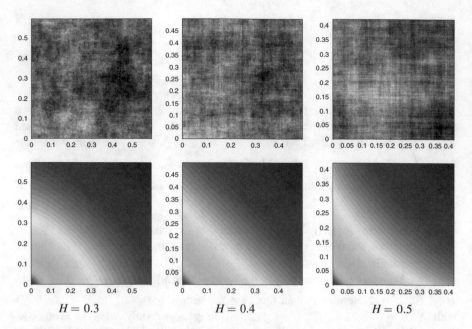

Fig. 4.13 Top: realizations of anisotropic self-similar fields $X_{H,E}$ with $E = \alpha^{-1} I_2$ and $H\alpha = 0.2$ for $n = 2^{10}$; Bottom: corresponding variograms $v_{H,E}$ (see [7])

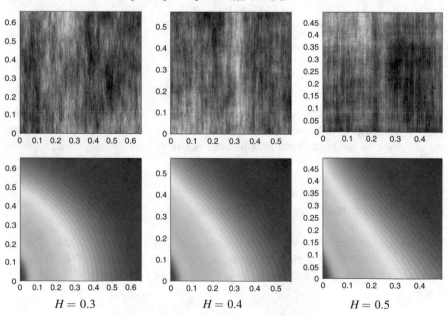

Fig. 4.14 Top: realizations of operator scaling fields $X_{H,E}$ with $E = \mathrm{diag}(\alpha_1^{-1}, \alpha_2^{-1})$ and $H\alpha_1 = 0.2$, $H\alpha_2 = 0.3$ for $n = 2^{10}$; Bottom: corresponding variograms $v_{H,E}$ (see [7])

Note that the sequence $(\Delta_u B_H(k))_{k \in \mathbb{Z}}$ is a stationary Gaussian centered sequence, with common variance given by $v_H(u) = c_H |u|^{2H}$. Hence, considering the statistics

$$V_n(u) = \frac{1}{n-u} \sum_{k=0}^{n-1-u} \Delta_u B_H(k)^2, \qquad (4.5)$$

we immediately see that $\mathbb{E}(V_n(u)) = v_H(u)$. Actually, since $\Delta_u B_H$ is Gaussian and $\mathrm{Cov}(\Delta_u B_H(k), \Delta_u B_H(0)) \xrightarrow[k \to +\infty]{} 0$, this sequence is ergodic (see [45]) so that $V_n(u)$ is a strongly consistent estimator of $v_H(u)$ meaning that $V_n(u) \xrightarrow[n \to +\infty]{} v_H(u)$ a.s. This naturally leads us to consider for two different steps $u \neq v$, the statistic

$$\widehat{H}_n = \frac{1}{2} \log\left(\frac{V_n(u)}{V_n(v)}\right) / \log\left(\frac{u}{v}\right). \qquad (4.6)$$

Theorem 4.6 *For $H \in (0, 1)$, the statistic $\widehat{H}_n$ is a strongly consistent estimator of H. Moreover for $H \in (0, 3/4)$ it is also asymptotically normal.*

Proof The consistency is immediate once remarked that $H = \frac{1}{2} \log\left(\frac{v_H(u)}{v_H(v)}\right) / \log\left(\frac{u}{v}\right)$. To prove asymptotic normality, the first step is a central limit theorem for

$$\frac{V_n(u)}{v_H(u)} = \frac{1}{n-u} \sum_{k=0}^{n-1-u} X_u(k)^2 \text{ with } X_u(k) = \frac{\Delta_u B_H(k)}{\sqrt{v_H(u)}},$$

so that $(X_u(k))_k$ is a centered stationary Gaussian sequence with unit variance. Then, denoting $H_2(x) = x^2 - 1$ the Hermite polynomial of order 2 we consider

$$Q_n(u) := \sqrt{n-u} \left(\frac{V_n(u)}{v_H(u)} - 1\right) = \frac{1}{\sqrt{n-u}} \sum_{k=0}^{n-1-u} H_2(X_u(k)). \qquad (4.7)$$

We will use a general result of Breuer Major [17] giving a summability condition on the covariance sequence to get asymptotic normality.

Proposition 4.13 ([17]) *If $(\rho_u(k))_k = (\mathrm{Cov}(X_u(k), X_u(0)))_k$ satisfies*

$$\sigma_u^2 = \sum_{k \in \mathbb{Z}} \rho_u(k)^2 < +\infty, \text{ then}$$

i) $Var(Q_n(u)) \to 2\sigma_u^2$;

ii) $\dfrac{Q_n(u)}{\sqrt{Var(Q_n(u))}} \xrightarrow{d} N$, with $N \sim \mathcal{N}(0, 1)$.

But since $X_u(k) = \frac{\Delta_u B_H(k)}{\sqrt{v_H(u)}}$ we obtain that

$$\rho_u(k) = \frac{1}{2} \left(|k+u|^{2H} - 2|k|^{2H} + |k-u|^{2H} \right) = O_{|k|\to+\infty}(|k|^{-2(1-H)}).$$

Hence, we check that $\sum_{k\in\mathbb{Z}} \rho_u(k)^2 < +\infty$ for $H < 3/4$, and by Slutsky's theorem $Q_n(u) \xrightarrow{d} \mathcal{N}(0, \sigma_u^2)$. By Delta-method (see [58] for instance), asymptotic normality of $\widehat{H}_n$ will follow from asymptotic normality of the couple $(Q_n(u), Q_n(v))$. Note that we already have checked it for each marginal. However since $Q_n(u)$ and $Q_n(v)$ are in the same Wiener chaos of order two we can use a very strong result of [50] saying that if $\mathrm{Cov}(Q_n(u), Q_n(v)) \to \sigma_{uv}$, asymptotic normality of marginals imply asymptotic normality of the couple namely

$$(Q_n(u), Q_n(v)) \xrightarrow{d} \mathcal{N}\left(0, \begin{pmatrix} \sigma_u^2 & \sigma_{uv} \\ \sigma_{uv} & \sigma_v^2 \end{pmatrix}\right),$$

that concludes the proof.   □

In order to get rid of the upper bound $H < 3/4$ for asymptotic normality, Istas and Lang [36] have proposed to consider generalized quadratic variations. In particular we can replace $\Delta_u B_H(k)$ by second order increments

$$\Delta_u^{(2)} B_H(k) = B_H(k+2u) - 2B_H(k+u) + B_H(k)$$

so that we keep a centered Gaussian stationary sequence with a similar variance given by $v_H^{(2)}(u) = \mathrm{Var}(\Delta_u^{(2)} B_H(k)) = c_H^{(2)}|u|^{2H}$ but now the covariance sequence is $O_{k\to+\infty}(|k|^{-2(2-H)})$ ensuring asymptotic normality for $Q_n^{(2)}(u)$, obtained by replacing X_u by $X_u^{(2)} = \frac{\Delta_u^{(2)} B_H}{\sqrt{v_H^{(2)}(u)}}$ in (4.7) and, as a consequence for $\widehat{H}_n^{(2)}$ for all $H \in (0, 1)$. This way of estimation is particularly robust and consistency as well as asymptotic normality still hold for infill or high frequency estimation where we now assume to observe $(B_H(k/n))_{0\leq k\leq n}$ instead of $(B_H(k))_{0\leq k\leq n}$. In this framework we have to replace $\Delta_u^{(2)} B_H(k)$ by

$$\Delta_{u/n}^{(2)} B_H(k/n) = B_H\left(\frac{k+2u}{n}\right) - 2B_H\left(\frac{k+u}{n}\right) + B_H\left(\frac{k}{n}\right).$$

Let us note that by self-similarity $(\Delta_{u/n}^{(2)} B_H(k/n))_k \overset{d}{=} (n^{-H}\Delta_u^{(2)} B_H(k))_n$ but we can also replace the self-similar process B_H by Y a centered Gaussian process with stationary increments such that

$$v_Y(u) = \mathbb{E}\left((Y(t+u) - Y(t))^2\right) = c_Y|u|^{2H} + O_{|u|\to0}\left(|u|^{2H+\varepsilon}\right).$$

Fig. 4.15 A realization of a 2-dimensional random field and two line processes. In green for the horizontal direction $\theta = (1, 0)$. In red for the vertical one $\theta = (0, 1)$

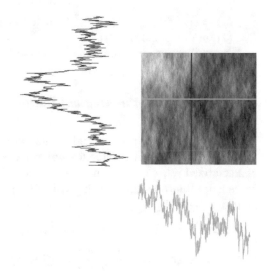

Asymptotic normality still hold assuming that $\varepsilon > 1/2$ but now, under our assumption, we get an estimator of Hölder regularity of Y, given by H in view of Proposition 4.9. We refer to [12] for more details and to [6] for complements.

4.3.2.2 Application to Random Fields by Line Processes

This framework may be used when considering random fields $(X(x))_{x\in\mathbb{R}^d}$ instead of one-dimensional processes. Actually, for $x_0 \in \mathbb{R}^d$ and $\theta \in S^{d-1}$ we can consider the line process $L_{x_0,\theta}(X) = \{X(x_0 + t\theta); t \in \mathbb{R}\}$ already defined in Sect. 4.2.1.3 (Fig. 4.15).

Recall that when X is a centered stationary Gaussian random field with stationary increments and variogram v_X, the process $L_{x_0,\theta}(X)$ is centered Gaussian with stationary increments and variogram

$$v_\theta(t) = \mathbb{E}\left((X(x_0 + t\theta) - X(x_0))^2\right) = v_X(t\theta), \ t \in \mathbb{R}.$$

When moreover X is self-similar of order $H \in (0, 1)$, we clearly get that $v_\theta(t) = v_X(\theta)|t|^{2H}$, ensuring that $L_{x_0,\theta}(X) - L_{x_0,\theta}(X)(0)$ is also self-similar of order $H \in (0, 1)$ and therefore it is a (non-standard) fractional Brownian motion with Hurst parameter H. Hence estimators set up in previous section may be used to estimate H. Note also that when X is isotropic we must have $\theta \in S^{d-1} \mapsto v_X(\theta)$ a constant function. Finally, let us remark that considering (E, H)-operator scaling fields with $\alpha_1, \ldots, \alpha_d \in (0, 1]$, $H \in (0, 1)$ and E the diagonal matrix $\operatorname{diag}(\alpha_1^{-1}, \ldots, \alpha_d^{-1})$, for any $1 \le i \le d$, the line process $L_{x_0,e_i}(X) - L_{x_0,e_i}(X)(0)$ is a fractional Brownian motion with Hurst parameter $H\alpha_i$, where $(e_i)_{1\le i\le d}$ is the canonical basis of $\mathbb{R}^d$. This follows from the fact that $v_{e_i}(t) = v_{H,E}(te_i) = |t|^{\alpha_i H}$ since

$v_{H,E}(x) = (|x_1|^{\alpha_1} + \ldots + |x_d|^{\alpha_d})^{2H}$. Hence we are able to estimate $\alpha_1 H, \ldots, \alpha_d H$ (see [7] for numerical results).

We can illustrate this with some applications in medical imaging.

4.3.3 Application in Medical Imaging Analysis

There are numerous methods and studies around what is called fractal analysis in medical imaging. We refer to [46] for a good review. The main goal is to characterized self-similarity of images with a fractal index $H \in (0, 1)$ to extract some helpful information for diagnosis. Our point of view consists in considering an image $(I(k_1, k_2))_{1 \leq k_1, k_2 \leq n}$ as a realization of a random field. Then,

- **Extract** a line from the image $(L_\theta(k))_{1 \leq k \leq n_\theta}$ for θ a direction.
- **Compute** $v_\theta(u) = \dfrac{1}{n_\theta - u} \displaystyle\sum_{k=1}^{n_\theta - u} (L_\theta(k + u) - L_\theta(k))^2$.
- **Average** along several lines of the same direction $\overline{v_\theta}(u)$ and compute $\widehat{H_\theta}(u, v) = \frac{1}{2} \log \left(\frac{\overline{v_\theta}(u)}{\overline{v_\theta}(v)} \right) / \log \left(\frac{u}{v} \right)$.

Of course there are several implementation issues according to the chosen direction. It follows that considering estimation on oriented lines without interpolation does not allow to reach any direction. Moreover precision is not the same in all directions (Fig.4.16). However the procedure is successful when considering horizontal and vertical direction and may also be compared with diagonal directions. We present in the following some results obtained with two kind of medical images: bone radiographs and mammograms.

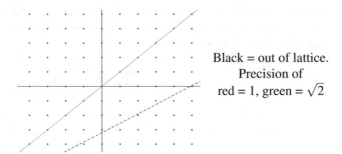

Black = out of lattice.
Precision of
red = 1, green = $\sqrt{2}$

Fig. 4.16 Available directions on a square lattice

4.3.3.1 Osteoporosis and Bone Radiographs

According to experts from the World Health Organization, osteoporosis is a disease affecting many millions of people around the world. It is characterized by low bone mass and micro-architectural deterioration of bone tissue, leading to bone fragility and a consequent increase in risk of fracture. Bone mineral density allows to measure low bone mass and is used for diagnosis while micro-architectural deterioration is not quantify. Several medical research teams have been working on micro-architectural deterioration assessment from bone radiographs, an easiest and cheapest clinical exam. Several authors have proposed to use fractal analysis with different methods of analysis for instance for calcaneous bone in [31], and cancellous bone in [20]. In [5], it allows to discriminate between osteoporotic cases $H_{mean} = 0.679 \pm 0.053$ and control cases $H_{mean} = 0.696 \pm 0.030$, by coupling with bone mineral density.

In [10], we have considered a data set composed of 211 numeric radiographs high-resolution of calcaneum (bone heel) with standardized acquisition of region of interest (ROI) 400×400 of Inserm unit U658 [43] (see Fig. 4.17). Figure 4.18 gives the results we obtained for horizontal and vertical directions. Log-log plots are linear for small scales in adequation with a self-similarity property valid for small scales. Estimated values are greater in the vertical direction than in the horizontal

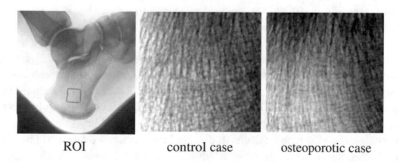

ROI   control case   osteoporotic case

Fig. 4.17 Standardized acquisition of region of interest (ROI) of Inserm unit U658 [43]

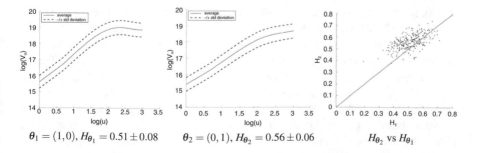

$\theta_1 = (1,0), H_{\theta_1} = 0.51 \pm 0.08$ $\theta_2 = (0,1), H_{\theta_2} = 0.56 \pm 0.06$ H_{θ_2} vs H_{θ_1}

Fig. 4.18 Mean of log-log plot of mean quadratic variations in horizontal and vertical direction. The last plot indicates estimated values of couple $(H_{\theta_1}, H_{\theta_2})$ for each 211 images (see [10])

Fig. 4.19 Region of interest
extracted from real
mammograms

 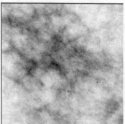

dense breast tissue   fatty breast tissue

direction contradicting an isotropic or self-similar model. Similar results were
recently obtained using different kinds of estimators in [32]. Moreover, comparisons
with diagonal direction lead us to think that operator scaling random fields could be
used for modeling.

4.3.3.2 Mammograms and Density Analysis

Fractal analysis has also been used in mammograms analysis. In particular, it
was used for the characterization and classification of mammogram density [19].
Actually, breast tissues are mainly composed of two kinds of tissues called dense
and fatty (see Fig. 4.19) and the amount of dense tissues is believed to be a risk
factor for developing breast cancer [19, 33].

In [33], the hypothesis of a self-similar behavior is validated using a power
spectrum method with an estimated fractal index range $H \in [0.33, 0.42]$. Based on
the variogram method presented above we also found a local self-similar behavior
with similar values $H = 0.31 \pm 0.05$ on a data set of 58 cases with 2 mammograms
(left and right) ROI of size 512×512 in [8]. Note that, contrarily to bones data,
we do not have a standardized procedure to extract ROI. Very interesting results
were obtained in [40] who manages to discriminate between dense and fatty breast
tissues using the Wavelet Transform Modulus Maxima method with respective
fractal indices given by $H \in [0.55, 0.75]$ and $H \in [0.2, 0.35]$. Fractal analysis is
also linked with lesion detectability in mammogram textures. Actually, as observed
in [18], it may be more difficult to detect lesion in dense tissues than in fatty tissues.
This was mathematically proven in [30], using a-contrario model, for textures like
isotropic fractional Brownian fields, showing that size and contrast of lesions are
linearly linked in log-log plot with a slope depending on the H index. This is
illustrated in Fig. 4.20, where we have added a spot with an increasing radius on
two simulations of fractional Brownian fields of size 512×512 for $H = 0.3$,
corresponding to values of fatty tissues, and for $H = 0.7$, corresponding to values
of dense tissues. Contrarily to white noise images, obtained with independent
identically distributed Gaussian variables on each pixel, in fractal images the more
the radius increases, the less the spot is observable.

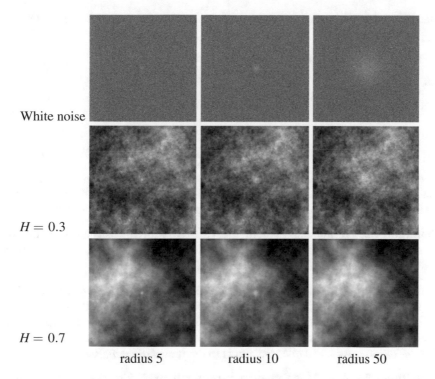

Fig. 4.20 Examples of simulated spots with various sizes (radius 5, 10, and 50) but similar contrast in a white noise texture (top row) and in fractional Brownian fields $H = 0.3$ and $H = 0.7$ (bottom row)

4.4 Geometric Construction

We present here geometric construction for some fractional Brownian fields based on Chentsov's representation of random fields using a random measure and a particular class of Borel sets indexed by points of $\mathbb{R}^d$. This is particularly interesting as it yields possible extensions, in particular beyond the Gaussian framework.

4.4.1 Random Measures

A random measure will be defined as a stochastic process indexed by some Borel set. We still consider $(\Omega, \mathscr{A}, \mathbb{P})$ a probability space. Let $k \geq 1$ and μ be a σ-finite non-negative measure on $(\mathbb{R}^k, \mathscr{B}(\mathbb{R}^k))$. Let set

$$\mathscr{E}_\mu = \{A \in \mathscr{B}(\mathbb{R}^k) \text{ such that } \mu(A) < +\infty\}.$$

Definition 4.22 A random measure M is a stochastic process $M = \{M(A); A \in \mathscr{E}_\mu\}$ satisfying

- For all $A \in \mathscr{E}_\mu$, $M(A)$ is a real random variable on $(\Omega, \mathscr{A})$;
- For $A_1, \ldots, A_n \in \mathscr{E}_\mu$ disjoint sets the random variables $M(A_1), \ldots, M(A_n)$ are independent;
- For $(A_n)_{n\in\mathbb{N}}$ disjoint sets such that $\underset{n\in\mathbb{N}}{\cup} A_n \in \mathscr{E}_\mu$,

$$M(\underset{n\in\mathbb{N}}{\cup} A_n) = \sum_{n\in\mathbb{N}} M(A_n) \text{ a.s.}$$

Let us emphasize that this definition does not ensure that almost surely M is a measure. However some random measures may be almost surely measures.

Definition 4.23 A Poisson random measure N with intensity μ is a random measure such that for any $A \in \mathscr{E}_\mu$, the random variable $N(A)$ follows a Poisson distribution of intensity $\mu(A)$ ie $N(A) \sim \mathscr{P}(\mu(A))$.

In this case N is a random discrete measure given by $N = \sum_{i\in I} \delta_{T_i}$, where $\Phi = (T_i)_{i\in I}$ is a countable family of random variables with values in $\mathbb{R}^k$ called Poisson point process on $\mathbb{R}^k$ with intensity μ (see [37] for instance). For example, when $k = 1$ and $\mu = \lambda \times$Lebesgue for some $\lambda > 0$, $(N([0, t]))_{t\geq 0}$ is the classical Poisson process of intensity λ and Φ corresponds to the jumps of the Poisson process.

Definition 4.24 A Gaussian random measure W with intensity μ is a random measure such that for any $A \in \mathscr{E}_\mu$, the random variable $W(A)$ follows a normal distribution with mean 0 and variance $\mu(A)$, ie $W(A) \sim \mathscr{N}(0, \mu(A))$.

In this case, W is not an a.s. measure. It is a centered Gaussian stochastic process (sometimes called set-indexed process, see [34] for instance) with covariance given by

$$\text{Cov}(W(A), W(B)) = \mu(A \cap B) = \frac{1}{2}(\mu(A) + \mu(B) - \mu(A \Delta B)),$$

for all $A, B \in \mathscr{E}_\mu$, with $A \Delta B = (A \cap B^c) \bigcup (B \cap A^c)$. Let us note that this is also the covariance function of any second order random measure M satisfying $\text{Var}(M(A)) = \mu(A)$ and so for N a Poisson random measure of intensity μ. For example, when $k = 1$ and $\mu = \lambda \times$ Lebesgue for some $\lambda > 0$, $(W([0, t]))_{t\geq 0}$ is the classical (non-standard) Brownian motion with diffusion λ, up to continuity of sample paths. Conversely, considering a Brownian motion $(B_t)_{t\in\mathbb{R}}$ one can define a Gaussian random measure on $\mathbb{R}$ given by $W(A) = \int_{-\infty}^{+\infty} \mathbf{1}_A(t) dB_t$.

The link between Poisson and Gaussian measures is given by the central limit theorem. Actually, if $N^{(1)}, \ldots, N^{(n)}$ are independent Poisson random measures with the same intensity μ, by superposition principle $\sum_{i=1}^{n} N^{(i)}$ is a Poisson random measure with intensity $n \times \mu$. By the central limit theorem we immediately deduce that for $A \in \mathscr{E}_\mu$

$$\frac{1}{\sqrt{n}} \left(\sum_{i=1}^{n} N^{(i)}(A) - n\mu(A) \right) \xrightarrow[n \to +\infty]{d} W(A).$$

More generally we have the following normal approximation for Poisson measures in high intensity.

Proposition 4.14 *If N_λ is a Poisson random measure with intensity $\lambda \times \mu$ and W is a Gaussian random measure with the same intensity μ, then*

$$\left(\lambda^{-1/2} \left(N_\lambda(A) - \lambda\mu(A) \right) \right)_{A \in \mathscr{E}_\mu} \xrightarrow[\lambda \to +\infty]{fdd} \left(W(A) \right)_{A \in \mathscr{E}_\mu}.$$

4.4.2 Chentsov's Representation: Lévy and Takenaka Constructions

Chentsov's type representation (see [56]) consists in constructing a random field X with M a random measure with intensity μ on $\mathbb{R}^k$ and $\mathscr{V} = \{V_x; x \in \mathbb{R}^d\}$ a class of sets of $\mathscr{E}_\mu$ indexed by $\mathbb{R}^d$ for $d \geq 1$, by setting

$$X_x = M(V_x), \quad x \in \mathbb{R}^d.$$

Then X is called Chentsov random field associated with M and $\mathscr{V}$. If M is a second order random measure satisfying $\text{Var}(M(A)) = \mu(A)$ then X is a second order random field with

$$\forall x, y \in \mathbb{R}^d, \quad \text{Var}(X_x - X_y) = \mu(V_x \Delta V_y).$$

Then invariance properties of X imply several relations on μ and $\mathscr{V}$. If X has stationary increments then we must have $\mu(V_x \Delta V_y) = \mu(V_{x-y} \Delta V_0)$ for all $x, y \in \mathbb{R}^d$; If X is isotropic and $X_0 = 0$ a.s. then $\mu(V_{Rx}) = \mu(V_x)$, for all vectorial rotations R; Finally if X is H-self-similar and $X_0 = 0$ a.s. then we obtain $\mu(V_{cx}) = c^{2H}\mu(V_x)$, for all $c > 0$. It follows that for X to be isotropic, H self-similar with stationary increments, we necessarily have $\mu(V_x \Delta V_y) = \mu(V_{x-y}) = c\|x - y\|^{2H}$, $t, s \in \mathbb{R}^d$, for some constant $c > 0$. This is only possible when $H \in (0, 1/2]$. This comes from the fact that $V_{2x} \subset (V_{2x} \Delta V_x) \cup V_x$. Hence, by increments stationarity, $\mu(V_{2x}) \leq \mu(V_x \Delta V_0) + \mu(V_x) \leq 2\mu(V_x)$ since $\mu(V_0) = \text{Var}(X_0) = 0$. By self-similarity we obtain that $2^{2H}\mu(V_x) \leq 2\mu(V_x)$ for all $x \in \mathbb{R}^d$, implying $H \leq 1/2$. We describe in the following the different constructions given by Lévy and Chentsov (1948 & 1957) for $H = 1/2$ and Takenaka (1987) for $H \in (0, 1/2)$.

Proposition 4.15 *Let μ and $\mathcal{V}$ be defined on $(\mathbb{R}^d, \mathcal{B}(\mathbb{R}^d))$ by*

- $\mu(dz) = \|z\|^{-d+1} dz$
- $\mathcal{V} = \{V_x, x \in \mathbb{R}^d\}$ *with* $V_x = B\left(\frac{x}{2}, \frac{\|x\|}{2}\right) = \left\{z \in \mathbb{R}^d : \left\|z - \frac{x}{2}\right\| < \frac{\|x\|}{2}\right\}$, *the ball of diameter $[0, x]$, for all $x \in \mathbb{R}^d$.*

Then, $\mu(V_x \Delta V_y) = \mu(V_{x-y}) = c_d \|x - y\|$, for all $x, y \in \mathbb{R}^d$.

Proof For $x \in \mathbb{R}^d$, we may used polar coordinates to identify V_x with $\left\{(r, \theta) \in \mathbb{R}_+ \times S^{d-1} : 0 < r < \theta \cdot x\right\}$ Then,

$$\mu(V_x) = \int_{S^{d-1}} \int_{\mathbb{R}_+} 1_{\{r < \theta \cdot x\}} dr d\theta = \frac{1}{2} \int_{S^{d-1}} |\theta \cdot x| d\theta = \frac{c_d}{2} \|x\|,$$

with $c_d = \int_{S^{d-1}} |e_1 \cdot x| d\theta$. Moreover, for $y \neq x$,

$$\mu(V_x \cap V_y^c) = \int_{S^{d-1}} \int_{\mathbb{R}_+} 1_{\{\theta \cdot y \leq r < \theta \cdot x\}} dr d\theta$$

$$= \int_{0 < \theta \cdot y < \theta \cdot x} \theta \cdot (x - y) d\theta + \int_{\theta \cdot y < 0 < \theta \cdot x} \theta \cdot x d\theta.$$

Similarly, by a change of variables,

$$\mu(V_y \cap V_x^c) = \int_{\theta \cdot y < \theta \cdot x < 0} |\theta \cdot (x - y)| d\theta + \int_{\theta \cdot y < 0 < \theta \cdot x} (-\theta \cdot y) \, d\theta,$$

so that

$$\mu(V_x \Delta V_y) = \frac{1}{2} \int_{S^{d-1}} |\theta \cdot (x - y)| d\theta = \frac{c_d}{2} \|x - y\|.$$

$\square$

One can therefore check that the Chentsov random field associated with a Gaussian measure W of intensity μ and $\mathcal{V}$, given in Proposition 4.15 is a (non-standard) Levy Chentsov field, or equivalently, a fractional Brownian field of index $H = 1/2$. The construction for $H \in (0, 1/2)$ has been given by Takenaka and relies on the following proposition.

Proposition 4.16 *Let $H \in (0, 1/2)$, μ and $\mathcal{V}$ be defined on $(\mathbb{R}^{d+1}, \mathcal{B}(\mathbb{R}^{d+1}))$ by*

- $\mu_H(dz, dr) = r^{2H-d-1} 1_{r>0} dz dr$ *for* $(z, r) \in \mathbb{R}^d \times \mathbb{R}$;
- $\mathcal{V} = \{V_x, x \in \mathbb{R}^d\}$ *with* $V_x = \mathcal{C}_x \Delta \mathcal{C}_0$ *where* $\mathcal{C}_x = \left\{(z, r) \in \mathbb{R}^d \times \mathbb{R} : \|z - x\| \leq r\right\}$, *for all $x \in \mathbb{R}^d$.*

Then, $\mu_H(V_x \Delta V_y) = \mu_H(V_{x-y}) = c_{H,d} \|x - y\|^{2H}$, for all $x, y \in \mathbb{R}^d$.

Proof Let $x \in \mathbb{R}^d$ with $x \neq 0$. Let us note that $\mu_H(\mathscr{C}_x) = +\infty$ but, integrating first with respect to r,

$$\mu_H(\mathscr{C}_x \cap \mathscr{C}_0^c) = \frac{1}{d - 2H} \int_{\|z - x\| < \|z\|} \left(\|z - x\|^{2H-d} - \|z\|^{2H-d} \right) dz$$

$$= c_{H,d} \|x\|^{2H} = \mu_H(\mathscr{C}_0 \cap \mathscr{C}_x^c),$$

using translation invariance of Lebesgue's measure, where

$$c_{H,d} = \frac{1}{d - 2H} \int_{\|z - e_1\| < \|z\|} \left(\|z - e_1\|^{2H-d} - \|z\|^{2H-d} \right) dz \in (0, +\infty).$$

Again by translation invariance of Lebesgue's measure, for $y \neq x$, we get $\mu_H(\mathscr{C}_x \Delta \mathscr{C}_y) = \mu_H(\mathscr{C}_{x-y} \Delta \mathscr{C}_0) = c_{H,d} \|x - y\|^{2H}$. The result follows once remarked that $V_x \Delta V_y = \mathscr{C}_x \Delta \mathscr{C}_y$. $\qquad\square$

Of course, considering an associated Gaussian random measure we obtain the Chentsov's representation of fractional Brownian fields for $H \in (0, 1/2)$. Let us remark that it also allows to define self-similar symmetric α-stable fields considering an $S\alpha S$ random measure (see [56]) but we leave our second order framework in this way! However, considering instead a Poisson random measure, we can define a Poisson analogous of fractional Brownian fields when $H \in (0, 1/2)$.

4.4.3 Fractional Poisson Fields

When $N_{\lambda,H}$ is a Poisson random measure on $\mathbb{R}^d \times \mathbb{R}$ with intensity $\lambda \times \mu_H$ for $\lambda > 0$, and μ_H given by Proposition 4.16, we can define a Chentsov's field by

$$N_{\lambda,H}(\mathscr{C}_x \Delta \mathscr{C}_0) = N_{\lambda,H}(\mathscr{C}_x \cap \mathscr{C}_0^c) + N_{\lambda,H}(\mathscr{C}_x^c \cap \mathscr{C}_0), \ \forall x \in \mathbb{R}^d.$$

However, since $N_{\lambda,H}(\mathscr{C}_x \Delta \mathscr{C}_0)$ is a Poisson random variable of parameter $\lambda \mu_H(\mathscr{C}_x \Delta \mathscr{C}_0)$, this field is not a centered. But remarking that $\mu_H\left(\mathscr{C}_x \cap \mathscr{C}_0^c\right) = \mu_H\left(\mathscr{C}_x^c \cap \mathscr{C}_0\right)$, we may define the centered fractional Poisson field on $\mathbb{R}^d$ by

$$F_{\lambda,H}(x) = N_{\lambda,H}(\mathscr{C}_x \cap \mathscr{C}_0^c) - N_{\lambda,H}(\mathscr{C}_x^c \cap \mathscr{C}_0), \ \forall x \in \mathbb{R}^d.$$

Actually, $F_{\lambda,H}$ may also be defined as the stochastic integral with respect to the Poisson random measure $N_{\lambda,H}$ as

$$F_{\lambda,H}(x) = \int_{\mathbb{R}^d \times \mathbb{R}} \left(\mathbf{1}_{\mathscr{C}_x \cap \mathscr{C}_0^c}(z, r) - \mathbf{1}_{\mathscr{C}_x^c \cap \mathscr{C}_0}(z, r) \right) N_{\lambda,H}(dz, dr)$$

$$= \int_{\mathbb{R}^d \times \mathbb{R}} \left(\mathbf{1}_{B(z,r)}(x) - \mathbf{1}_{B(z,r)}(0) \right) N_{\lambda,H}(dz, dr).$$

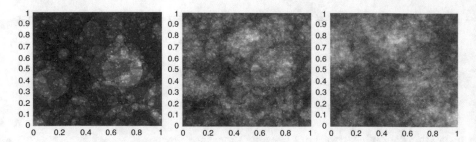

Fig. 4.21 Convergence of a fractional Poisson field to a fractional Brownian field as the intensity increases

Heuristically, we can throw centers and radius with respect to a Poisson point process on $\mathbb{R}^d \times \mathbb{R}$ of intensity $\lambda \times \mu_H$, meaning that centers are thrown uniformly in $\mathbb{R}^d$ with intensity λ and independently marked with a radius. Then $F_{\lambda,H}(x)$ will count the number of balls falling on x minus the number of balls falling on 0. It is close to a shot noise random field obtained as the sum of randomly dilated and translated contributions. We refer to [11] and [13] for more details. Then $(F_{\lambda,H}(x))_{x \in \mathbb{R}^d}$ is centered, with stationary increments, isotropic with covariance

$$\mathrm{Cov}(F_{\lambda,H}(x), F_{\lambda,H}(y)) = \frac{\lambda c_{H,d}}{2}\left(\|x\|^{2H} + \|y\|^{2H} - \|x - y\|^{2H}\right).$$

This field is not self-similar but

$$(F_{\lambda,H}(cx))_{x \in \mathbb{R}^d} \stackrel{fdd}{=} (F_{\lambda c^{2H},H}(x))_{x \in \mathbb{R}^d}, \ \forall c > 0.$$

Moreover, according to normal approximation of Poisson measures for high intensity we can prove that $(\lambda^{-1/2} F_{\lambda,H}(x))_{x \in \mathbb{R}^d} \xrightarrow[\lambda \to +\infty]{fdd} (\sqrt{C_{H,d}} B_H(x))_{x \in \mathbb{R}^d}$. This is illustrated in Fig. 4.21. Note that sharp contours of fractional Poisson fields disappear in the asymptotic Gaussian limits, that are Hölder continuous. Another interesting property of this field is that its distribution is also preserved along lines (and more generally along affine subspaces). More precisely, for $x_0 \in \mathbb{R}^d$ and $\theta \in S^{d-1}$, defining the line process $L_{x_0,\theta}(F_{\lambda,H}) = \left(F_{\lambda,H}(x_0 + t\theta)\right)_{t \in \mathbb{R}}$, computing characteristic functions we can prove that $\left(L_{x_0,\theta}(t) - L_{x_0,\theta}(0)\right)_{t \in \mathbb{R}}$ has the same distribution than a fractional Poisson process (defined for $d = 1$), with the same H index and intensity given by $c_{H,d}\lambda$ where $c_{H,d} = \int_{\mathbb{R}^{d-1}}(1 - \|y\|^2)^{1/2-H}\mathbf{1}_{\|y\| \le 1}dy$ (see [11]). Hence, we can also use estimation based on variograms to build estimators of H. Consistency has been proven in [13]. Figure 4.22 presents sample paths comparison between Poisson and Gaussian cases. To conclude, let us note that, contrarily to the Gaussian case, one can prove that the increments $\mathbb{E}(|F_{\lambda,H}(x) - F_{\lambda,H}(x)|^q)$ behave like $\|x - y\|^{2H}$ for any $q \ge 2$ as $\|x - y\| \to 0$. Such a feature still holds allowing some interactions for the radii as done in [54].

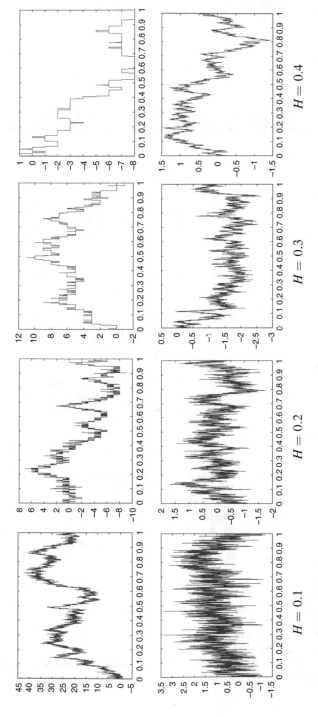

Fig. 4.22 Sample paths of fractional Poisson process (top) vs fractional Brownian motion (bottom)

Acknowledgements I would like to warmly thanks all my co-authors for the different works partially presented here, especially Clément Chesseboeuf and Olivier Durieu for their careful reading.

References

1. R.J. Adler, *The Geometry of Random Field* (Wiley, Hoboken, 1981)
2. D. Allard, R. Senoussi, E. Porcu, Anisotropy models for spatial data. Math. Geosci. **48**(3), 305–328 (2016)
3. A. Ayache, F. Roueff, A Fourier formulation of the Frostman criterion for random graphs and its applications to wavelet series. Appl. Comput. Harmon. Anal. **14**, 75–82 (2003)
4. A. Benassi, S. Cohen, J. Istas, Local self-similarity and the Hausdorff dimension. C. R. Acad. Sci. **336**(3), 267–272 (2003)
5. C.L. Benhamou, S. Poupon, E. Lespessailles, S. Loiseau, R. Jennane, V. Siroux, W. Ohley, L. Pothuaud, Fractal analysis of radiographic trabecular bone texture and bone mineral density: two complementary parameters related to osteoporotic fractures. J. Bone Miner. Res. **16**(4), 697–704 (2001)
6. C. Berzin, A. Latour, J.R. León, *Inference on the Hurst Parameter and the Variance of Diffusions Driven by Fractional Brownian Motion. Lecture Notes in Statistics*, vol. 216 (Springer, Cham, 2014). With a foreword by Aline Bonami
7. H. Biermé, C. Lacaux, Fast and exact synthesis of some operator scaling Gaussian random fields. Appl. Comput. Harmon. Anal. (2018). https://doi.org/10.1016/j.acha.2018.05.004
8. H. Biermé, F. Richard, Statistical tests of anisotropy for fractional brownian textures: application to full-field digital mammography. J. Math. Imaging Vision **36**(3), 227–240 (2010)
9. H. Biermé, M.M. Meerschaert, H.P. Scheffler, Operator scaling stable random fields. Stoch. Process. Appl. **117**(3), 312–332 (2007)
10. H. Biermé, C.L. Benhamou, F. Richard, Parametric estimation for gaussian operator scaling random fields and anisotropy analysis of bone radiograph textures, in *Proceedings of the International Conference on Medical Image Computing and Computer Assisted Intervention (MICCAI'09), Workshop on Probabilistic Models for Medical Imaging*, ed. by K. Pohl, London, UK, September 2009, pp. 13–24
11. H. Biermé, A. Estrade, I. Kaj, Self-similar random fields and rescaled random balls models. J. Theor. Probab. **23**(4), 1110–1141 (2010)
12. H. Biermé, A. Bonami, J.R. León, Central limit theorems and quadratic variations in terms of spectral density. Electron. J. Probab. **16**(3), 362–395 (2011)
13. H. Biermé, Y. Demichel, A. Estrade, Fractional Poisson field and fractional Brownian field: why are they resembling but different? Electron. Commun. Probab. **18**, 11–13 (2013)
14. H. Biermé, L. Moisan, F. Richard, A turning-band method for the simulation of anisotropic fractional Brownian fields. J. Comput. Graph. Stat. **24**(3), 885–904 (2015)
15. M. Bilodeau, D. Brenner, *Theory of Multivariate Statistics*. Springer Texts in Statistics (Springer, New York, 1999)
16. A. Bonami, A. Estrade, Anisotropic analysis of some Gaussian models. J. Fourier Anal. Appl. **9**(3), 215–236 (2003)
17. P. Breuer, P. Major, Central limit theorems for nonlinear functionals of Gaussian fields. J. Multivar. Anal. **13**(3), 425–441 (1983)
18. A. Burgess, F. Jacobson, P. Judy, Human observer detection experiments with mammograms and power-law noise. Med. Phys. **28**(4), 419–437 (2001)
19. C. Caldwell, S. Stapleton, D. Holdsworth, et al., On the statistical nature of characterisation of mammographic parenchymal patterns by fractal dimension. Phys. Med. Biol. **35**(2), 235–247 (1990)

20. C.B. Caldwell, J. Rosson, J. Surowiak, T. Hearn, Use of fractal dimension to characterize the structure of cancellous bone in radiographs of the proximal femur, in *Fractals in Biology and Medicine* (Birkhäuser, Basel, 1994), pp. 300–306

21. G. Chan, An effective method for simulating Gaussian random fields, in *Proceedings of the Statistical Computing Section* (American Statistical Association, Boston, 1999), pp. 133–138. www.stat.uiowa.edu/~grchan/

22. S. Cohen, J. Istas, *Fractional Fields and Applications. Mathématiques & Applications (Berlin) [Mathematics & Applications]*, vol. 73 (Springer, Heidelberg, 2013). With a foreword by Stéphane Jaffard

23. P.F. Craigmile, Simulating a class of stationary Gaussian processes using the Davies-Harte algorithm, with application to long memory processes. J. Time Ser. Anal. **24**(5), 505–511 (2003)

24. R. Dalang, D. Khoshnevisan, C. Mueller, D. Nualart, Y. Xiao, *A Minicourse on Stochastic Partial Differential Equations. Lecture Notes in Mathematics*, ed. by D. Khoshnevisan, F. Rassoul-Agha, vol. 1962 (Springer, Berlin, 2009). Held at the University of Utah, Salt Lake City, UT, May 8–19, 2006

25. S. Davies, P. Hall, Fractal analysis of surface roughness by using spatial data. J. R. Stat. Soc. Ser. B **61**, 3–37 (1999)

26. C.R. Dietrich, G.N. Newsam, Fast and exact simulation of stationary gaussian processes through circulant embedding of the covariance matrix. SIAM J. Sci. Comput. **18**(4), 1088–1107 (1997)

27. K.J. Falconer, *Fractal Geometry* (Wiley, Hoboken, 1990)

28. W. Feller, *An Introduction to Probability Theory and Its Applications. Vol. II.* 2nd edn. (Wiley, New York, 1971)

29. T. Gneiting, H. Sevciková, D.B. Percivala, M. Schlather, Y. Jianga, Fast and exact simulation of large gaussian lattice systems in $\mathbb{R}^2$: exploring the limits. J. Comput. Graph. Stat. **15**, 483–501 (1996)

30. B. Grosjean, L. Moisan, A-contrario detectability of spots in textured backgrounds. J. Math. Imaging Vision **33**(3), 313–337 (2009)

31. R. Harba, G. Jacquet, R. Jennane, T. Loussot, C.L. Benhamou, E. Lespessailles, D. Tourlière, Determination of fractal scales on trabecular bone X-ray images. Fractals **2**(3), 451–456 (1994)

32. K. Harrar, R. Jennane, K. Zaouchi, T. Janvier, H. Toumi, E. Lespessailles, Oriented fractal analysis for improved bone microarchitecture characterization. Biomed. Signal Process. Control **39**, 474–485 (2018)

33. J. Heine, R. Velthuizen, Spectral analysis of full field digital mammography data. Med. Phys. **29**(5), 647–661 (2002)

34. E. Herbin, E. Merzbach, The set-indexed Lévy process: stationarity, Markov and sample paths properties. Stoch. Process. Appl. **123**(5), 1638–1670 (2013)

35. J. Istas, On fractional stable fields indexed by metric spaces. Electron. Commun. Probab. **11**, 242–251 (2006)

36. J. Istas, G. Lang, Quadratic variations and estimation of the local Hölder index of a Gaussian process. Ann. Inst. Henri Poincaré Probab. Stat. **33**(4), 407–436 (1997)

37. O. Kallenberg, *Foundations of Modern Probability*. Probability and Its Applications (New York) (Springer, New York, 1997)

38. L.M. Kaplan, C.C.J. Kuo, An improved method for 2-d self-similar image synthesis. IEEE Trans. Image Process. **5**(5), 754–761 (1996)

39. I. Karatzas, E. Shreve, *Brownian Motion and Stochastic Calculus* (Springer, New York, 1998)

40. P. Kesterner, J.M. Lina, P. Saint-Jean, A. Arneodo, Waveled-based multifractal formalism to assist in diagnosis in digitized mammograms. Image Anal. Stereol. **20**, 169–174 (2001)

41. A.N. Kolmogorov, The local structure of turbulence in an incompressible viscous fluid for very large reynolds number. Dokl. Akad. Nauk SSSR **30**, 301–305 (1941)

42. R. Leipus, A. Philippe, D. Puplinskaitė, D. Surgailis, Aggregation and long memory: recent developments. J. Indian Stat. Assoc. **52**(1), 81–111 (2014)

43. E. Lespessailles, C. Gadois, I. Kousignian, J.P. Neveu, P. Fardellone, S. Kolta, C. Roux, J.P. Do-Huu, C.L. Benhamou, Clinical interest of bone texture analysis in osteoporosis: a case control multicenter study. Osteoporos. Int. **19**, 1019–1028 (2008)

44. Y. Li, W. Wang, Y. Xiao, Exact moduli of continuity for operator-scaling Gaussian random fields. Bernoulli **21**(2), 930–956 (2015)

45. G. Lindgren, *Stationary Stochastic Processes: Theory and Applications.* Chapman & Hall/CRC Texts in Statistical Science Series (CRC Press, Boca Raton, 2013)

46. R. Lopes, N. Betrouni, Fractal and multifractal analysis: a review. Med. Image Anal. **13**, 634–649 (2009)

47. B.B. Mandelbrot, J. Van Ness, Fractional Brownian motion, fractionnal noises and applications. SIAM Rev. **10**, 422–437 (1968)

48. G. Matheron, The intrinsic random functions and their application. Adv. Appl. Probab. **5**, 439–468 (1973)

49. I. Molchanov, K. Ralchenko, A generalisation of the fractional Brownian field based on non-Euclidean norms. J. Math. Anal. Appl. **430**(1), 262–278 (2015)

50. G. Peccati, C. Tudor, Gaussian limits for vector-valued multiple stochastic integrals. Séminaire de Probabilités **XXXVIII**, 247–262 (2004)

51. E. Perrin, R. Harba, R. Jennane, I. Iribarren, Fast and exact synthesis for 1-D fractional Brownian motion a nd fractional gaussian noises. IEEE Signal Process. Lett. **9**(11), 382–384 (2002)

52. V. Pilipauskaitė, D. Surgailis, Scaling transition for nonlinear random fields with long-range dependence. Stochastic Process. Appl. **127**(8), 2751–2779 (2017)

53. C.E. Powell, Generating realisations of stationary gaussian random fields by circulant embedding (2014). Technical report

54. N. Privault, Poisson sphere counting processes with random radii. ESAIM Probab. Stat. **20**, 417–431 (2016)

55. W. Rudin, *Real and Complex Analysis* (McGraw-Hill, New York, 1986)

56. G. Samorodnitsky, M.S. Taqqu, *Stable Non-Gaussian Random Processes: Stochastic Models with Infinite Variance.* Stochastic Modeling (Chapman & Hall, New York, 1994)

57. M.L. Stein, Fast and exact simulation of fractional Brownian surfaces. J. Comput. Graph. Stat. **11**(3), 587–599 (2002)

58. A.W. van der Vaart, *Asymptotic Statistics. Cambridge Series in Statistical and Probabilistic Mathematics*, vol. 3 (Cambridge University Press, Cambridge, 1998)

Chapter 5
Introduction to the Theory of Gibbs Point Processes

David Dereudre

Abstract The Gibbs point processes (GPP) constitute a large class of point processes with interaction between the points. The interaction can be attractive, repulsive, depending on geometrical features whereas the null interaction is associated with the so-called Poisson point process. In a first part of this mini-course, we present several aspects of finite volume GPP defined on a bounded window in $\mathbb{R}^d$. In a second part, we introduce the more complicated formalism of infinite volume GPP defined on the full space $\mathbb{R}^d$. Existence, uniqueness and non-uniqueness of GPP are non-trivial questions which we treat here with completely self-contained proofs. The DLR equations, the GNZ equations and the variational principle are presented as well. Finally we investigate the estimation of parameters. The main standard estimators (MLE, MPLE, Takacs-Fiksel and variational estimators) are presented and we prove their consistency. For sake of simplicity, during all the mini-course, we consider only the case of finite range interaction and the setting of marked points is not presented.

5.1 Introduction

The spatial point processes are well studied objects in probability theory and statistics for modelling and analysing spatial data which appear in several disciplines as statistical mechanics, material science, astronomy, epidemiology, plant ecology, seismology, telecommunication, and others [4, 10]. There exist many models of such random points configurations in space and the most popular one is surely the Poisson point process. It corresponds to the natural way of producing independent locations of points in space without interaction. For dependent random structures, we can mention for instance the Cox processes, determinantal point processes, Gibbs point processes, etc. None of them is established as the most relevant model

D. Dereudre (✉)
University Lille, Villeneuve-d'Ascq, France
e-mail: david.dereudre@univ-lille.fr

© Springer Nature Switzerland AG 2019   181
D. Coupier (ed.), *Stochastic Geometry*, Lecture Notes in Mathematics 2237,
https://doi.org/10.1007/978-3-030-13547-8_5

for applications. In fact the choice of the model depends on the nature of the dataset, the knowledge of (physical or biological) mechanisms producing the pattern, the aim of the study (theoretical, applied or numerical).

In this mini-course, we focus on Gibbs point processes (GPP) which constitute a large class of points processes, able to fit several kinds of patterns and which provide a clear interpretation of the interaction between the points, such as attraction or repulsion depending on their relative position. Note that this class is particularly large since several point processes can be represented as GPP (see [24, 33] for instance). The main disadvantage of GPP is the complexity of the model due to an intractable normalizing constant which appears in the local conditional densities. Therefore their analytical studies are in general based on implicit equilibrium equations which lead to complicated and delicate analysis. Moreover, the theoretical results which are needed to investigate the Gibbs point process theory are scattered across several publications or books. The aim of this mini-course is to provide a solid and self-contained theoretical basis for understanding deeply the Gibbs point process theory. The results are in general not exhaustive but the main ideas and tools are presented in accordance with modern and recent developments. The main strong restriction here involves the range of the interaction, which is assumed to be finite. The infinite range interaction requires the introduction of tempered configuration spaces and for sake of simplicity we decided to avoid this level of complexity. The mini-course is addressed for Master and Phd students and also for researchers who want to discover or investigate the domain. The manuscript is based on a mini-course given during the conference of GDR 3477 géométrie stochastique, at university of Nantes in April 2016.

In a first section, we introduce the finite volume GPP on a bounded window $\Lambda \subset \mathbb{R}^d$. They are simply defined as point processes in Λ whose the distributions are absolutely continuous with respect to the Poisson point process distribution. The unnormalized densities are of form $z^N e^{-\beta H}$, where z and β are positive parameters (called respectively activity and inverse temperature), N is the number of points and H an energy function. Clearly, these distributions favour (or penalize) configurations with low (or high) energy E. This distortion strengthens as β is large. The parameter z allows to tune the mean number of points. This setting is relatively simple since all the objects are defined explicitly. However, the intractable normalization constant is ever a problem and most of quantities are not computable. Several standard notions (DLR and GNZ equations, Ruelle's estimates, etc.) are treated in this first section as a preparation for the more complicated setting of infinite volume GPP developed in the second section. Note that we do not present the setting of marked Gibbs point processes in order to keep the notations as simple as possible. However, all the results can be easily extended in this case.

In a second section, we present the theory of infinite volume GPP in $\mathbb{R}^d$. There are several motivations for studying such infinite volume regime. Firstly, the GPP are the standard models in statistical physics for modelling systems with a large number of interacting particles (around 10^{23} according to the Avogadro's number). Therefore, the case where the number of particles is infinite is an idealization of this setting and furnishes microscopic descriptions of gas, liquid or solid. Macroscopic quantities like the density of particles, the pressure and the mean energy are consequently easily defined by mean values or laws of large numbers. Secondly, in the spatial statistic context, the asymptotic properties of estimators or tests are obtained when the observation window tends to the full space $\mathbb{R}^d$. This strategy requires the existence of infinite volume models. Finally, since the infinite volume GPP are stationary (shift invariant) in $\mathbb{R}^d$, several powerful tools, as the ergodic theorem or the central limit Theorem for mixing field, are available in this infinite volume regime.

The infinite volume Gibbs measures are defined by a collection of implicit DLR equations (Dobrushin, Lanford and Ruelle). The existence, uniqueness and non-uniqueness are non trivial questions which we treat in depth with self-contained proofs in this second section. The phase transition between uniqueness and non uniqueness is one of the most difficult conjectures in statistical physics. This phenomenon is expected to occur for all standard interactions although it is proved rigorously only for few models. The area interaction is one of such models and the complete proof of its phase transition is given here. The GNZ equations, the variational principle are discussed as well.

In the last section, we investigate the estimation of parameters which appear in the distribution of GPP. For sake of simplicity we deal only with the activity parameter z and the inverse temperature β. We present several standard procedures (MLE, MPLE, Takacs-Fiksel procedure) and a new variational procedure. We show the consistency of estimators, which highlights that many theoretical results are possible in spite of lack of explicit computations. We will see that the GNZ equations play a crucial role in this task. For sake of simplicity the asymptotic normality is not presented but some references are given.

Let us finish this introduction by giving standard references. Historically, the GPP have been introduced for statistical mechanics considerations and an unavoidable reference is the book by Ruelle [47]. Important theoretical contributions are also developed in two Lecture Notes [20, 46] by Georgii and Preston. For the relations between GPP and stochastic geometry, we can mention the book [8] by Chiu et al. and for spatial statistic and numerical considerations, the book by Møller and Waagepetersen [42] is the standard reference. Let us mention also the book [51] by van Lieshout on the applications of GPP.

5.2 Finite Volume Gibbs Point Processes

In this first section we present the theory of Gibbs point process on a bounded set $\Lambda \subset \mathbb{R}^d$. A Gibbs point process (GPP) is a point process with interactions between the points defined via an energy functional on the space of configurations. Roughly speaking, the GPP produces random configurations for which the configurations with low energy have more chance to appear than the configurations with high energy (see Definition 5.2). In Sect. 5.2.1 we recall succinctly some definitions of point process theory and we introduce the reference Poisson point process. The energy functions are discussed in Sect. 5.2.2 and the definition of finite volume GPP is given in Sect. 5.2.3. Some first properties are presented as well. The central DLR equations and GNZ equations are treated in Sects. 5.2.4 and 5.2.5. Finally we finish the first section by giving Ruelle estimates in the setting of superstable and lower regular energy functions.

5.2.1 Poisson Point Process

In this first section, we describe briefly the setting of point process theory and we introduce the reference Poisson point process. We only give the main definitions and concepts and we suggest [10, 36] for a general presentation.

The space of configurations $\mathscr{C}$ is defined as the set of locally finite subsets in $\mathbb{R}^d$:

$$\mathscr{C} = \{\gamma \subset \mathbb{R}^d, \gamma_\Lambda := \gamma \cap \Lambda \text{ is finite for any bounded set } \Lambda \subset \mathbb{R}^d\}.$$

Note that we consider only the simple point configurations, which means that the points do not overlap. We denote by $\mathscr{C}_f$ the space of finite configurations in $\mathscr{C}$ and by $\mathscr{C}_\Lambda$ the space of finite configurations inside $\Lambda \subset \mathbb{R}^d$.

The space $\mathscr{C}$ is equipped with the sigma-field $\mathscr{F}_\mathscr{C}$ generated by the counting functions N_Λ for all bounded measurable $\Lambda \subset \mathbb{R}^d$, where $N_\Lambda : \gamma \mapsto \#\gamma_\Lambda$. A point process Γ is then simply a measurable function from any probability space $(\Omega, \mathscr{F}, P)$ to $(\mathscr{C}, \mathscr{F}_\mathscr{C})$. As usual, the distribution (or the law) of a point process Γ is defined by the image of P to $(\mathscr{C}, \mathscr{F}_\mathscr{C})$ by the application Γ. We say that Γ has finite intensity if, for any bounded set Λ, the expectation $\mu(\Lambda) := E(N_\Lambda(\Gamma))$ is finite. In this case, μ is a sigma-finite measure called intensity measure of Γ. When $\mu = \zeta\lambda^d$, where λ^d is the Lebesgue measure on $\mathbb{R}^d$ and $\zeta \geq 0$ a positive real, we simply say that Γ has finite intensity ζ.

The main class of point processes is the family of Poisson point processes, which furnish the natural way of producing independent points in space. Let μ be a sigma-finite measure in $\mathbb{R}^d$. A Poisson point process with intensity μ is a point process Γ such that, for any bounded Λ in $\mathbb{R}^d$, these properties both occur

- The random variable $N_\Lambda(\Gamma)$ is distributed following a Poisson distribution with parameter $\mu(\Lambda)$.
- Given the event $\{N_\Lambda(\Gamma) = n\}$, the n points in Γ_Λ are independent and distributed following the distribution $\mu_\Lambda/\mu(\Lambda)$.

The distribution of such a Poisson point process is denoted by π^μ. When the intensity is $\mu = \zeta\lambda^d$, we say that the Poisson point process is stationary (or homogeneous) with intensity $\zeta > 0$, and denote its distribution π^ζ. For any measurable set $\Lambda \subset \mathbb{R}^d$, we denote by π_Λ^ζ the distribution of a Poisson point process with intensity $\zeta\lambda_\Lambda^d$ which is also the distribution of a stationary Poisson point process with intensity ζ restricted to Λ. For sake of brevity, π and π_Λ denote the distribution of Poisson point processes with intensity $\zeta = 1$.

5.2.2   Energy Functions

In this section, we present the energy functions with the standard assumptions which we assume in this mini-course. The choices of energy functions come from two main motivations. First, the GPP are natural models in statistical physics for modelling continuum interacting particles systems. In general, in this setting the energy function is a sum of the energy contribution of all pairs of points (see expression (5.1)). The GPP are also used in spatial statistics to fit as best as possible the real datasets. So, in a first step, the energy function is chosen by the user with respect to the characteristics of the dataset. Then the parameters are estimated in a second step.

Definition 5.1 An energy function is a measurable function

$$H : \mathscr{C}_f \mapsto \mathbb{R} \cup \{+\infty\}$$

such that the following assumptions hold

- H is **non-degenerate**:

$$H(\emptyset) < +\infty.$$

- H is **hereditary**: for any $\gamma \in \mathscr{C}_f$ and $x \in \gamma$ then

$$H(\gamma) < +\infty \Rightarrow H(\gamma \setminus \{x\}) < +\infty.$$

- H is **stable**: there exists a constant A such that for any $\gamma \in \mathscr{C}_f$

$$H(\gamma) \geq AN_{\mathbb{R}^d}(\gamma).$$

The stability implies that the energy is superlinear. If the energy function H is positive then the choice $A = 0$ works but in the interesting cases, the constant A is negative. The hereditary means that the set of allowed configurations (configurations with finite energy) is stable when points are removed. The non-degeneracy is very natural. Without this assumption, the energy would be equal to infinity everywhere (by hereditary).

(1) Pairwise Interaction Let us start with the most popular energy function which is based on a function (called pair potential)

$$\varphi : \mathbb{R}^+ \to \mathbb{R} \cup \{+\infty\}.$$

The pairwise energy function is defined for any $\gamma \in \mathscr{C}_f$ by

$$H(\gamma) = \sum_{\{x,y\} \subset \gamma} \varphi(|x - y|). \tag{5.1}$$

Note that such an energy function is trivially hereditary and non-degenerate. The stability is more delicate and we refer to general results in [47]. However if φ is positive the result is obvious.

A standard example coming from statistical physics is the so-called Lennard-Jones pair potential where $\varphi(r) = ar^{-12} + br^{-6}$ with $a > 0$ and $b \in \mathbb{R}$. In the interesting case $b < 0$, the pair potential $\varphi(r)$ is positive (repulsive) for small r and negative (attractive) for large r. The stability is not obvious and is proved in Proposition 3.2.8 in [47].

The Strauss interaction corresponds to the pair potential $\varphi(r) = \mathbf{1}_{[0,R]}(r)$ where $R > 0$ is a support parameter. This interaction exhibits a constant repulsion between the particles at distance smaller than R. This simple model is very popular in spatial statistics.

The multi-Strauss interaction corresponds to the pair potential

$$\varphi(r) = \sum_{i=1}^{k} a_i \mathbf{1}_{]R_{i-1}, R_i]},$$

where $(a_i)_{1 \leq i \leq k}$ is a sequence of real numbers and $0 = R_0 < R_1 < \ldots < R_k$ a sequence of increasing real numbers. Clearly, the pair potential exhibits a constant attraction or repulsion at different scales. The stability occurs provided that the parameter a_1 is large enough (see Section 3.2 in [47]).

(2) Energy Functions Coming from Geometrical Objects Several energy functions are based on local geometrical characteristics. The main motivation is to provide random configurations such that special geometrical features appear with higher probability under the Gibbs processes than the original Poisson point process. In this paragraph we give examples related to the Delaunay-Voronoi diagram. Obviously other geometrical graph structures could be considered.

Let us recall that for any $x \in \gamma \in \mathscr{C}_f$ the Voronoi cell $C(x, \gamma)$ is defined by

$$C(x, \gamma) = \left\{ w \in \mathbb{R}^d, \text{ such that } \forall y \in \gamma \ |x - w| \leq |x - y| \right\}.$$

The Delaunay graph with vertices γ is defined by considering the edges

$$D(\gamma) = \left\{ \{x, y\} \subset \gamma \text{ such that } C(x, \gamma) \cap C(y, \gamma) \neq \emptyset \right\}.$$

See [40] for a general presentation on the Delauany-Voronoi tessellations.

A first geometric energy function can be defined by

$$H(\gamma) = \sum_{x \in \gamma} \mathbf{1}_{C(x,\gamma) \text{ is bounded}} \varphi(C(x, \gamma)), \tag{5.2}$$

where φ is any function from the space of polytopes in $\mathbb{R}^d$ to $\mathbb{R}$. Examples of such functions φ are the Area, the $(d-1)$-Hausdorff measure of the boundary, the number of faces, etc. Clearly these energy functions are non-degenerate and hereditary. The stability holds as soon as the function φ is bounded from below.

Another kind of geometric energy function can be constructed via a pairwise interaction along the edges of the Delaunay graph. Let us consider a finite pair potential $\varphi : \mathbb{R}^+ \mapsto \mathbb{R}$. Then the energy function is defined by

$$H(\gamma) = \sum_{\{x,y\} \subset D(\gamma)} \varphi(|x - y|) \tag{5.3}$$

which is again clearly non-degenerate and hereditary. The stability occurs in dimension $d = 2$ thanks to Euler's formula. Indeed the number of edges in the Delaunay graph is linear with respect to the number of vertices. Therefore the energy function is stable as soon as the pair potential φ is bounded from below. In higher dimension $d > 2$, the stability is more complicated and not really understood. Obviously, if φ is positive, the stability occurs.

Let us give a last example of geometric energy function which is not based on the Delaunay-Voronoi diagram but on a germ-grain structure. For any radius $R > 0$ we define the germ-grain structure of $\gamma \in \mathscr{C}$ by

$$L_R(\gamma) = \bigcup_{x \in \gamma} B(x, R),$$

where $B(x, R)$ is the closed ball centred at x with radius R. Several interesting energy functions are built from this germ-grain structure. First the Widom-Rowlinson interaction is simply defined by

$$H(\gamma) = \text{Area}(L_R(\gamma)), \tag{5.4}$$

where the "Area" is simply the Lebesgue measure λ^d. This model is very popular since it is one of a few models for which the phase transition result is proved (see Sect. 5.3.8). This energy function is sometimes called Area-interaction [3, 52]. If the Area functional is replaced by any linear combination of the Minkowski functionals we obtain the Quermass interaction [12].

Another example is the random cluster interaction defined by

$$H(\gamma) = \text{Ncc}(L_R(\gamma)), \tag{5.5}$$

where Ncc denotes the functional which counts the number of connected components. This energy function is introduced first in [7] for its relations with the Widom-Rowlinson model. See also [14] for a general study in the infinite volume regime.

5.2.3   Finite Volume GPP

Let $\Lambda \subset \mathbb{R}^d$ such that $0 < \lambda^d(\Lambda) < +\infty$. In this section we define the finite volume GPP on Λ and we give its first properties.

Definition 5.2 The finite volume Gibbs measure on Λ with activity $z > 0$, inverse temperature $\beta \geq 0$ and energy function H is the distribution

$$P_\Lambda^{z,\beta} = \frac{1}{Z_\Lambda^{z,\beta}} z^{N_\Lambda} e^{-\beta H} \pi_\Lambda, \tag{5.6}$$

where $Z_\Lambda^{z,\beta}$, called partition function, is the normalization constant $\int z^{N_\Lambda} e^{-\beta H} d\pi_\Lambda$. A finite volume Gibbs point process (GPP) on Λ with activity $z > 0$, inverse temperature $\beta \geq 0$ and energy function H is a point process on Λ with distribution $P_\Lambda^{z,\beta}$.

Note that $P_\Lambda^{z,\beta}$ is well-defined since the partition function $Z_\Lambda^{z,\beta}$ is positive and finite. Indeed, thanks to the non degeneracy of H

$$Z_\Lambda^{z,\beta} \geq \pi_\Lambda(\emptyset) e^{-\beta H(\{\emptyset\})} = e^{-\lambda^d(\Lambda)} e^{-\beta H(\{\emptyset\})} > 0$$

and thanks to the stability of H

$$Z_\Lambda^{z,\beta} \leq e^{-\lambda^d(\Lambda)} \sum_{n=0}^{+\infty} \frac{(z e^{-\beta A} \lambda^d(\Lambda))^n}{n!} = e^{\lambda^d(\Lambda)(z e^{-\beta A}-1)} < +\infty.$$

In the case $\beta = 0$, we recover that $P_\Lambda^{z,\beta}$ is the Poisson point process π_Λ^z. So the activity parameter z is the mean number of points per unit volume when the interaction is null. When the interaction is active ($\beta > 0$), $P_\Lambda^{z,\beta}$ favours the configurations with low energy and penalizes the configurations with high energy. This distortion strengthens as β is large.

There are many motivations for the exponential form of the density in (5.6). Historically, it is due to the fact that the finite volume GPP solves the variational principle of statistical physics. Indeed, $P_\Lambda^{z,\beta}$ is the unique probability measure which realizes the minimum of the free excess energy, equal to the mean energy plus the entropy. It expresses the common idea that the equilibrium states in statistical physics minimize the energy and maximize the "disorder". This result is presented in the following proposition. Recall first that the relative entropy of a probability measure P on $\mathscr{C}_\Lambda$ with respect to the Poisson point process π_Λ^ζ is defined by

$$I(P|\pi_\Lambda^\zeta) = \begin{cases} \int \log(f)dP & \text{if } P \preccurlyeq \pi_\Lambda^z \text{ with } f = \dfrac{dP}{d\pi_\Lambda^\zeta} \\ +\infty & \text{otherwise.} \end{cases} \tag{5.7}$$

Proposition 5.1 (Variational Principle) *Let H be an energy function, $z > 0$, $\beta \geq 0$. Then*

$$\{P_\Lambda^{z,\beta}\} = argmin_{P \in \mathscr{P}_\Lambda} \beta E_P(H) - \log(z)E_P(N_\Lambda) + I(P|\pi_\Lambda),$$

where $\mathscr{P}_\Lambda$ is the space of probability measures on $\mathscr{C}_\Lambda$ with finite intensity and $E_P(H)$ is the expectation of H under P, which is always defined (maybe equal to infinity) since H is stable.

Proof First we note that

$$\beta E_{P_\Lambda^{z,\beta}}(H) - \log(z)E_{P_\Lambda^{z,\beta}}(N_\Lambda) + I(P_\Lambda^{z,\beta}|\pi_\Lambda)$$

$$= \beta \int H dP_\Lambda^{z,\beta} - \log(z)E_{P_\Lambda^{z,\beta}}(N_\Lambda) + \int \log\left(z^{N_\Lambda}\frac{e^{-\beta H}}{Z_\Lambda^{z,\beta}}\right)dP_\Lambda^{z,\beta}$$

$$= -\log(Z_\Lambda^{z,\beta}). \tag{5.8}$$

This equality implies that the minimum of $\beta E_P(H) - \log(z)E_P(N_\Lambda) + I(P|\pi_\Lambda)$ should be equal to $-\log(Z_\Lambda^{z,\beta})$. So for any $P \in \mathscr{P}_\Lambda$ such that $E_P(H) < +\infty$ and $I(P|\pi_\Lambda) < +\infty$ let us show that $\beta E_P(H) - \log(z)E_P(N_\Lambda) + I(P|\pi_\Lambda) \geq -\log(Z_\Lambda^{z,\beta})$ with equality if and only if $P = P_\Lambda^{z,\beta}$. Let f be the density of P with respect to π_Λ.

$$\log(Z_\Lambda^{z,\beta}) \geq \log\left(\int_{\{f>0\}} z^{N_\Lambda}e^{-\beta H}d\pi_\Lambda\right)$$

$$= \log\left(\int z^{N_\Lambda}e^{-\beta H}f^{-1}dP\right)$$

$$\geq \int \log\left(z^{N_\Lambda}e^{-\beta H}f^{-1}\right)dP$$

$$= -\beta E_P(H) - \log(z)E_P(N_\Lambda) - \log(f)dP.$$

The second inequality, due to the Jensen's inequality, is an equality if and only if $z^{N_\Lambda}e^{-\beta H}f^{-1}$ is P a.s. constant which is equivalent to $P = P_\Lambda^{z,\beta}$. The proposition is proved.

The parameters z and β allow to fit the mean number of points and the mean value of the energy under the GPP. Indeed when z increases, the mean number of points increases as well and similarly when β increases, the mean energy decreases. This phenomenon is expressed in the following proposition. The proof is a simple computation of derivatives.

Let us note that it is not easy to tune both parameters simultaneously since the mean number of points changes when β is modified (and vice versa). The estimation of the parameters z and β is discussed in the last Sect. 5.4.

Proposition 5.2 *The function* $z \mapsto E_{P_\Lambda^{z,\beta}}(N_\Lambda)$ *is continuous and differentiable, with derivative* $z \mapsto Var_{P_\Lambda^{z,\beta}}(N_\Lambda)/z$ *on* $(0, +\infty)$. *Similarly the function* $\beta \mapsto E_{P_\Lambda^{z,\beta}}(H)$ *is continuous and differentiable with derivative* $\beta \mapsto -Var_{P_\Lambda^{z,\beta}}(H)$ *on* $\mathbb{R}^+$.

Let us finish this section by explaining succinctly how to simulate such finite volume GPP. There are essentially two algorithms. The first one is based on a MCMC procedure where GPP are viewed as equilibrium states of Markov chains. The simulation is obtained by letting run for a long enough time the Markov chain. The simulation is not exact and the error is essentially controlled via a monitoring approach (see [42]). The second one is a coupling from the past algorithm which provided exact simulations. However, the computation time is often very long and these algorithms are not really that used in practice (see [32]).

5.2.4 DLR Equations

The DLR equations are due to Dobrushin, Lanford and Ruelle and give the local conditional distributions of GPP in any bounded window Δ given the configuration outside Δ. We need to define a family of local energy functions $(H_\Delta)_{\Delta \subset \mathbb{R}^d}$.

Definition 5.3 For any bounded set Δ and any finite configuration $\gamma \in \mathscr{C}_f$ we define

$$H_\Delta(\gamma) := H(\gamma) - H(\gamma_{\Delta^c}),$$

with the convention $\infty - \infty = 0$.

The quantity $H_\Delta(\gamma)$ gives the energetic contribution of points in γ_Δ towards the computation of the energy of γ. As an example, let us compute these quantities in the setting of pairwise interaction introduced in (5.1);

$$H_\Delta(\gamma) = \sum_{\{x,y\} \subset \gamma} \varphi(|x - y|) - \sum_{\{x,y\} \subset \gamma_{\Delta^c}} \varphi(|x - y|) = \sum_{\substack{\{x, y\} \subset \gamma \\ \{x, y\} \cap \Delta \neq \emptyset}} \varphi(|x - y|).$$

Note that $H_\Delta(\gamma)$ does not depend only on points in Δ. However, trivially we have $H(\gamma) = H_\Delta(\gamma) + H(\gamma_{\Delta^c})$, which shows that the energy of γ is the sum of the energy $H_\Delta(\gamma)$ plus something which does not depends on γ_Δ.

Proposition 5.3 (DLR Equations for Finite Volume GPP) *Let $\Delta \subset \Lambda$ be two bounded sets in $\mathbb{R}^d$ with $\lambda^d(\Delta) > 0$. Then for $P_\Lambda^{z,\beta}$-a.s. all γ_{Δ^c}*

$$P_\Lambda^{z,\beta}(d\gamma_\Delta | \gamma_{\Delta^c}) = \frac{1}{Z_\Delta^{z,\beta}(\gamma_{\Delta^c})} z^{N_\Delta(\gamma)} e^{-\beta H_\Delta(\gamma)} \pi_\Delta(d\gamma_\Delta), \tag{5.9}$$

where $Z_\Delta^{z,\beta}(\gamma_{\Delta^c})$ is the normalizing constant $\int z^{N_\Delta(\gamma)} e^{-\beta H_\Delta(\gamma)} \pi_\Delta(d\gamma_\Delta)$. In particular the right term in (5.9) does not depend on Λ.

Proof From the definition of H_Δ and the stochastic properties of the Poisson point process we have

$$P_\Lambda^{z,\beta}(d\gamma) = \frac{1}{Z_\Lambda^{z,\beta}} z^{N_\Lambda(\gamma)} e^{-\beta H(\gamma)} \pi_\Lambda(d\gamma)$$

$$= \frac{1}{Z_\Lambda^{z,\beta}} z^{N_\Delta(\gamma)} e^{-\beta H_\Delta(\gamma)} z^{N_{\Lambda\setminus\Delta}(\gamma)} e^{-\beta H(\gamma_{\Lambda\setminus\Delta})} \pi_\Delta(d\gamma_\Delta) \pi_{\Lambda\setminus\Delta}(d\gamma_{\Lambda\setminus\Delta}).$$

This expression ensures that the unnormalized conditional density of $P_\Lambda^{z,\beta}(d\gamma_\Delta | \gamma_{\Delta^c})$ with respect to $\pi_\Delta(d\gamma_\Delta)$ is $\gamma_\Delta \mapsto z^{N_\Delta(\gamma)} e^{-\beta H_\Delta(\gamma)}$. The normalization is necessary $Z_\Delta^{z,\beta}(\gamma_{\Delta^c})$ and the proposition is proved.

The DLR equations give the local conditional marginal distributions of GPP. They are the main tool to understand the local description of $P_\Lambda^{z,\beta}$, in particular when Λ is large. Note that the local marginal distributions (not conditional) are in general not accessible. It is a difficult point of the theory of GPP. This fact will be reinforced in the infinite volume regime, where the local distributions can be non-unique.

The DLR equations have a major issue due the intractable normalization constant $Z_\Delta^{z,\beta}(\gamma_{\Delta^c})$. In the next section the problem is partially solved via the GNZ equations.

5.2.5   GNZ Equations

The GNZ equations are due to Georgii, Nguyen and Zessin and have been introduced first in [43]. They generalize the Slivnyak-Mecke formulas for Poisson point processes. In this section we present and prove these equations. We need first to define the energy of a point inside a configuration.

Definition 5.4 Let $\gamma \in \mathcal{C}_f$ be a finite configuration and $x \in \mathbb{R}^d$. Then the local energy of x in γ is defined by

$$h(x, \gamma) = H(\{x\} \cup \gamma) - H(\gamma),$$

with the convention $+\infty - (+\infty) = 0$. Note that if $x \in \gamma$ then $h(x, \gamma) = 0$.

Proposition 5.4 (GNZ Equations) *For any positive measurable function f from $\mathbb{R}^d \times \mathcal{C}_f$ to $\mathbb{R}$,*

$$\int \sum_{x \in \gamma} f(x, \gamma \setminus \{x\}) P_\Lambda^{z,\beta}(d\gamma) = z \int \int_\Lambda f(x, \gamma) e^{-\beta h(x,\gamma)} dx \, P_\Lambda^{z,\beta}(d\gamma). \quad (5.10)$$

Proof Let us decompose the left term in (5.10).

$$\int \sum_{x \in \gamma} f(x, \gamma \setminus \{x\}) P_\Lambda^{z,\beta}(d\gamma)$$

$$= \frac{1}{Z_\Lambda^{z,\beta}} \int \sum_{x \in \gamma} f(x, \gamma \setminus \{x\}) z^{N_\Lambda(\gamma)} e^{-\beta H(\gamma)} \pi_\Lambda(d\gamma)$$

$$= \frac{e^{-\lambda^d(\Lambda)}}{Z_\Lambda^{z,\beta}} \sum_{n=1}^{+\infty} \frac{z^n}{n!} \sum_{k=1}^n \int_{\Lambda^k} f(x_k, \{x_1, \ldots, x_n\} \setminus \{x_k\}) e^{-\beta H(\{x_1, \ldots, x_n\})} dx_1 \ldots dx_n$$

$$= \frac{e^{-\lambda^d(\Lambda)}}{Z_\Lambda^{z,\beta}} \sum_{n=1}^{+\infty} \frac{z^n}{(n-1)!} \int_{\Lambda^k} f(x, \{x_1, \ldots, x_{n-1}\}) e^{-\beta H(\{x_1, \ldots, x_{n-1}\})}$$

$$e^{-\beta h(x, \{x_1, \ldots, x_{n-1}\})} dx_1 \ldots dx_{n-1} dx$$

$$= \frac{z}{Z_\Lambda^{z,\beta}} \int_\Lambda \int f(x, \gamma) z^{N_\Lambda(\gamma)} e^{-\beta H(\gamma)} e^{-\beta h(x,\gamma)} \pi_\Lambda(d\gamma) dx$$

$$= z \int \int_\Lambda f(x, \gamma) e^{-\beta h(x,\gamma)} dx \, P_\Lambda^{z,\beta}(d\gamma).$$

As usual the function f in (5.10) can be chosen without a constant sign. We just need to check that both terms in (5.10) are integrable.

In the following proposition we show that the equations GNZ (5.10) characterize the probability measure $P_\Lambda^{z,\beta}$.

Proposition 5.5 *Let $\Lambda \subset \mathbb{R}^d$ bounded such that $\lambda^d(\Lambda) > 0$. Let P be a probability measure on $\mathcal{C}_\Lambda$ such that for any positive measurable function f from $\mathbb{R}^d \times \mathcal{C}_f$ to $\mathbb{R}$*

$$\int \sum_{x \in \gamma} f(x, \gamma \setminus \{x\}) P(d\gamma) = z \int \int_\Lambda f(x, \gamma) e^{-\beta h(x,\gamma)} dx \, P(d\gamma).$$

Then it holds that $P = P_\Lambda^{z,\beta}$.

Proof Let us consider the measure $Q = \mathbf{1}_{\{H < +\infty\}} z^{-N_\Lambda} e^{\beta H} P$. Then

$$\int \sum_{x \in \gamma} f(x, \gamma \setminus \{x\}) Q(d\gamma)$$

$$= \int \sum_{x \in \gamma} f(x, \gamma \setminus \{x\}) \mathbf{1}_{\{H(\gamma) < +\infty\}} z^{-N_\Lambda(\gamma)} e^{\beta H(\gamma)} P(d\gamma)$$

$$= z^{-1} \int \sum_{x \in \gamma} f(x, \gamma \setminus \{x\}) \mathbf{1}_{\{H(\gamma \setminus \{x\}) < +\infty\}} \mathbf{1}_{\{h(x, \gamma \setminus \{x\}) < +\infty\}}$$

$$z^{-N_\Lambda(\gamma \setminus \{x\})} e^{\beta H(\gamma \setminus \{x\})} e^{\beta h(x, \gamma \setminus \{x\})} P(d\gamma)$$

$$= \int \int_\Lambda f(x, \gamma) \mathbf{1}_{\{H(\gamma) < +\infty\}} \mathbf{1}_{\{h(x, \gamma) < +\infty\}} e^{-\beta h(x, \gamma)}$$

$$z^{-N_\Lambda(\gamma)} e^{\beta H(\gamma)} e^{\beta h(x, \gamma)} dx \, P(d\gamma)$$

$$= \int \int_\Lambda f(x, \gamma) \mathbf{1}_{\{h(x, \gamma) < +\infty\}} dx \, Q(d\gamma).$$

We deduce that Q satisfies the Slivnyak-Mecke formula on $\{\gamma \in \mathscr{C}_\Lambda, H(\gamma) < +\infty\}$. It is well-known (see [36] for instance) that it implies that the measure Q (after normalization) is the Poisson point process π_Λ restricted to $\{\gamma \in \mathscr{C}_\Lambda, H(\gamma) < +\infty\}$. The proposition is proved.

These last two propositions show that the GNZ equations contain completely the informations on $P_\Lambda^{z,\beta}$. Note again that the normalization constant $Z_\Lambda^{z,\beta}$ is not present in the equations.

5.2.6 Ruelle Estimates

In this section we present Ruelle estimates in the context of superstable and lower regular energy functions. These estimates are technical and we refer to the original paper [48] for the proofs.

Definition 5.5 An energy function H is said superstable if $H = H_1 + H_2$ where H_1 is an energy function (see Definition 5.1) and H_2 is a pairwise energy function defined in (5.1) with a non-negative continuous pair potential φ such that $\varphi(0) > 0$. The energy function H is said lower regular if there exists a summable decreasing sequence of positive reals $(\psi_k)_{k \geq 0}$ (i.e. $\sum_{k=0}^{+\infty} \psi_k < +\infty$) such that for any finite configurations γ^1 and γ^2

$$H(\gamma^1 \cup \gamma^2) - H(\gamma^1) - H(\gamma^2)$$

$$\geq - \sum_{k,k' \in \mathbb{Z}^d} \psi_{\|k-k'\|} \left(N^2_{[k+[0,1]^d]}(\gamma^1) + N^2_{[k'+[0,1]^d]}(\gamma^2) \right). \qquad (5.11)$$

Let us give the main example of superstable and lower regular energy function.

Proposition 5.6 (Proposition 1.3 [48]) *Let H be a pairwise energy function with a pair potential $\varphi = \varphi_1 + \varphi_2$ where φ_1 is stable and φ_2 is non-negative continuous with $\varphi_2(0) > 0$. Moreover, we assume that there exists a positive decreasing function ψ from $\mathbb{R}^+$ to $\mathbb{R}$ such that*

$$\int_0^{+\infty} r^{d-1} \psi(r) dr < +\infty$$

and such that for any $x \in \mathbb{R}$, $\varphi(x) \geq -\psi(\|x\|)$. Then the energy function H is superstable and lower regular.

In particular, the Lennard-Jones pair potential or the Strauss pair potential defined in Sect. 5.2.2 are superstable and lower regular. Note also that all geometric energy functions presented in Sect. 5.2.2 are not superstable.

Proposition 5.7 (Corollary 2.9 [48]) *Let H be a superstable and lower regular energy function. Let $z > 0$ and $\beta > 0$ be fixed. Then for any bounded subset $\Delta \subset \mathbb{R}^d$ with $\lambda^d(\Delta) > 0$ there exist two positive constants c_1, c_2 such that for any bounded set Λ and $k \geq 0$*

$$P_\Lambda^{z,\beta}(N_\Delta \geq k) \leq c_1 e^{-c_2 k^2}. \qquad (5.12)$$

In particular, Ruelle estimates (5.12) ensure that the random variable N_Δ admits exponential moments for all orders under $P_\Lambda^{z,\beta}$. Surprisingly, the variate N_Δ^2 admits exponential moments for small orders. This last fact is not true under the Poisson point process $\pi_\Lambda^z = P_\Lambda^{z,0}$. The interaction between the points improves the integrability properties of the GPP with respect to the Poisson point process.

5.3   Infinite Volume Gibbs Point Processes

In this section we present the theory of infinite volume GPP corresponding to the case "$\Lambda = \mathbb{R}^d$" of the previous section. Obviously, a definition inspired by (5.6) does not work since the energy of an infinite configuration γ is meaningless. A natural construction would be to consider a sequence of finite volume GPP $(P_{\Lambda_n}^{z,\beta})_{n \geq 1}$ on bounded windows $\Lambda_n = [-n, n]^d$ and let n tend to infinity. It is more or less what we do in the following Sects. 5.3.1 and 5.3.2, except that the convergence occurs only for a subsequence and that the field is stationarized (see Eq. (5.14)). As far as we know, there does not exist a general proof of the convergence of the sequence $(P_{\Lambda_n}^{z,\beta})_{n\geq 1}$ without extracted a subsequence. The stationarization is a convenient setting here in order to use the tightness entropy tools. In Sects. 5.3.3 and 5.3.4 we prove that the accumulation points $P^{z,\beta}$ satisfy the DLR equations which is the standard definition of infinite volume GPP (see Definition 5.8). We make precise that the main new assumption in this section is the finite range property (see Definition 5.7). It means that the points interact with each other only if their distance is smaller than a fixed constant $R > 0$. The GNZ equations in the infinite volume regime are discussed in Sect. 5.3.5. The variational characterisation of GPP, in the spirit of Proposition 5.1, is presented in Sect. 5.3.6. Uniqueness and non-uniqueness results of infinite volume GPP are treated in Sects. 5.3.7 and 5.3.8. These results, whose proofs are completely self contained here, ensure the existence of a phase transition for the Area energy function presented in (5.4). It means that the associated infinite volume Gibbs measures are unique for some parameters (z, β) and non-unique for other parameters.

5.3.1   The Local Convergence Setting

In this section we define the topology of local convergence which is the setting we use to prove the existence of an accumulation point for the sequence of finite volume Gibbs measures.

First, we say that a function from $\mathscr{C}$ to $\mathbb{R}$ is local if there exists a bounded set $\Delta \subset \mathbb{R}^d$ such that for all $\gamma \in \mathscr{C}$, $f(\gamma) = f(\gamma_\Delta)$.

Definition 5.6 The local convergence topology on the space of probability measures on $\mathscr{C}$ is the smallest topology such that for any local bounded function f from $\mathscr{C}$ to $\mathbb{R}$ the function $P \mapsto \int f dP$ is continuous. We denote by $\tau_{\mathscr{L}}$ this topology.

Let us note that the continuity of functions f in the previous definition is not required. For instance the function $\gamma \mapsto f(\gamma) = \mathbf{1}_{N_\Delta(\gamma)\geq k}$, where Δ is a bounded set in $\mathbb{R}^d$ and k any integer, is a bounded local function. For any vector $u \in \mathbb{R}^d$ we denote by τ_u the translation by the vector u acting on $\mathbb{R}^d$ or $\mathscr{C}$. A probability P on $\mathscr{C}$ is said stationary (or shift invariant) if for any vector $u \in \mathbb{R}^d$ $P = P \circ \tau_u^{-1}$.

Our tightness tool is based on the specific entropy which is defined for any stationary probability P on $\mathscr{C}$ by

$$I_\zeta(P) = \lim_{n\to+\infty} \frac{1}{\lambda^d(\Lambda_n)} I(P_{\Lambda_n}|\pi_{\Lambda_n}^\zeta), \tag{5.13}$$

where $I(P_{\Lambda_n}|\pi_{\Lambda_n}^\zeta)$ is the relative entropy of P_{Λ_n}, the projection of P on Λ_n, with respect to $\pi_{\Lambda_n}^\zeta$ (see Definition 5.7). Note that the specific entropy $I_\zeta(P)$ always exists (i.e. the limit in (5.13) exists); see chapter 15 in [22]. The tightness tool presented in Lemma 5.1 below is a consequence of the following proposition.

Proposition 5.8 (Proposition 15.14 [22]) *For any $\zeta > 0$ and any value $K \geq 0$, the set*

$$\{P \in \mathscr{P} \text{ such that } I_\zeta(P) \leq K\}$$

is sequentially compact for the topology $\tau_{\mathscr{L}}$, where $\mathscr{P}$ is the space of stationary probability measures on $\mathscr{C}$ with finite intensity.

5.3.2 An Accumulation Point $P^{z,\beta}$

In this section we prove the existence of an accumulation point for a sequence of stationarized finite volume GPP. To the end we consider the Gibbs measures $(P_{\Lambda_n}^{z,\beta})_{n\geq 1}$ on $\Lambda_n := [-n, n]^d$, where $(P_{\Lambda}^{z,\beta})$ is defined in (5.6) for any $z > 0$, $\beta \geq 0$ and energy function H. We assume that H is **stationary**, which means that for any vector $u \in \mathbb{R}^d$ and any finite configuration $\gamma \in \mathscr{C}_f$

$$H(\tau_u(\gamma)) = H(\gamma).$$

For any $n \geq 1$, the empirical field $\bar{P}_{\Lambda_n}^{z,\beta}$ is defined by the probability measure on $\mathscr{C}$ such that for any test function f

$$\int f(\gamma) \bar{P}_{\Lambda_n}^{z,\beta}(d\gamma) = \frac{1}{\lambda^d(\Lambda_n)} \int_{\Lambda_n} \int f(\tau_u(\gamma)) P_{\Lambda_n}^{z,\beta}(d\gamma) du. \tag{5.14}$$

The probability measure $\bar{P}_{\Lambda_n}^{z,\beta}$ can be interpreted as the Gibbs measure $P_{\Lambda_n}^{z,\beta}$ where the origin of the space $\mathbb{R}^d$ (i.e. the point $\{0\}$) is replaced by a random point chosen uniformly inside Λ_n. It is a kind of stationarization of $P_{\Lambda_n}^{z,\beta}$ and any accumulation point of the sequence $(\bar{P}_{\Lambda_n}^{z,\beta})_{n\geq 1}$ is necessary stationary.

Proposition 5.9 *The sequence $(\bar{P}_{\Lambda_n}^{z,\beta})_{n\geq 1}$ is tight for the $\tau_{\mathscr{L}}$ topology. We denote by $P^{z,\beta}$ any of its accumulation points.*

Proof Our tightness tool is the following lemma whose the proof is a consequence of Proposition 5.8 (See also Proposition 15.52 in [22]).

Lemma 5.1 *The sequence $(\bar{P}_{\Lambda_n}^{z,\beta})_{n\geq 1}$ is tight for the $\tau_{\mathscr{L}}$ topology if there exits $\zeta > 0$ such that*

$$\sup_{n\geq 1} \frac{1}{\lambda^d(\Lambda_n)} I(P_{\Lambda_n}^{z,\beta} | \pi_{\Lambda_n}^\zeta) < +\infty. \tag{5.15}$$

So, let us compute $I(P_{\Lambda_n}^{z,\beta} | \pi_{\Lambda_n}^{\zeta})$ and check that we can find $\zeta > 0$ such that (5.15) holds.

$$I(P_{\Lambda_n}^{z,\beta} | \pi_{\Lambda_n}^{\zeta}) = \int \log\left(\frac{dP_{\Lambda_n}^{z,\beta}}{d\pi_{\Lambda_n}^{\zeta}}\right) dP_{\Lambda_n}^{z,\beta}$$

$$= \int \left[\log\left(\frac{dP_{\Lambda_n}^{z,\beta}}{d\pi_{\Lambda_n}}\right) + \log\left(\frac{d\pi_{\Lambda_n}}{d\pi_{\Lambda_n}^{\zeta}}\right)\right] dP_{\Lambda_n}^{z,\beta}$$

$$= \int \left[\log\left(z^{N_{\Lambda_n}} \frac{e^{-\beta H}}{Z_{\Lambda_n}^{z,\beta}}\right) + \log\left(e^{(\zeta-1)\lambda^d(\Lambda_n)}\left(\frac{1}{\zeta}\right)^{N_{\Lambda_n}}\right)\right] dP_{\Lambda_n}^{z,\beta}$$

$$= \int \left[-\beta H + \log\left(\frac{z}{\zeta}\right) N_{\Lambda_n}\right] dP_{\Lambda_n}^{z,\beta} + (\zeta-1)\lambda^d(\Lambda_n) - \log(Z_{\Lambda_n}^{z,\beta}).$$

Thanks to the non degeneracy and the stability of H we find that

$$I(P_{\Lambda_n}^{z,\beta} | \pi_{\Lambda_n}^{\zeta}) \leq \int \left(-A\beta + \log\left(\frac{z}{\zeta}\right)\right) N_{\Lambda_n} dP_{\Lambda_n}^{z,\beta}$$

$$+ \lambda^d(\Lambda_n)\Big((\zeta-1) + 1 + \beta H(\{\emptyset\})\Big).$$

Choosing $\zeta > 0$ such that $-A\beta + \log(z/\zeta) \leq 0$ we obtain

$$I(P_{\Lambda_n}^{z,\beta} | \pi_{\Lambda_n}^{\zeta}) \leq \lambda^d(\Lambda_n)(\zeta + \beta H(\{\emptyset\}))$$

and (5.15) holds. Proposition 5.9 is proved.

In the following, for sake of simplicity, we say that $\bar{P}_{\Lambda_n}^{z,\beta}$ converges to $P^{z,\beta}$ although it occurs only for a subsequence.

Note that the existence of an accumulation points holds under very weak assumptions on the energy function H. Indeed the two major assumptions are the stability and the stationarity. The superstability or the lower regularity presented in Definition 5.5 are not required here. However, if the energy function H is superstable and lower regular, then the accumulation points $P^{z,\beta}$ inherits Ruelle estimates (5.12). This fact is obvious since the function $\gamma \mapsto \mathbf{1}_{\{N_\Delta(\gamma) \geq k\}}$ is locally bounded.

Corollary 5.1 *Let H be a superstable and lower regular energy function (see Definition 5.5). Let $z > 0$ and $\beta > 0$ be fixed. Then for any bounded subset $\Delta \subset \mathbb{R}^d$ with $\lambda^d(\Delta) > 0$, there exists c_1 and c_2 two positive constants such that for any $k \geq 0$*

$$P^{z,\beta}(N_\Delta \geq k) \leq c_1 e^{-c_2 k^2}. \tag{5.16}$$

The important point now is to prove that $P^{z,\beta}$ satisfies good stochastic properties as for instance the DLR or GNZ equations. At this stage, without extra assumptions, these equations are not necessarily satisfied. Indeed it is possible to build energy functions H such that the accumulation point $P^{z,\beta}$ is degenerated and charges only the empty configuration. In this mini-course our extra assumption is the finite range property presented in the following section. More general settings have been investigated for instance in [16] or [47].

5.3.3 The Finite Range Property

The finite range property expresses that further a certain distance $R > 0$ the points do not interact each other. Let us recall the Minkoswki $\oplus$ operator acting on sets in $\mathbb{R}^d$. For any two sets $A, B \subset \mathbb{R}^d$, the set $A \oplus B$ is defined by $\{x + y, x \in A$ and $y \in B\}$.

Definition 5.7 The energy function H has a finite range $R > 0$ if for every bounded Δ, the local energy H_Δ (see Definition 5.3) is a local function on $\Delta \oplus B(0, R)$. It means that for any finite configuration $\gamma \in \mathscr{C}_f$

$$H_\Delta(\gamma) := H(\gamma) - H(\gamma_{\Delta^c}) = H(\gamma_{\Delta \oplus B(0,R)}) - H(\gamma_{\Delta \oplus B(0,R) \setminus \Delta^c}).$$

Let us illustrate the finite range property in the setting of pairwise interaction defined in (5.1). Assume that the interaction potential $\varphi : \mathbb{R}^+ \to \mathbb{R} \cup \{+\infty\}$ has a support included in $[0, R]$. Then the associated energy function has a finite R;

$$H_\Delta(\gamma) = \sum_{\substack{\{x, y\} \subset \gamma \\ \{x, y\} \cap \Delta \neq \emptyset \\ |x - y| \leq R}} \varphi(|x - y|)$$

$$= \sum_{\substack{\{x, y\} \subset \gamma_{\Delta \oplus B(0,R)} \\ \{x, y\} \cap \Delta \neq \emptyset}} \varphi(|x - y|).$$

Also the area energy function (5.4) inherits the finite range property. A simple computation gives

$$H_\Delta(\gamma) = \text{Area}\left(\bigcup_{x \in \gamma_\Delta} B(x, R) \setminus \bigcup_{x \in \gamma_{\Delta \oplus B(0,2R)} \setminus \Delta} B(x, R) \right) \tag{5.17}$$

which provides a range of interaction equals to $2R$.

Let us note that the energy functions defined in (5.2),(5.3) and (5.5) do not have the finite range property. Similarly the pairwise energy function (5.1) with the Lennard-Jones potential is not finite range since the support of the pair potential is not bounded. A truncated version of such potential is sometimes considered.

Let us finish this section by noting that the finite range property allows to extend the domain of definition of H_Δ from the space $\mathscr{C}_f$ to the set $\mathscr{C}$. Indeed, since $H_\Delta(\gamma) = H_\Delta(\gamma_{\Delta \oplus B(0,R)})$, this equality provides a definition of $H_\Delta(\gamma)$ when γ is in $\mathscr{C}$. This point is crucial in order to correctly define the DLR equations in the infinite volume regime.

5.3.4 DLR Equations

In Sect. 5.2 on the finite volume GPP, the DLR equations are presented as properties for $P_\Lambda^{z,\beta}$ (see Sect. 5.2.4). In the setting of infinite volume GPP, the DLR equations are the main points of the definition of GPP.

Definition 5.8 (Infinite Volume GPP) Let H be a stationary and finite range energy function. A stationary probability P on $\mathscr{C}$ is an infinite volume Gibbs measure with activity $z > 0$, inverse temperature $\beta \geq 0$ and energy function H if for any bounded $\Delta \subset \mathbb{R}^d$ such that $\lambda^d(\Delta) > 0$ then for P-a.s. all γ_{Δ^c}

$$P(d\gamma_\Delta | \gamma_{\Delta^c}) = \frac{1}{Z_\Delta^{z,\beta}(\gamma_{\Delta^c})} z^{N_\Delta(\gamma)} e^{-\beta H_\Delta(\gamma)} \pi_\Delta(d\gamma_\Delta), \tag{5.18}$$

where $Z_\Delta^{z,\beta}(\gamma_{\Delta^c})$ is the normalizing constant $\int z^{N_\Delta(\gamma)} e^{-\beta H_\Delta(\gamma)} \pi_\Delta(d\gamma_\Delta)$. As usual, an infinite volume GPP is a point process whose distribution is an infinite volume Gibbs measure.

Note that the DLR equations (5.18) make sense since $H_\Delta(\gamma)$ is well defined for any configuration $\gamma \in \mathscr{C}$ (see the end of Sect. 5.3.3). Note also that the DLR equations (5.18) can be reformulated in an integral form. Indeed P satisfies (5.18) if and only if for any local bounded function f from $\mathscr{C}$ to $\mathbb{R}$

$$\int f dP = \int f(\gamma_\Delta' \cup \gamma_{\Delta^c}) \frac{1}{Z_\Delta^{z,\beta}(\gamma_{\Delta^c})} z^{N_\Delta(\gamma_\Delta')} e^{-\beta H_\Delta(\gamma_\Delta' \cup \gamma_{\Delta^c})} \pi_\Delta(d\gamma_\Delta') P(d\gamma).$$

$$\tag{5.19}$$

The term "equation" is now highlighted by the formulation (5.19) since the unknown variate P appears in both left and right sides. The existence, uniqueness and non-uniqueness of solutions of such DLR equations are non trivial questions. In the next theorem, we show that the accumulation point $P^{z,\beta}$ obtained in Sect. 5.3.2 is such a solution. Infinite volume Gibbs measure exist and the question of existence is solved. The uniqueness and non-uniqueness are discussed in Sects. 5.3.7 and 5.3.8.

Theorem 5.1 *Let H be a stationary and finite range energy function. Then for any* $z > 0$ *and* $\beta \geq 0$ *the probability measure* $P^{z,\beta}$ *defined in Proposition 5.9 is an infinite volume Gibbs measure.*

Proof We have just to check that $P^{z,\beta}$ satisfies, for any bounded Δ and any positive local bounded function f, the Eq. (5.19). Let us define the function f_Δ by

$$f_\Delta : \gamma \mapsto \int f(\gamma'_\Delta \cup \gamma_{\Delta^c}) \frac{1}{Z^{z,\beta}_\Delta(\gamma_{\Delta^c})} z^{N_\Delta(\gamma'_\Delta)} e^{-\beta H_\Delta(\gamma'_\Delta \cup \gamma_{\Delta^c})} \pi_\Delta(d\gamma'_\Delta).$$

Since f is local and bounded and since H is finite range, the function f_Δ is bounded and local as well. From the convergence of the sequence $(\bar{P}^{z,\beta}_{\Lambda_n})_{n\geq 1}$ to $P^{z,\beta}$ with respect to the $\tau_{\mathscr{L}}$ topology, we have

$$\int f_\Delta d P^{z,\beta}$$

$$= \lim_{n\to\infty} \int f_\Delta d \bar{P}^{z,\beta}_{\Lambda_n}$$

$$= \lim_{n\to\infty} \frac{1}{\lambda^d(\Lambda_n)} \int_{\Lambda_n} \int f_\Delta(\tau_u(\gamma)) P^{z,\beta}_{\Lambda_n}(d\gamma) du.$$

$$= \lim_{n\to\infty} \frac{1}{\lambda^d(\Lambda_n)} \int_{\Lambda_n} \int \int f(\gamma'_\Delta \cup \tau_u(\gamma)_{\Delta^c}) \frac{z^{N_\Delta(\gamma'_\Delta)}}{Z^{z,\beta}_\Delta(\tau_u(\gamma)_{\Delta^c})} e^{-\beta H_\Delta(\gamma'_\Delta \cup \tau_u(\gamma)_{\Delta^c})}$$

$$\pi_\Delta(d\gamma'_\Delta) P^{z,\beta}_{\Lambda_n}(d\gamma) du$$

$$= \lim_{n\to\infty} \frac{1}{\lambda^d(\Lambda_n)} \int_{\Lambda_n} \int \int f\Big(\tau_u\big(\gamma'_{\tau_{-u}(\Delta)} \cup \gamma_{\tau_{-u}(\Delta)^c}\big)\Big) \frac{z^{N_{\tau_{-u}(\Delta)}(\gamma'_{\tau_{-u}(\Delta)})}}{Z^{z,\beta}_{\tau_{-u}(\Delta)}(\gamma_{\tau_{-u}(\Delta)^c})}$$

$$e^{-\beta H_{\tau_{-u}(\Delta)}\big(\gamma'_{\tau_{-u}(\Delta)} \cup \gamma_{\tau_{-u}(\Delta)^c}\big)} \pi_{\tau_{-u}(\Delta)}(d\gamma'_{\tau_{-u}(\Delta)}) P^{z,\beta}_{\Lambda_n}(d\gamma) du. \tag{5.20}$$

Denoting by Λ_n^* the set of $u \in \Lambda_n$ such that $\tau_{-u}(\Delta) \subset \Lambda_n$, by Proposition 5.3, $P^{z,\beta}_{\Lambda_n}$ satisfies the DLR equation on $\tau_{-u}(\Delta)$ as soon as $\tau_{-u}(\Delta) \subset \Lambda_n$ (i.e. $u \in \Lambda_n^*$). It follows that for any $u \in \Lambda_n^*$

$$\int f(\tau_u \gamma) P^{z,\beta}_{\Lambda_n}(d\gamma)$$

$$= \int \int f\Big(\tau_u\big(\gamma'_{\tau_{-u}(\Delta)} \cup \gamma_{\tau_{-u}(\Delta)^c}\big)\Big) \frac{z^{N_{\tau_{-u}(\Delta)}(\gamma'_{\tau_{-u}(\Delta)})}}{Z^{z,\beta}_{\tau_{-u}(\Delta)}(\gamma_{\tau_{-u}(\Delta)^c})} e^{-\beta H_{\tau_{-u}(\Delta)}\big(\gamma'_{\tau_{-u}(\Delta)} \cup \gamma_{\tau_{-u}(\Delta)^c}\big)}$$

$$\pi_{\tau_{-u}(\Delta)}(d\gamma'_{\tau_{-u}(\Delta)}) P^{z,\beta}_{\Lambda_n}(d\gamma). \tag{5.21}$$

By noting that $\lambda^d(\Lambda_n^*)$ is equivalent to $\lambda^d(\Lambda_n)$ when n goes to infinity, we obtain in compiling (5.20) and (5.21)

$$
\int f_\Delta dP^{z,\beta} = \lim_{n\to\infty} \frac{1}{\lambda^d(\Lambda_n)} \int_{\Lambda_n^*} \int \int f(\tau_u\gamma) P_{\Lambda_n}^{z,\beta}(d\gamma) du
$$

$$
= \lim_{n\to\infty} \int f(\gamma) \bar{P}_{\Lambda_n}^{z,\beta}(d\gamma)
$$

$$
= \int f dP^{z,\beta}
$$

which gives the expected integral DLR equation on Δ with test function f.

5.3.5   GNZ Equations

In this section we deal with the GNZ equations in the infinite volume regime. As in the finite volume case, the main advantage of such equations is that the intractable normalization factor $Z_\Lambda^{z,\beta}$ is not present.

Note first that, in the setting of finite range interaction $R > 0$, the local energy $h(x, \gamma)$ defined in Definition 5.4 is well-defined for any configuration $\gamma \in \mathscr{C}$ even if γ is infinite. Indeed, we clearly have $h(x, \gamma) = h(x, \gamma_{B(x,R)})$.

Theorem 5.2 *Let P be a probability measure on $\mathscr{C}$. Let H be a finite range energy function and $z > 0$, $\beta \geq 0$ be two parameters. Then P is an infinite volume Gibbs measure with energy function H, activity $z > 0$ and inverse temperature β if and only if for any positive measurable function f from $\mathbb{R}^d \times \mathscr{C}$ to $\mathbb{R}$*

$$
\int \sum_{x\in\gamma} f(x, \gamma\setminus\{x\}) P(d\gamma) = z \int \int_{\mathbb{R}^d} f(x, \gamma) e^{-\beta h(x,\gamma)} dx\, P(d\gamma). \qquad (5.22)
$$

Proof Let us start with the proof of the "only if" part. Let P be an infinite volume Gibbs measure. By standard monotonicity arguments it is sufficient to prove (5.22) for any local positive measurable function f. So let $\Delta \subset \mathbb{R}^d$ be a bounded set such that $f(x, \gamma) = 1_\Delta(x) f(x, \gamma_\Delta)$. Applying now the DLR equation (5.19) on the set Δ we find

$$
\int \sum_{x\in\gamma} f(x, \gamma\setminus\{x\}) P(d\gamma)
$$

$$
= \int \int \sum_{x\in\gamma_\Delta'} f(x, \gamma_\Delta'\setminus\{x\}) \frac{1}{Z_\Delta^{z,\beta}(\gamma_{\Delta^c})} z^{N_\Delta(\gamma_\Delta')} e^{-\beta H_\Delta(\gamma_\Delta'\cup\gamma_{\Delta^c})} \pi_\Delta(d\gamma_\Delta') P(d\gamma).
$$

By computations similar to those developed in the proof of Proposition 5.4, we obtain

$$\int \sum_{x \in \gamma} f(x, \gamma \setminus \{x\}) P(d\gamma) = z \int \int_{\Delta} \int f(x, \gamma'_{\Delta}) \frac{1}{Z_{\Delta}^{z,\beta}(\gamma_{\Delta^c})} e^{-\beta h(x, \gamma'_{\Delta} \cup \gamma_{\Delta^c})}$$

$$z^{N_{\Delta}(\gamma'_{\Delta})} e^{-\beta H_{\Delta}(\gamma'_{\Delta} \cup \gamma_{\Delta^c})} \pi_{\Delta}(d\gamma'_{\Delta}) dx \, P(d\gamma)$$

$$= z \int \int_{\mathbb{R}^d} f(x, \gamma) e^{-\beta h(x, \gamma)} dx \, P(d\gamma).$$

Let us now turn to the "if part". Applying Eq. (5.22) to the function $\tilde{f}(x, \gamma) = \psi(\gamma_{\Delta^c}) f(x, \gamma)$ where f is a local positive function with support Δ and ψ a positive test function we find

$$\int \psi(\gamma_{\Delta^c}) \sum_{x \in \gamma_{\Delta}} f(x, \gamma \setminus \{x\}) P(d\gamma) = z \int \psi(\gamma_{\Delta^c}) \int_{\mathbb{R}^d} f(x, \gamma) e^{-\beta h(x, \gamma)} dx \, P(d\gamma).$$

This implies that for P almost all γ_{Δ^c} the conditional probability measure $P(d\gamma_{\Delta}|\gamma_{\Delta^c})$ solves the GNZ equations on Δ with local energy function $\gamma_{\Delta} \mapsto h(x, \gamma_{\Delta} \cup \gamma_{\Delta^c})$. Following an adaptation of the proof of Proposition 5.5, we get that

$$P(d\gamma_{\Delta}|\gamma_{\Delta^c}) = \frac{1}{Z_{\Delta}^{z,\beta}(\gamma_{\Delta^c})} z^{N_{\Delta}(\gamma)} e^{-\beta H_{\Delta}(\gamma)} \pi_{\Delta}(d\gamma_{\Delta}),$$

which is exactly the DLR equation (5.18) on Δ. The theorem is proved.

Let us finish this section with an application of the GNZ equations which highlights that some properties of infinite volume GPP can be extracted from the implicit GNZ equations.

Proposition 5.10 *Let Γ be a infinite volume GPP for the hardcore pairwise interaction $\varphi(r) = +\infty \mathbf{1}_{[0,R]}(r)$ (see Definition 5.1) and the activity $z > 0$. Then*

$$\frac{z}{1 + z v_d R^d} \leq E\left(N_{[0,1]^d}(\Gamma)\right) \leq z, \tag{5.23}$$

where v_d is the volume of the unit ball in $\mathbb{R}^d$.

Note that the inverse temperature β does not play any role here and that $E_P\left(N_{[0,1]^d}(\Gamma)\right)$ is simply the intensity of Γ.

Proof The local energy of such harcore pairwise interaction is given by

$$h(x, \gamma) = \sum_{y \in \gamma_{B(x,R)}} \varphi(|x - y|) = +\infty \mathbf{1}_{\gamma_{B(x,R)} \neq \emptyset}.$$

So the GNZ equation (5.22) with the function $f(x, \gamma) = \mathbf{1}_{[0,1]^d}(x)$ gives

$$E\left(N_{[0,1]^d}(\Gamma)\right) = z \int_{[0,1]^d} P(\Gamma_{B(x,R)} = \emptyset)dx = z\, P(\Gamma_{B(0,R)} = \emptyset),$$

which provides a relation between the intensity and the spherical contact distribution of Γ. The upper bound in (5.23) follows. For the lower bound we have

$$
\begin{aligned}
E_P\left(N_{[0,1]^d}(\Gamma)\right) &= z\, P(\Gamma_{B(0,R)} = \emptyset) \\
&\geq z\left(1 - E_P\left(N_{B(0,R)}(\Gamma)\right)\right) \\
&= z\left(1 - v_d R^r E_P\left(N_{[0,1]^d}(\Gamma)\right)\right).
\end{aligned}
$$

Note also that a natural upper bound for $E_P\left(N_{[0,1]^d}\right)$ is obtained via the closed packing configuration. For instance, in dimension $d = 2$, it gives the upper bound $\pi/(2\sqrt{3}R^2)$.

5.3.6   Variational Principle

In this section, we extend the variational principle for finite volume GPP presented in Proposition 5.1 to the setting of infinite volume GPP. For brevity we present only the result without the proof which can be found in [13].

The variational principle claims that the Gibbs measures are the minimizers of the free excess energy defined by the sum of the mean energy and the specific entropy. Moreover, the minimum is equal to minus the pressure. Let us first define all these macroscopic quantities.

Let us start by introducing the pressure with free boundary condition. It is defined as the following limit

$$p^{z,\beta} := \lim_{n \to +\infty} \frac{1}{|\Lambda_n|} \ln(Z_{\Lambda_n}^{z,\beta}), \tag{5.24}$$

The existence of such limit is proved for instance in Lemma 1 in [13].

The second macroscopic quantity involves the mean energy of a stationary probability measure P. It is also defined by a limit but, in opposition to the pressure, we have to assume that it exists. The proof of such existence is generally based on stationary arguments and nice representations of the energy contribution per unit volume. It depends strongly on the expression of the energy function H. Examples are given below. So for any stationary probability measure P on $\mathscr{C}$ we assume that the following limit exists in $\mathbb{R} \cup \{+\infty\}$,

$$H(P) := \lim_{n \to \infty} \frac{1}{|\Lambda_n|} \int H(\gamma_{\Lambda_n}) dP(\gamma), \qquad (5.25)$$

and we call the limit mean energy of P.

We need to introduce a technical assumption on the boundary effects of H. We assume that for any infinite volume Gibbs measure P

$$\lim_{n \to \infty} \frac{1}{|\Lambda_n|} \int \partial H_{\Lambda_n}(\gamma) dP(\gamma) = 0, \qquad (5.26)$$

where $\partial H_{\Lambda_n}(\gamma) = H_{\Lambda_n}(\gamma) - H(\gamma_{\Lambda_n})$.

Theorem 5.3 (Variational Principle, Theorem 1, [13]) *We assume that H is stationary and finite range. Moreover, we assume that the mean energy exists for any stationary probability measure P (i.e. the limit (5.25) exists) and that the boundary effects assumption (5.26) holds. Let $z > 0$ and $\beta \geq 0$ two parameters. Then for any stationary probability measure P on $\mathscr{C}$ with finite intensity*

$$I_1(P) + \beta H(P) - \log(z) E_P(N_{[0,1]^d}) \geq -p^{z,\beta}, \qquad (5.27)$$

with equality if and only if P is a Gibbs measure with activity $z > 0$, inverse temperature β and energy function H.

Let us finish this section by presenting the two fundamental examples of energy functions satisfying the assumptions of Theorem 5.3.

Proposition 5.11 *Let H be the Area energy function defined in (5.4). Then both limits (5.25) and (5.26) exist. In particular, the assumptions of Theorem 5.3 are satisfied and the variational principle holds.*

Proof Let us prove only that the limit (5.25) exists. The existence of limit (5.26) can be shown in the same way. By definition of H and the stationarity of P,

$$\int H(\gamma_{\Lambda_n}) P(d\gamma) = \int \text{Area}(L_R(\gamma_{\Lambda_n})) P(d\gamma)$$

$$= \lambda^d(\Lambda_n) \int \text{Area}(L_R(\gamma) \cap [0,1]^d) P(d\gamma)$$

$$+ \int \left(\text{Area}(L_R(\gamma_{\Lambda_n})) - \text{Area}(L_R(\gamma) \cap [-n,n]^d) \right) P(d\gamma).$$

$$(5.28)$$

By geometric arguments, we get that

$$\left| \text{Area}(L_R(\gamma_{\Lambda_n})) - \text{Area}(L_R(\gamma) \cap [-n,n]^d) \right| \leq C n^{d-1},$$

for some constant $C > 0$. We deduce that the limit (5.25) exists with

$$H(P) = \int \text{Area}(L_R(\gamma) \cap [0, 1]^d) P(d\gamma).$$

Proposition 5.12 *Let H be the pairwise energy function defined in (5.1) with a superstable, lower regular pair potential with compact support. Then the both limits (5.25) and (5.26) exist. In particular the assumptions of Theorem 5.3 are satisfied and the variational principle holds.*

Proof Since the potential φ is stable with compact support, we deduce that $\varphi \geq 2A$ and H is finite range and lower regular. In this setting, the existence of the limit (5.25) is proved in [21], Theorem 1 with

$$H(P) = \begin{cases} \frac{1}{2} \int \sum_{0 \neq x \in \gamma} \varphi(x) P^0(d\gamma) & \text{if } E_P(N_{[0,1]^d}^2) < \infty \\ +\infty & \text{otherwise} \end{cases} \tag{5.29}$$

where P^0 is the Palm measure of P. Recall that P^0 can be viewed as the natural version of the conditional probability $P(.|0 \in \gamma)$ (see [36] for more details). It remains to prove the existence of the limit (5.26) for any Gibbs measure P on $\mathscr{C}$. A simple computation gives that, for any $\gamma \in \mathscr{C}$,

$$\partial H_{\Lambda_n}(\gamma) = \sum_{x \in \gamma_{\Lambda_n^\oplus \setminus \Lambda_n}} \sum_{y \in \gamma_{\Lambda_n \setminus \Lambda_n^\ominus}} \varphi(x - y),$$

where $\Lambda_n^\oplus = \Lambda_{n+R_0}$ and $\Lambda_n^\ominus = \Lambda_{n-R_0}$ with R_0 an integer larger than the range of the interaction R.

Therefore thanks to the stationarity of P and the GNZ equations (5.22), we obtain

$$\left| \int \partial H_{\Lambda_n}(\gamma) dP(\gamma) \right| \leq \int \sum_{x \in \gamma_{\Lambda_n^\oplus \setminus \Lambda_n}} \sum_{y \in \gamma \setminus \{x\}} |\varphi(x - y)| dP(\gamma)$$

$$= z \int \int_{\Lambda_n^\oplus \setminus \Lambda_n} e^{-\beta \sum_{y \in \gamma} \varphi(x-y)} \sum_{y \in \gamma} |\phi(x - y)| dx dP(\gamma)$$

$$= z |\Lambda_n^\oplus \setminus \Lambda_n| \int e^{-\beta \sum_{y \in \gamma_{B(0, R_0)}} \varphi(y)} \sum_{y \in \gamma_{B(0, R_0)}} |\varphi(y)| dP(\gamma).$$

Since $\varphi \geq 2A$, denoting by $C := \sup_{c \in [2A; +\infty)} |c| e^{-\beta c} < \infty$ we find that

$$\left| \int \partial H_{\Lambda_n}(\gamma) dP(\gamma) \right| \leq z C |\Lambda_n^\oplus \setminus \Lambda_n| \int N_{B(0, R_0)}(\gamma) e^{-2\beta A N_{B(0, R_0)}(\gamma)} dP(d\gamma).$$

$$\tag{5.30}$$

Using Ruelle estimates (5.16), the integral in the right term of (5.30) is finite. The boundary assumption (5.26) follows.

5.3.7 A Uniqueness Result

In this section we investigate the uniqueness of infinite volume Gibbs measures. The common belief claims that the Gibbs measures are unique when the activity z or (and) the inverse temperature β are small enough (low activity, high temperature regime). The non-uniqueness phenomenon (discussed in the next section) are in general related to some issues with the energy part in the variational principle (see Theorem 5.3). Indeed, either the mean energy has several minimizers or there is a conflict between the energy and the entropy. Therefore it is natural to expect the Gibbs measures are unique when β is small enough. When z is small, the mean number of points per unit volume is low and so the energy is in general low as well.

As far as we know, there do not exist general results which prove the uniqueness for small β or small z. In the case of pairwise energy functions (5.1), the uniqueness for any $\beta > 0$ and $z > 0$ small enough is proved via the Kirkwood-Salsburg equations (see Theorem 5.7 [48]). An extension of the Dobrushin uniqueness criterium in the continuum is developed as well [18]. The uniqueness of GPP can also be obtained via the cluster expansion machinery which provides a power series expansion of the partition function when z and β are small enough. This approach has been introduced first by Mayer and Montroll [37] and we refer to [45] for a general presentation.

In this section we give a simple and self-contained proof of the uniqueness of GPP for all $\beta \geq 0$ and any $z > 0$ small enough. We just assume that the energy function H has a local energy h uniformly bounded from below. This setting covers for instance the case of pairwise energy function (5.1) with non-negative pair potential or the Area energy function (5.4).

Let us start by recalling the existence of a percolation threshold for the Poisson Boolean model. For any configuration $\gamma \in \mathscr{C}$ the percolation of $L_R(\gamma) = \cup_{x \in \gamma} B(x, R)$ means the existence of an unbounded connected component in $L_R(\gamma)$.

Proposition 5.13 (Theorem 1 [28]) *For any* $d \geq 2$, *there exists* $0 < z_d < +\infty$ *such that for* $z < z_d$, $\pi^z(L_{1/2}$ *percolates* $) = 0$ *and for* $z > z_d$, $\pi^z(L_{1/2}$ *percolates* $) = 1$.

The value z_d is called the percolation threshold of the Poisson Boolean model with radius $1/2$. By scale invariance, the percolation threshold for any other radius R is simply $z_d/(2R)^d$. The exact value of z_d is unknown but numerical studies provide for instance the approximation $z_2 \simeq 1.4$ in dimension $d = 2$.

Theorem 5.4 *Let H be an energy function with finite range $R > 0$ such that the local energy h is uniformly bounded from below by a constant C. Then for any $\beta \geq 0$*

and $z < z_d e^{C\beta}/R^d$, there exists an unique Gibbs measure with energy function H, activity $z > 0$ and inverse temperature β.

Proof The proof is based on two main ingredients. The first one is the stochastic domination of Gibbs measures, with uniformly bounded from below local energy function h, by Poisson processes. This result is given in the following lemma, whose proof can be found in [23]. The second ingredient is a disagreement percolation result presented in Lemma 5.3 below.

Lemma 5.2 *Let H be an energy function such that the local energy h is uniformly bounded from below by a constant C. Then for any bounded set Δ and any outside configuration γ_{Δ^c} the Gibbs distribution inside Δ given by*

$$P^{z,\beta}(d\gamma_\Delta|\gamma_{\Delta^c}) = \frac{1}{Z_\Delta^{z,\beta}(\gamma_{\Delta^c})} z^{N_\Delta(\gamma)} e^{-\beta H_\Delta(\gamma)} \pi_\Delta(d\gamma_\Delta)$$

is stochastically dominated by the Poisson point distribution $\pi_\Delta^{ze^{-C\beta}}(d\gamma_\Delta)$.

Thanks to Strassen's Theorem, this stochastic domination can be interpreted via the following coupling (which could be the definition of the stochastic domination): There exist two point processes Γ and Γ' on Δ such that $\Gamma \subset \Gamma'$, $\Gamma \sim P^{z,\beta}(d\gamma_\Delta|\gamma_{\Delta^c})$ and $\Gamma' \sim \pi_\Delta^{ze^{-C\beta}}(d\gamma_\Delta)$.

Now the rest of the proof of Theorem 5.4 consists in showing that the Gibbs measure is unique as soon as $\pi^{ze^{-C\beta}}(L_{R/2}\text{percolates}) = 0$. Roughly speaking, if the dominating process does not percolate, the information coming from the boundary condition does not propagate in the heart of the model and the Gibbs measure is unique. To prove rigorously this phenomenon, we need a disagreement percolation argument introduced first in [50]. For any sets $A, B \in \mathbb{R}^d$, we denote by $A \ominus B$ the set $(A^c \oplus B)^c$.

Lemma 5.3 *Let $\gamma_{\Delta^c}^1$ and $\gamma_{\Delta^c}^2$ be two configurations on Δ^c. For any $R' > R$, there exist three point processes Γ^1, Γ^2 and Γ' on Δ such that $\Gamma^1 \subset \Gamma'$, $\Gamma^2 \subset \Gamma'$, $\Gamma^1 \sim P^{z,\beta}(d\gamma_\Delta|\gamma_{\Delta^c}^1)$, $\Gamma^2 \sim P^{z,\beta}(d\gamma_\Delta|\gamma_{\Delta^c}^2)$ and $\Gamma' \sim \pi_\Delta^{ze^{-C\beta}}(d\gamma_\Delta)$. Moreover, denoting by $L_{R'/2}^\Delta(\Gamma')$ the connected components of $L_{R'/2}(\Gamma')$ which are inside $\Delta \ominus B(0, R'/2)$, then $\Gamma^1 = \Gamma^2$ on the set $L_{R'/2}^\Delta(\Gamma')$.*

Proof Let us note first that, by Lemma 5.2, there exist three point processes Γ^1, Γ^2 and Γ' on Δ such that $\Gamma^1 \subset \Gamma'$, $\Gamma^2 \subset \Gamma'$, $\Gamma^1 \sim P^{z,\beta}(d\gamma_\Delta|\gamma_{\Delta^c}^1)$, $\Gamma^2 \sim P^{z,\beta}(d\gamma_\Delta|\gamma_{\Delta^c}^2)$ and $\Gamma' \sim \pi_\Delta^{ze^{-C\beta}}(d\gamma_\Delta)$. The main difficulty is now to show that we can build Γ^1 and Γ^2 such that $\Gamma^1 = \Gamma^2$ on the set $L_{R'/2}^\Delta(\Gamma')$.

Let us decompose Δ via a grid of small cubes where each cube has a diameter smaller than $\epsilon = (R' - R)/2$. We define an arbitrary numeration of these cubes $(C_i)_{1 \leq i \leq m}$ and we construct progressively the processes Γ^1, Γ^2 and Γ' on each cube C_i. Assume that they are already constructed on $C_I := \cup_{i \in I} C_i$ with all the expected properties: $\Gamma_{C_I}^1 \subset \Gamma'_{C_I}$, $\Gamma_{C_I}^2 \subset \Gamma'_{C_I}$, $\Gamma_{C_I}^1 \sim P^{z,\beta}(d\gamma_{C_I}|\gamma_{\Delta^c}^1)$, $\Gamma_{C_I}^2 \sim$

$P^{z,\beta}(d\gamma_{C_I}|\gamma_{\Delta^c}^2)$, $\Gamma_{C_I}' \sim \pi_{C_I}^{ze^{-C\beta}}(d\gamma_{C_I})$ and $\Gamma_{C_I}^1 = \Gamma_{C_I}^2$ on the set $L_{R'/2}^\Delta(\Gamma_{C_I}')$. Let us consider the smaller index $j \in \{1, \ldots m\}\backslash I$ such that either the distances $d(C_j, \gamma_{\Delta^c}^1)$ or $d(C_j, \gamma_{\Delta^c}^2)$ or $d(C_j, \Gamma_{C_I}')$ is smaller than $R' - \epsilon$.

- If such an index j does not exist, by the finite range property the following Gibbs distributions coincide on $\Delta^I = \Delta\backslash C_I$;

$$P^{z,\beta}(d\gamma_{\Delta^I}|\gamma_{\Delta^c}^1 \cup \Gamma_{C_I}^1) = P^{z,\beta}(d\gamma_{\Delta^I}|\gamma_{\Delta^c}^2 \cup \Gamma_{C_I}^2).$$

  Therefore we define Γ^1, Γ^2 and Γ' on Δ^I by considering $\Gamma_{\Delta^I}^1$ and Γ_{Δ^I}' as in Lemma 5.2 and by putting $\Gamma_{\Delta^I}^2 = \Gamma_{\Delta^I}^1$. We can easily check that all expected properties hold and the full construction of Γ^1, Γ^2 and Γ' is over.

- If such an index j does exist, we consider the double coupling construction of Γ^1, Γ^2 and Γ' on Δ^I. It means that $\Gamma_{\Delta^I}^1 \subset \Gamma_{\Delta^I}'$, $\Gamma_{\Delta^I}^2 \subset \Gamma_{\Delta^I}'$, $\Gamma_{\Delta^I}^1 \sim P^{z,\beta}(d\gamma_{\Delta^I}|\gamma_{\Delta^c}^1 \cup \Gamma_{C_I}^1)$, $\Gamma_{\Delta^I}^2 \sim P^{z,\beta}(d\gamma_{\Delta^I}|\gamma_{\Delta^c}^2 \cup \Gamma_{C_I}^2)$ and $\Gamma_{\Delta^I}' \sim \pi_{\Delta^I}^{ze^{C\beta}}(d\gamma_\Delta)$. Now we keep these processes $\Gamma_{\Delta^I}^1$, $\Gamma_{\Delta^I}^2$ and Γ_{Δ^I}' only on the window C_j. The construction of the processes Γ^1, Γ^2 and Γ' is now over $C_I \cup C_j$ and we can check again that all expected properties hold. We go on to the construction of the processes on a new cube in $(C_i)_{i\in\{1,\ldots n\}\backslash\{I,j\}}$ and so on.

Let us now finish the proof of Theorem 5.4 by considering two infinite volume GPP $\tilde{\Gamma}^1$ and $\tilde{\Gamma}^2$ with distribution P^1 and P^2. We have to show that for any local event A $P^1(A) = P^2(A)$. We denote by Δ_0 the support of such an event A. Let us consider a bounded subset $\Delta \supset \Delta_0$ and three new processes Γ_Δ^1, Γ_Δ^2 and Γ_Δ' on Δ constructed as in Lemma 5.3. Precisely, for any $i = 1, 2$ $\Gamma_\Delta^i \subset \Gamma_\Delta'$, $\Gamma' \sim \pi_\Delta^{ze^{-C\beta}}$, the conditional distribution of Γ_Δ^i given $\tilde{\Gamma}_{\Delta^c}^i$ is $P^{z,\beta}(|\tilde{\Gamma}_{\Delta^c}^i)$ and $\Gamma_\Delta^1 = \Gamma_\Delta^2$ on the set $L_{R'/2}^\Delta(\Gamma_\Delta')$. The parameter $R' > R$ is chosen such that

$$ze^{-C\beta}R'^d < z_d \qquad (5.31)$$

which is possible by assumption on z.

Thanks to the DLR equations (5.18), for any $i = 1, 2$ the processes Γ_Δ^i and $\tilde{\Gamma}_\Delta^i$ have the same distributions and therefore $P^i(A) = P(\Gamma_\Delta^i \in A)$. Denoting by $\{\Delta \leftrightarrow \Delta_0\}$ the event that there exists a connected component in $L_{R'/2}(\Gamma')$ which intersects $(\Delta \ominus B(0, R'/2))^c$ and Δ_0, we obtain that

$$|P^1(A) - P^2(A)| = |P(\Gamma_\Delta^1 \in A) - P(\Gamma_\Delta^2 \in A)|$$

$$\leq E\left(\mathbf{1}_{\{\Delta\leftrightarrow\Delta_0\}}\left|\mathbf{1}_{\Gamma_\Delta^1\in A} - \mathbf{1}_{\Gamma_\Delta^2\in A}\right|\right)$$

$$+ E\left(\mathbf{1}_{\{\Delta\leftrightarrow\Delta_0\}^c}\left|\mathbf{1}_{\Gamma_\Delta^1\in A} - \mathbf{1}_{\Gamma_\Delta^2\in A}\right|\right)$$

$$\leq P(\{\Delta \leftrightarrow \Delta_0\}) + E\left(\mathbf{1}_{\{\Delta \leftrightarrow \Delta_0\}^c}\left|\mathbf{1}_{\Gamma^1_{\Delta} \in A} - \mathbf{1}_{\Gamma^1_{\Delta} \in A}\right|\right)$$

$$= P(\{\Delta \leftrightarrow \Delta_0\}). \tag{5.32}$$

By the choice of R' in inequality (5.31) and Proposition 5.13, it follows that

$$\pi^{ze^{-C\beta}}\left(L_{R'/2}\text{percolates}\right) = 0$$

and we deduce, by a monotonicity argument, the probability $P(\{\Delta \leftrightarrow \Delta_0\})$ tends to 0 when Δ tends to $\mathbb{R}^d$ (see [38] for details on equivalent characterizations of continuum percolation). The left term in (5.32) does not depend on Δ and therefore it is null. Theorem 5.4 is proved.

5.3.8 A Non-uniqueness Result

In this section we discuss the non-uniqueness phenomenon of infinite volume Gibbs measures. It is believed to occur for almost all models provided that the activity z or the inverse temperature β is large enough. However, in the present continuous setting without spin, it is only proved for few models and several old conjectures are still valid. For instance, for the pairwise Lennard-Jones interaction defined in (5.1), it is conjectured that for β large (but not too large) there exists an unique z such that the Gibbs measures are not unique. It would correspond to a liquid-vapour phase transition. Similarly for β very large, it is conjectured that the non-uniqueness occurs as soon as z is larger than a threshold z_β. It would correspond to a crystallization phenomenon for which a symmetry breaking may occur. Indeed, it is expected, but not proved at all, that some continuum Gibbs measures would be not invariant under symmetries like translations, rotations, etc. This conjecture is probably one of the most important and difficult challenges in statistical physics. In all cases, the non-uniqueness appear when the local distribution of infinite volume Gibbs measures depend on the boundary conditions "at infinity".

In this section we give a complete proof of such non-uniqueness result for the Area energy interaction presented in (5.4). This result has been first proved in [52] but our proof is inspired by the one given in [7]. Roughly speaking, we build two different Gibbs measures which depend, via a percolation phenomenon, on the boundary conditions "at infinity". In one case, the boundary condition "at infinity" is empty and in the other case the boundary condition is full of particles. We show that the intensity of both infinite volume Gibbs measures are different.

Let us cite another famous non-uniqueness result for attractive pair and repulsive four-body potentials [34]. As far as we know, this result and the one presented below on the Area interaction, are the only rigorous proofs of non-uniqueness results for continuum particles systems without spin.

Theorem 5.5 *For $z = \beta$ large enough, the infinite volume Gibbs measures for the Area energy function H presented in (5.4), the activity z and the inverse temperature β are not unique.*

Proof In all the proof we fix $z = \beta$. Let us consider following finite volume Gibbs measures on $\Lambda_n = [-n, n]^d$ with different boundary conditions:

$$dP_{\Lambda_n}(\gamma) = \frac{1}{Z_{\Lambda_n}} \mathbf{1}_{\left\{ \gamma_{\Lambda_n \setminus \Lambda_n^\ominus} = \emptyset \right\}} z^{N_{\Lambda_n}(\gamma)} e^{-z\text{Area}\left(\Lambda_n \cap L_R(\gamma)\right)} d\pi_{\Lambda_n}(\gamma),$$

and

$$dQ_{\Lambda_n}(\gamma) = \frac{1}{Z'_{\Lambda_n}} z^{N_{\Lambda_n}(\gamma)} e^{-z\text{Area}\left(\Lambda_n^\ominus \cap L_R(\gamma)\right)} d\pi_{\Lambda_n}(\gamma),$$

where $\Lambda_n^\ominus = \Lambda_n \ominus B(0, R/2)$. Recall that R is the radius of balls in $L_R(\gamma) = \cup_{x \in \gamma} B(x, R)$ and that the range of the interaction is $2R$. As in Sect. 5.3.2 we consider the associated empirical fields $\bar{P}_{\Lambda_n}$ and $\bar{Q}_{\Lambda_n}$ defined by

$$\int f(\gamma) d\bar{P}_{\Lambda_n}(\gamma) = \frac{1}{\lambda^d(\Lambda_n)} \int_{\Lambda_n} f(\tau_u(\gamma)) dP_{\Lambda_n}(\gamma) du$$

and

$$\int f(\gamma) d\bar{Q}_{\Lambda_n}(\gamma) = \frac{1}{\lambda^d(\Lambda_n)} \int_{\Lambda_n} f(\tau_u(\gamma)) dQ_{\Lambda_n}(\gamma) du,$$

where f is any measurable bounded test function. Following the proof of Proposition 5.9 we get the existence of an accumulation point $\bar{P}$ (respectively $\bar{Q}$) for $(\bar{P}_{\Lambda_n})$ (respectively $(\bar{Q}_{\Lambda_n})$). As in Theorem 5.1, we show that $\bar{P}$ and $\bar{Q}$ satisfy the DLR equations and therefore they are both infinite volume Gibbs measures for the Area energy function, the activity z and the inverse temperature $\beta = z$. Now it remains to prove that $\bar{P}$ and $\bar{Q}$ are different when z is large enough. Note that the difference between $\bar{P}$ and $\bar{Q}$ comes only from their boundary conditions "at infinity" (i.e. the boundary conditions of P_{Λ_n} and Q_{Λ_n} when n goes to infinity).

Let us start with a representation of P_{Λ_n} and Q_{Λ_n} via the two type Widom-Rowlinson model on Λ_n. Consider the following event of allowed configurations on $\mathscr{C}_{\Lambda_n}^2$

$$\mathscr{A} = \left\{ (\gamma^1, \gamma^2) \in \mathscr{C}_{\Lambda_n}^2, \text{ s.t. } \begin{array}{l} a) \; L_{R/2}(\gamma^1) \cap L_{R/2}(\gamma^2) = \emptyset \\ b) \; L_{R/2}(\gamma^1) \cap \Lambda_n^c = \emptyset \end{array} \right\} \tag{5.33}$$

which assumes first that the balls with radii $R/2$ centred at γ^1 and γ^2 do not overlap and secondly that the balls centred at γ^1 are completely inside Λ_n.

The two type Widom-Rowlinson model on Λ_n with boundary condition b) is the probability measure $\tilde{P}_{\Lambda_n}$ on $\mathscr{C}_{\Lambda_n}^2$ which is absolutely continuous with respect to the product $(\pi_{\Lambda_n}^z)^{\otimes 2}$ with density

$$\frac{1}{\tilde{Z}_n} \mathbf{1}_{\mathscr{A}}(\gamma^1, \gamma^2) z^{N_{\Lambda_n}(\gamma^1)} z^{N_{\Lambda_n}(\gamma^2)} d\pi_{\Lambda_n}(\gamma^1) d\pi_{\Lambda_n}(\gamma^2),$$

where $\tilde{Z}_{\Lambda_n}$ is a normalization factor.

Lemma 5.4 *The first marginal (respectively the second marginal) distribution of $\tilde{P}_{\Lambda_n}$ is P_{Λ_n} (respectively Q_{Λ_n}).*

Proof By definition of $\tilde{P}_{\Lambda_n}$, its first marginal admits the following unnormalized density with respect to $\pi_{\Lambda_n}(d\gamma^1)$

$$f(\gamma^1) = \int \mathbf{1}_{\mathscr{A}}(\gamma^1, \gamma^2) z^{N_{\Lambda_n}(\gamma^1)} z^{N_{\Lambda_n}(\gamma^2)} d\pi_{\Lambda_n}(\gamma^2)$$

$$= e^{(z-1)\lambda^d(\Lambda_n)} z^{N_{\Lambda_n}(\gamma^1)} \int \mathbf{1}_{\mathscr{A}}(\gamma^1, \gamma^2) d\pi_{\Lambda_n}^z(\gamma^2)$$

$$= e^{(z-1)\lambda^d(\Lambda_n)} z^{N_{\Lambda_n}(\gamma^1)} \mathbf{1}_{\left\{\gamma^1_{\Lambda_n \setminus \Lambda_n^{\ominus}} = \emptyset\right\}} e^{-z \text{Area}\left(\Lambda_n \cap L_R(\gamma^1)\right)}$$

which is proportional to the density of P_{Λ_n}. A similar computation gives the same result for Q_{Λ_n}. ∎

Now let us give a representation of the two type Widom-Rowlinson model via the random cluster model. The random cluster process R_{Λ_n} is a point process on Λ_n distributed by

$$\frac{1}{\hat{Z}_n} z^{N_{\Lambda_n}(\gamma)} 2^{N_{cc}^{\Lambda_n}(\gamma)} d\pi_{\Lambda_n}(\gamma),$$

where $N_{cc}^{\Lambda_n}(\gamma)$ is the number of connected components of $L_{R/2}(\gamma)$ which are completely included in Λ_n. Then we build two new point processes $\hat{\Gamma}_{\Lambda_n}^1$ and $\hat{\Gamma}_{\Lambda_n}^2$ by splitting randomly and uniformly the connected component of R_{Λ_n}. Each connected component inside Λ_n is given to $\hat{\Gamma}_{\Lambda_n}^1$ or $\hat{\Gamma}_{\Lambda_n}^2$ with probability an half each. The connected components hitting Λ_n^c are given to $\hat{\Gamma}_{\Lambda_n}^2$. Rigorously this construction is done by the following way. Let us consider $(C_i(\gamma))_{1 \le i \le N_{cc}^{\Lambda_n}(\gamma)}$ the collection of connected components of $L_{R/2}(\gamma)$ inside Λ_n. Let $(\epsilon_i)_{i \ge 1}$ be a sequence of independent Bernoulli random variables with parameter $1/2$. The processes $\hat{\Gamma}_{\Lambda_n}^1$ and $\hat{\Gamma}_{\Lambda_n}^2$ are defined by

$$\hat{\Gamma}_{\Lambda_n}^1 = \bigcup_{1 \le i \le N_{cc}^{\Lambda_n}(R_{\Lambda_n}), \, \epsilon_i = 1} R_{\Lambda_n} \cap C_i(R_{\Lambda_n}) \quad \text{and} \quad \hat{\Gamma}_{\Lambda_n}^2 = R_{\Lambda_n} \setminus \hat{\Gamma}_{\Lambda_n}^1.$$

Lemma 5.5 *The distribution of* $(\hat{\Gamma}^1_{\Lambda_n}, \hat{\Gamma}^2_{\Lambda_n})$ *is the two-type Widom-Rowlinson model with boundary condition* b). *In particular,* $\hat{\Gamma}^1_{\Lambda_n} \sim P_{\Lambda_n}$ *and* $\hat{\Gamma}^2_{\Lambda_n} \sim Q_{\Lambda_n}$.

Proof For any bounded measurable test function f we have

$$E(f(\hat{\Gamma}^1_{\Lambda_n}, \hat{\Gamma}^2_{\Lambda_n}))$$

$$= E\left[f\left(\bigcup_{1 \le i \le N^{\Lambda_n}_{cc}(R_{\Lambda_n}), \, \epsilon_i=1} R_{\Lambda_n} \cap C_i(R_{\Lambda_n}), \right.\right.$$

$$\left.\left. R_{\Lambda_n} \cap \left(\bigcup_{1 \le i \le N^{\Lambda_n}_{cc}(R_{\Lambda_n}), \, \epsilon_i=1} C_i(R_{\Lambda_n}) \right)^c \right) \right]$$

$$= \frac{1}{\hat{Z}_n} \int \sum_{(\epsilon_i) \in \{0,1\}^{N^{\Lambda_n}_{cc}(\gamma)}} \frac{1}{2^{N^{\Lambda_n}_{cc}(\gamma)}}$$

$$f\left(\bigcup_{1 \le i \le N^{\Lambda_n}_{cc}(\gamma), \, \epsilon_i=1} \gamma \cap C_i(\gamma), \gamma \cap \left(\bigcup_{1 \le i \le N^{\Lambda_n}_{cc}(\gamma), \, \epsilon_i=1} C_i(\gamma) \right)^c \right)$$

$$z^{N_{\Lambda_n}(\gamma)} 2^{N^{\Lambda_n}_{cc}(\gamma)} d\pi_{\Lambda_n}(\gamma)$$

$$= \frac{1}{\hat{Z}_n} \int \sum_{(\epsilon_x) \in \{0,1\}^\gamma} (\mathbf{1}_{\mathscr{A}} f) \left(\bigcup_{x \in \gamma, \, \epsilon_x=1} \{x\}, \gamma \setminus \bigcup_{x \in \gamma, \, \epsilon_x=1} \{x\} \right) z^{N_{\Lambda_n}(\gamma)} d\pi_{\Lambda_n}(\gamma)$$

$$= \frac{1}{\hat{Z}_n} \int (\mathbf{1}_{\mathscr{A}} f) \left(\bigcup_{(x,\epsilon_x) \in \tilde{\gamma}, \, \epsilon_x=1} \{x\}, \bigcup_{(x,\epsilon_x) \in \tilde{\gamma}, \, \epsilon_x=0} \{x\} \right) (2z)^{N_{\Lambda_n}(\gamma)} d\tilde{\pi}_{\Lambda_n}(\tilde{\gamma})$$

where $\tilde{\pi}_{\Lambda_n}$ is a marked Poisson point process on $\Lambda_n \times \{0, 1\}$. It means that the points are distributed by π_{Λ_n} and that each point x is marked independently by a Bernoulli variable ϵ_x with parameter $1/2$. We obtain

$$E(f(\hat{\Gamma}^1_{\Lambda_n}, \hat{\Gamma}^2_{\Lambda_n}))$$

$$= \frac{e^{|\Lambda_n|}}{\hat{Z}_n} \int (\mathbf{1}_{\mathscr{A}} f) \left(\bigcup_{(x,\epsilon_x) \in \tilde{\gamma}, \, \epsilon_x=1} \{x\}, \bigcup_{(x,\epsilon_x) \in \tilde{\gamma}, \, \epsilon_x=0} \{x\} \right) z^{N_{\Lambda_n}(\gamma)} d\tilde{\pi}^2_{\Lambda_n}(\tilde{\gamma})$$

$$= \frac{e^{|\Lambda_n|}}{\hat{Z}_n} \int \int (\mathbf{1}_{\mathscr{A}} f) \left(\gamma^1, \gamma^2 \right) z^{N_{\Lambda_n}(\gamma^1)} z^{N_{\Lambda_n}(\gamma^2)} d\pi_{\Lambda_n}(\gamma^1) d\pi_{\Lambda_n}(\gamma^2),$$

which proves the lemma.

Note that the random cluster process R_{Λ_n} is a finite volume GPP with energy function $\hat{H} = -N_{cc}^{\Lambda_n}$, activity z and inverse temperature $\log(2)$. Its local energy $\hat{h}$ is defined by

$$\hat{h}(x, \gamma) = N_{cc}^{\Lambda_n}(\gamma) - N_{cc}^{\Lambda_n}(\gamma \cup \{x\}).$$

Thanks to a geometrical argument, it is not difficult to note that $\hat{h}$ is uniformly bounded from above by a constant c_d (depending only on the dimension d). For instance, in the case $d = 2$, a ball with radius $R/2$ can overlap at most five disjoints balls with radius $R/2$ and therefore $c_2 = 5 - 1 = 4$ is suitable.

By Lemma 5.2, we deduce that the distribution of R_{Λ_n} dominates the Poisson point distribution $\pi_{\Lambda_n}^{2ze^{-c_d}}$. So we choose

$$z > \frac{z_d e^{c_d}}{2R^d}$$

which implies that the Boolean model with intensity $2ze^{-c_d}$ and radii $R/2$ percolates with probability one (see Proposition 5.13). For any $\gamma \in \mathscr{C}$, we denote by $C_\infty(\gamma)$ the unbounded connected components in $L_{R/2}(\gamma)$ (if it exists) and we define by α the intensity of points in $C_\infty(\gamma)$ under the distribution $\pi^{2ze^{-c_d}}$;

$$\alpha := \int N_{[0,1]^d}\left(\gamma \cap C_\infty(\gamma)\right) d\pi^{2ze^{-c_d}}(\gamma) > 0. \tag{5.34}$$

We are now in position to finish the proof of Theorem 5.5 by proving that the difference in intensities between $\bar{Q}$ and $\bar{P}$ is larger than α.

The local convergence topology $\tau_{\mathscr{L}}$ ensures that, for any local bounded function f, the evaluation $P \mapsto \int f dP$ is continuous. Actually, the continuity of such evaluation holds for the larger class of functions f satisfying: (1) f is local on some bounded set Δ, (2) there exists $A > 0$ such that $|f(\gamma)| \leq A(1 + \#(\gamma))$. In particular, the application $P \mapsto i(P) := \int N_{[0,1]^d}(\gamma) P(d\gamma)$ is continuous (see [25] for details). We deduce that

$$i(\bar{Q}) - i(\bar{P}) = \int N_{[0,1]^d}(\gamma) d\bar{Q}(\gamma) - \int N_{[0,1]^d}(\gamma) d\bar{P}(\gamma)$$

$$= \lim_{n\to\infty} \left(\int N_{[0,1]^d}(\gamma) d\bar{Q}_{\Lambda_n}(\gamma) - \int N_{[0,1]^d}(\gamma) d\bar{P}_{\Lambda_n}(\gamma)\right)$$

$$= \lim_{n\to\infty} \frac{1}{\lambda^d(\Lambda_n)} \int_{\Lambda_n} \left(\int N_{[0,1]^d}(\tau_u \gamma) d Q_{\Lambda_n}(\gamma)\right.$$

$$\left. - \int N_{[0,1]^d}(\tau_u \gamma) d P_{\Lambda_n}(\gamma)\right) du.$$

By the representation of P_{Λ_n} and Q_{λ_n} given in Lemma 5.5, we find

$$
i(\bar{Q}) - i(\bar{P}) = \lim_{n \to \infty} \frac{1}{\lambda^d(\Lambda_n)} \int_{\Lambda_n} E\left(N_{[0,1]^d}(\tau_u \hat{\Gamma}^2_{\Lambda_n}) - N_{[0,1]^d}(\tau_u \hat{\Gamma}^1_{\Lambda_n})\right) du
$$

$$
= \lim_{n \to \infty} \frac{1}{\lambda^d(\Lambda_n)} \int_{\Lambda_n} E\left(N_{\tau_u[0,1]^d}(R_{\Lambda_n} \cap C_b(R_{\Lambda_n}))\right) du,
$$

where $C_b(\gamma)$ are the connected components of $L_{R/2}(\gamma)$ hitting Λ_n^c. Since the distribution of R_{Λ_n} dominates $\pi_{\Lambda_n}^{2ze^{-c_d}}$,

$$
i(\bar{Q}) - i(\bar{P}) \geq \lim_{n \to \infty} \frac{1}{\lambda^d(\Lambda_n)} \int_{[-n,n-1]^d} \int N_{\tau_u[0,1]^d}\left(\gamma \cap C(\gamma)_\infty\right) d\pi_{\Lambda_n}^{2ze^{-c_d}}(\gamma) du,
$$

$$
\geq \lim_{n \to \infty} \frac{1}{\lambda^d(\Lambda_n)} \int_{[-n,n-1]^d} \alpha \, du = \alpha > 0.
$$

The theorem is proved.

5.4 Estimation of Parameters

In this section we investigate the parametric estimation of the activity z^* and the inverse temperature β^* of an infinite volume Gibbs point process Γ. As usual the star specifies that the parameters z^*, β^* are unknown whereas the variable z and β are used for the optimization procedures. Here the dataset is the observation of Γ trough the bounded window $\Lambda_n = [-n, n]^d$ (i.e. the process Γ_{Λ_n}). The asymptotic means that the window Λ_n increases to the whole space $\mathbb{R}^d$ (i.e. n goes to infinity) without changing the realization of Γ.

For sake of simplicity, we decide to treat only the case of two parameters (z, β) but it would be possible to consider energy functions depending on an extra parameter $\theta \in \mathbb{R}^p$. The case where H depends linearly on θ can be treated exactly as z and β. For the non linear case the setting is much more complicated and each procedure has to be adapted. References are given in each section.

In all the section, we assume that the energy function H is stationary and has a finite range $R > 0$. The existence of Γ is therefore guaranteed by Theorem 5.1. The procedures presented below are not affected by the uniqueness or non-uniqueness of the distribution of such GPP.

In Sect. 5.4.1, we start by presenting the natural maximum likelihood estimator. Afterwards, in Sect. 5.4.2, we introduce the general Takacs-Fiksel estimator which is a mean-square procedure based on the GNZ equations. The standard maximum pseudo-likelihood estimator is a particular case of such estimator and is presented in Sect. 5.4.3. An application to an unobservable issue is treated in Sect. 5.4.4. The last Sect. 5.4.5 is devoted to a new estimator based on a variational GNZ equation.

5.4.1 Maximum Likelihood Estimator

The natural method to estimate the parameters is the likelihood inference. However a practical issue is that the likelihood depends on the intractable partition function. In the case of sparse data, approximations were first proposed in [44], before simulation-based methods have been developed [26]. Here, we treat only the theoretical aspects of the MLE and these practical issues are not investigated.

Definition 5.9 The maximum likelihood estimator of (z^*, β^*) is given for any $n \geq 1$ by

$$(\hat{z}_n, \hat{\beta}_n) = \text{argmax}_{z>0, \beta \geq 0} \frac{1}{Z_{\Lambda_n}^{z,\beta}} z^{N_{\Lambda_n}(\Gamma)} e^{-\beta H(\Gamma_{\Lambda_n})}. \tag{5.35}$$

Note that the argmax is not necessarily unique and that the boundary effects are not considered in this version of MLE. Other choices could be considered.

In this section we show the consistency of such estimators. The next natural question concerns the asymptotic distribution of the MLE but this problem is more arduous and is still partially unsolved today. Indeed, Mase [35] and Jensen [29] proved that the MLE is asymptotically normal when the parameters z and β are small enough. Without these conditions, phase transition may occur and some long-range dependence phenomenon can appear. The MLE might then exhibit a non standard asymptotic behavior, in the sense that the rate of convergence might differ from the standard square root of the size of the window and the limiting law might be non-Gaussian.

The next theorem is based on a preprint by S. Mase (Asymptotic properties of MLEs of Gibbs models on Rd, 2002, unpublished preprint). See also [15] for general results on consistency.

Theorem 5.6 *We assume that the energy function H is stationary, finite range and not almost surely constant (i.e. there exists a subset $\Lambda \subset \mathbb{R}^d$ such that $H(\gamma_\Lambda)$ is not $\pi_\Lambda(d\gamma_\Lambda)$ almost surely constant). We assume also that the mean energy exists for any stationary probability measure P (i.e. the limit (5.25) exists) and that the boundary effects assumption (5.26) holds. Moreover we assume that for any ergodic Gibbs measure P, the following limit holds for P-almost every γ*

$$\lim_{n \mapsto \infty} \frac{1}{\lambda^d(\Lambda_n)} H(\gamma_{\Lambda_n}) = H(P). \tag{5.36}$$

Then, almost surely the parameters $(\hat{z}_n, \hat{\beta}_n)$ converge to (z^*, β^*) when n goes to infinity.

Proof Let us assume that the Gibbs distribution P of Γ is ergodic. Otherwise P can be represented as a mixture of ergodic stationary Gibbs measures (see [46, Theorem 2.2 and 4.1]). Therefore the proof of the consistency of the MLE reduces to the case when P is ergodic, which is assumed henceforth.

Let us consider the log-likelihood contrast function

$$K_n(\theta, \beta) = -\log(Z_{\Lambda_n}^{e^{-\theta}, \beta}) - \theta N_{\Lambda_n}(\Gamma) - \beta H(\Gamma_{\Lambda_n})$$

related to the parametrization $\theta = -\log(z)$. It is clear that $(\hat{z}_n, \hat{\beta}_n) = (e^{-\tilde{\theta}_n}, \tilde{\beta}_n)$ where $(\tilde{\theta}_n, \tilde{\beta}_n)$ is the argmax of $(\theta, \beta) \mapsto K_n(\theta, \beta)$. So it is sufficient to show that $(\tilde{\theta}_n, \tilde{\beta}_n)$ converges almost surely to $(-\log(z^*), \beta^*)$. The limit (5.24), the ergodic Theorem and the assumption (5.36) imply the existence of the following limit contrast function

$$K(\theta, \beta) := -p^{e^{-\theta}, \beta} - \theta E_P(N_{[0,1]^d}(\Gamma)) - \beta H(P) = \lim_{n \to \infty} \frac{K_n(\theta, \beta)}{\lambda^d(\Lambda_n)}.$$

The variational principle (Theorem 5.3) ensures that $(\theta, \beta) \mapsto K(\theta, \beta)$ is lower than $I_1(P)$ with equality if and only if P is a Gibbs measure with energy function H, activity z and inverse temperature β. Since H is not almost surely constant, it is easy to see that two Gibbs measures with different parameters z, β are different (this fact can be viewed used the DLR equations in a very large box Λ). Therefore $K(\theta, \beta)$ is maximal, equal to $I_1(P)$, if and only if $(\theta, \beta) = (\theta^*, \beta^*)$.

Therefore it remains to prove that the maximizers of $(\theta, \beta) \mapsto K_n(\theta, \beta)$ converge to the unique maximizer of $(\theta, \beta) \mapsto K(\theta, \beta)$. First note that the functions K_n are concave. Indeed, the Hessian of K_n is negative since

$$\frac{\partial^2 K_n(\theta, \beta)}{\partial^2 \theta} = -\text{Var}_{P_{\Lambda_n}^{e^{-\theta}, \beta}}(N_{\Lambda_n}), \quad \frac{\partial^2 K_n(\theta, \beta)}{\partial^2 \beta} = -\text{Var}_{P_{\Lambda_n}^{e^{-\theta}, \beta}}(H)$$

and

$$\frac{\partial^2 K_n(\theta, \beta)}{\partial \theta \partial \beta} = -\text{Cov}_{P_{\Lambda_n}^{e^{-\theta}, \beta}}(N_{\Lambda_n}, H).$$

The convergence result for the argmax follows since the function $(\theta, \beta) \mapsto K(\theta, \beta)$ is necessarily strictly concave at (θ^*, β^*) because $K(\theta, \beta)$ is maximal uniquely at (θ^*, β^*).

Let us finish this section with a discussion on the extra assumption (5.36) which claims that the empirical mean energy converges to the expected value energy. This assumption is in general proved via the ergodic theorem or a law of large numbers. In the case of the Area energy function H defined in (5.4), it is a direct consequence of a decomposition as in (5.28) and the ergodic Theorem. In the case of pairwise interaction, the verification follows essentially the proof of Proposition 5.12.

5.4.2   Takacs-Fiksel Estimator

In this section we present an estimator introduced in the eighties by Takacs and Fiksel [19, 49]. It is based on the GNZ equations presented in Sect. 5.3.5. Let us start by explaining briefly the procedure. Let f be a test function from $\mathbb{R}^d \times \mathscr{C}$ to $\mathbb{R}$. We define the following quantity for any $z > 0$, $\beta > 0$ and $\gamma \in \mathscr{C}$

$$C_{\Lambda_n}^{z,\beta}(f,\gamma) = \sum_{x \in \gamma_{\Lambda_n}} f(x, \gamma \setminus \{x\}) - z \int_{\Lambda_n} e^{-\beta h(x,\gamma)} f(x,\gamma) dx. \tag{5.37}$$

By the GNZ equation (5.22) we obtain

$$E\left(C_{\Lambda_n}^{z^*,\beta^*}(f,\Gamma)\right) = 0$$

where Γ is a GPP with parameter z^* and β^*. Thanks to the ergodic Theorem it follows that for n large enough

$$\frac{C_{\Lambda_n}^{z^*,\beta^*}(f,\Gamma)}{\lambda^d(\Lambda_n)} \approx 0.$$

Then the Takacs-Fiksel estimator is defined as a mean-square method based on functions $C_{\Lambda_n}^{z^*,\beta^*}(f_k, \Gamma)$ for a collection of test functions $(f_k)_{1 \le k \le K}$.

Definition 5.10 Let $K \ge 2$ be an integer and $(f_k)_{1 \le k \le K}$ a family of K functions from $\mathbb{R}^d \times \mathscr{C}$ to $\mathbb{R}$. The Takacs-Fiksel estimator $(\hat{z}_n, \hat{\beta}_n)$ of (z^*, β^*) is defined by

$$(\hat{z}_n, \hat{\beta}_n) = \operatorname{argmin}_{(z,\beta) \in \mathscr{D}} \sum_{k=1}^{K} \left(C_{\Lambda_n}^{z,\beta}(f_k, \Gamma)\right)^2,$$

where $\mathscr{D} \subset (0, +\infty) \times [0, +\infty)$ is a bounded domain containing (z^*, β^*).

In opposition to the MLE procedure, the contrast function does not depend on the partition function. This estimator is explicit except for the computation of integrals and the optimization procedure. In [9] the Takacs-Fiksel procedure is presented in a more general setting including the case where the functions f_k depend on parameters z and β. This generalization may lead to a simpler procedure in choosing f_k such that the integral term in (5.37) is explicitly computable.

In the rest of the section, we prove the consistency of the estimator. General results on consistency and asymptotic normality are developed in [9].

Theorem 5.7 (Consistency) *We make the following integrability assumption: for any $1 \le k \le K$*

$$E\left(|f_k(0,\Gamma)|(1 + |h(0,\Gamma)|) \sup_{(z,\beta) \in \mathscr{D}} e^{-\beta h(0,\Gamma)}\right) < +\infty. \tag{5.38}$$

We assume also the following identifiability condition: the equality

$$\sum_{k=1}^{K} E\left(f_k(0, \Gamma)\left(ze^{-\beta h(0,\Gamma)} - z^*e^{-\beta^* h(0,\Gamma)}\right)\right)^2 = 0 \tag{5.39}$$

holds if and only $(z, \beta) = (z^*, \beta^*)$. *Then the Takacs-Fiksel estimator* $(\hat{z}_n, \hat{\beta}_n)$ *presented in Definition 5.10 converges almost surely to* (z^*, β^*).

Proof As in the proof of Theorem 5.6, without loss of generality, we assume that the Gibbs distribution of Γ is ergodic. Therefore, thanks to the ergodic Theorem, almost surely for any $1 \le k \le K$

$$\lim_{n \mapsto \infty} \frac{C_{\Lambda_n}^{z,\beta}(f_k, \Gamma)}{\lambda^d(\Lambda_n)} = E\left[\sum_{x \in \Gamma_{[0,1]^d}} f_k(x, \Gamma \setminus x) \right]$$

$$- zE\left[\int_{[0,1]^d} e^{-\beta h(x,\Gamma)} f_k(x, \Gamma)dx \right]. \tag{5.40}$$

By the GNZ equation (5.22)

$$E\left[\sum_{x \in \Gamma_{[0,1]^d}} f_k(x, \Gamma \setminus x) \right] = z^* E\left[\int_{[0,1]^d} e^{-\beta^* h(x,\Gamma)} f_k(x, \Gamma)dx \right]. \tag{5.41}$$

Using the stationarity and compiling (5.40) and (5.41), we obtain that the contrast function

$$K_n(z, \beta) = \sum_{k=1}^{K} \left(\frac{C_{\Lambda_n}^{z,\beta}(f_k, \Gamma)}{\lambda^d(\Lambda_n)} \right)^2$$

admits almost surely the limit

$$\lim_{n \mapsto \infty} K_n(z, \beta) = K(z, \beta) := \sum_{k=1}^{K} E\left(f_k(0, \Gamma)\left(ze^{-\beta h(0,\Gamma)} - z^*e^{-\beta^* h(0,\Gamma)}\right)\right)^2,$$

which is null if and only if $(z, \beta) = (z^*, \beta^*)$. Therefore it remains to prove that the minimizers of the contrast function converge to the minimizer of the limit contrast function. In the previous section we solved a similar issue for the MLE procedure using the convexity of contrast functions. This argument does not work here and we need more sophisticated tools.

We define by $W_n(.)$ the modulus of continuity of the contrast function K_n; let η be a positive real

$$W_n(\eta) = \sup \left\{ |K_n(z, \beta) - K_n(z', \beta')|, \right.$$

$$\left. \text{with } (z, \beta), (z', \beta') \in \mathscr{D}, \ \|(z - z', \beta - \beta')\| \le \eta \right\}.$$

Lemma 5.6 (Theorem 3.4.3 [27]) *Assuming that there exists a sequence* $(\epsilon_l)_{l \ge 1}$, *which goes to zero when l goes to infinity, such that for any* $l \ge 1$

$$P \left(\limsup_{n \mapsto +\infty} \left\{ W_n \left(\frac{1}{l} \right) \ge \epsilon_l \right\} \right) = 0 \tag{5.42}$$

then almost surely the minimizers of $(z, \beta) \mapsto K_n(z, \beta)$ *converges to the minimizer of* $(z, \beta) \mapsto K(z, \beta)$.

Let us show that the assertion (5.42) holds. Thanks to equalities (5.40), (5.41) and assumption (5.38), there exists a constant C_1 such that for n large enough, any $1 \le k \le K$ and any $(z, \beta) \in \mathscr{D}$

$$\frac{|C_{\Lambda_n}^{z,\beta}(f_k, \Gamma)|}{\lambda^d(\Lambda_n)} \le C_1. \tag{5.43}$$

We deduce that for n large enough

$$|K_n(z, \beta) - K_n(z', \beta')|$$

$$\le \frac{C_1}{\lambda^d(\Lambda_n)} \sum_{k=1}^{K} \int_{\Lambda_n} |f_k(x, \Gamma)| \left| z e^{-\beta h(x, \Gamma)} - z' e^{-\beta' h(x, \Gamma)} \right| dx$$

$$\le \frac{C_1 |\beta - \beta'|}{\lambda^d(\Lambda_n)} \max_{1 \le k \le K} \int_{\Lambda_n} |f_k(x, \Gamma) h(x, \Gamma)| \sup_{(z, \beta'') \in \mathscr{D}} z e^{-\beta'' h(x, \Gamma)} dx$$

$$+ \frac{C_1 |z - z'|}{\lambda^d(\Lambda_n)} \max_{1 \le k \le K} \int_{\Lambda_n} |f_k(x, \Gamma)| \sup_{(z, \beta'') \in \mathscr{D}} e^{-\beta'' h(x, \Gamma)} dx.$$

By the ergodic Theorem, the following convergences hold almost surely

$$\lim_{n \mapsto +\infty} \frac{1}{\lambda^d(\Lambda_n)} \int_{\Lambda_n} |f_k(x, \Gamma) h(x, \Gamma)| \sup_{(z, \beta'') \in \mathscr{D}} z e^{-\beta'' h(x, \Gamma)} dx$$

$$= E \left(|f_k(0, \Gamma) h(0, \Gamma)| \sup_{(z, \beta'') \in \mathscr{D}} z e^{-\beta'' h(0, \Gamma)} \right) < +\infty,$$

and

$$\lim_{n \mapsto +\infty} \frac{1}{\lambda^d(\Lambda_n)} \int_{\Lambda_n} |f_k(x, \Gamma)| \sup_{(z, \beta'') \in \mathcal{D}} e^{-\beta'' h(x, \Gamma)} dx$$

$$= E\left(|f_k(0, \Gamma)| \sup_{(z, \beta'') \in \mathcal{D}} e^{-\beta'' h(0, \Gamma)}\right) < +\infty.$$

This implies the existence of a constant $C_2 > 0$ such that for n large enough, any $1 \leq k \leq K$ and any $(z, \beta) \in \mathcal{D}$

$$|K_n(z, \beta) - K_n(z', \beta')| < C_2 \|(z - z', \beta - \beta')\|.$$

The assumption (5.42) occurs with the sequence $\epsilon_l = C_2/l$ and Theorem 5.7 is proved.

Remark 5.1 (On the Integrability Assumption) The integrability assumption (5.38) is sometimes difficult to check, especially when the local energy $h(0, \gamma)$ is not bounded from below. For instance in the setting of pairwise energy function H defined in (5.1) with a pair potential φ having negative values, Ruelle estimates (5.16) are very useful. Indeed, by stability of the energy function, the potential φ is necessary bounded from below by $2A$ and therefore

$$E\left(e^{-\beta h(0, \Gamma)}\right) < E\left(e^{-2A\beta N_{B(0,R)}(\Gamma)}\right) < +\infty,$$

where R is the range of the interaction.

Remark 5.2 (On the Identifiability Assumption) In the identifiability assumption (5.39), the sum is null if and only if each term is null. Assuming that the functions are regular enough, each term is null as soon as (z, β) belongs to a 1-dimensional manifold embedded in $\mathbb{R}^2$ containing (z^*, β^*). Therefore, assumption (5.39) claims that (z^*, β^*) is the unique element of these K manifolds. If $K \leq 2$, there is no special geometric argument to ensure that K 1-dimensional manifolds in $\mathbb{R}^2$ have an unique intersection point. For this reason, it is recommended to choose $K \geq 3$. See Section 5 in [9] for more details and complements on this identifiability assumption.

5.4.3 Maximum Pseudo-Likelihood Estimator

In this section we present the maximum pseudo-likelihood estimator, which is a particular case of the Takacs-Fiksel estimator. This procedure has been first introduced by Besag in [5] and popularized by Jensen and Moller in [31] and Baddeley and Turner in [2].

Definition 5.11 The maximum pseudo-likelihood estimator $(\hat{z}_n, \hat{\beta}_n)$ is defined as a Takacs-Fiksel estimator (see Definition 5.10) with $K = 2$, $f_1(x, \gamma) = 1$ and $f_2(x, \gamma) = h(x, \gamma)$.

This particular choice of functions f_1, f_2 simplifies the identifiability assumption (5.39). The following theorem is an adaptation of Theorem 5.7 in the present setting of MPLE. The asymptotic normality is investigated first in [30] (see also [6] for more general results).

Theorem 5.8 (Consistency) *Assuming*

$$E\left((1 + h(0, \Gamma)^2) \sup_{(z,\beta)\in\mathscr{D}} e^{-\beta h(0,\Gamma)}\right) < +\infty \tag{5.44}$$

and

$$P\left(h(0, \Gamma) = h(0, \emptyset)\right) < 1, \tag{5.45}$$

then the maximum pseudo-likelihood estimator $(\hat{z}_n, \hat{\beta}_n)$ converges almost surely to (z^*, β^*).

Proof Let us check the assumptions of Theorem 5.7. Clearly, the integrability assumption (5.44) ensures the integrability assumptions (5.38) with $f_1 = 1$ and $f_2 = h$. So it remains to show that assumption (5.45) implies the identifiability assumption (5.39). Consider the parametrization $z = e^{-\theta}$ and ψ the function

$$\psi(\theta, \beta) = E\left(e^{-\theta^* - \beta^* h(0,\Gamma)}(e^U - U - 1)\right),$$

with

$$U = \beta^* h(0, \Gamma) + \theta^* - \beta h(0, \Gamma) - \theta.$$

The function ψ is convex, non negative and equal to zero if and only if U is almost surely equal to zero. By assumption (5.45) this fact occurs when $(z, \beta) = (z^*, \beta^*)$. Therefore the gradient $\nabla\psi = 0$ if and only if $(z, \beta) = (z^*, \beta^*)$. Noting that

$$\frac{\partial\psi(\theta, \beta)}{\partial\theta} = E\left(z^* e^{-\beta^* h(0,\Gamma)} - ze^{-\beta h(0,\Gamma)}\right)$$

and

$$\frac{\partial\psi(\theta, \beta)}{\partial\beta} = E\left(h(0, \Gamma)\left(z^* e^{-\beta^* h(0,\Gamma)} - ze^{-\beta h(0,\Gamma)}\right)\right),$$

the identification assumption (5.39) holds. The theorem is proved.

5.4.4   Solving an Unobservable Issue

In this section we give an application of the Takacs-Fiksel procedure in a setting of partially observable dataset. Let us consider a Gibbs point process Γ for which we observe only $L_R(\Gamma)$ in place of Γ. This setting appears when Gibbs point processes are used for producing random surfaces via germ-grain structures (see [41] for instance). Applications for modelling micro-structure in materials or micro-emulsion in statistical physics are developed in [8].

The goal is to furnish an estimator of z^* and β^* in spite of this unobservable issue. Note that the number of points (or balls) is not observable from $L_R(\Gamma)$ and therefore the MLE procedure is not achievable, since the likelihood is not computable. When β is known and fixed to zero, it corresponds to the estimation of the intensity of the Boolean model from its germ-grain structure (see [39] for instance).

In the following we assume that Γ a Gibbs point process for the Area energy function defined in (5.4), the activity z^* and the inverse temperature β^*. This choice is natural since the energy function depends on the observations $L_R(\Gamma)$. The more general setting of Quermass interaction is presented in [17] but for sake of simplicity, we treat only here the simpler case of Area interaction.

We opt for a Takacs-Fiksel estimator but the main problem is that the function

$$C_{\Lambda_n}^{z,\beta}(f, \gamma) = \sum_{x \in \gamma_{\Lambda_n}} f(x, \gamma \backslash \{x\}) - z \int_{\Lambda_n} e^{-\beta h(x,\gamma)} f(x, \gamma) dx,$$

which appears in the procedure, is not computable since the positions of points are not observable. The main idea is to choose the function f properly such that the sum is observable although each term of the sum is not. To this end, we define

$$f_1(x, \gamma) = \text{Surface}\Big(\partial B(x, R) \cap L_R^c(\gamma)\Big)$$

and

$$f_2(x, \gamma) = \mathbf{1}_{\{B(x,R) \cap L_R(\gamma) = \emptyset\}},$$

where $\partial B(x, R)$ is the boundary of the ball $B(x, R)$ (i.e. the sphere $S(x, R)$) and the "Surface" means the $(d-1)$-dimensional Hausdorff measure in $\mathbb{R}^d$. Clearly the function f_1 gives the surface of the portion of the sphere $S(x, R)$ outside the germ-grain structure $L_R(\gamma)$. The function f_2 indicates if the ball $B(x, R)$ hits the germ-grain structure $L_R(\gamma)$. Therefore we obtain that

$$\sum_{x \in \gamma_{\Lambda_n}} f_1(x, \gamma \backslash \{x\}) = \text{Surface}\Big(\partial L_R(\gamma_{\Lambda_n})\Big)$$

and

$$\sum_{x \in \gamma_{\Lambda_n}} f_2(x, \gamma \setminus \{x\}) = N_{\text{iso}}\Big(L_R(\gamma_{\Lambda_n})\Big),$$

where $N_{\text{iso}}(L_R(\gamma_{\Lambda_n}))$ is the number of isolated balls in the germ-grain structure $L_R(\gamma_{\Lambda_n})$. Let us note that these quantities are not exactly observable since, in practice, we observe $L_R(\gamma) \cap \Lambda_n$ rather than $L_R(\gamma_{\Lambda_n})$. However, if we omit this boundary effect, the values $C_{\Lambda_n}^{z,\beta}(f_1, \Gamma)$ and $C_{\Lambda_n}^{z,\beta}(f_2, \Gamma)$ are observable and the Takacs-Fiksel procedure is achievable. The consistency of the estimator is guaranteed by Theorem 5.7. The integrability assumption (5.38) is trivially satisfied since the functions f_1, f_2 and h are uniformly bounded. The verification of the identifiability assumption (5.39) is more delicate and we refer to [9], example 2 for a proof. Numerical estimations on simulated and real datasets can be found in [17].

5.4.5 A Variational Estimator

In this last section, we present a new estimator based on a variational GNZ equation which is a mix between the standard GNZ equation and an integration by parts formula. This equation has been first introduced in [11] for statistical mechanics issues and used recently in [1] for spatial statistic considerations. In the following, we present first this variational equation and afterwards we introduce its associated estimator of β^*. The estimation of z^* is not considered here.

Theorem 5.9 *Let Γ be a GPP for the energy function H, the activity z and the inverse temperature β. We assume that, for any $\gamma \in \mathscr{C}$, the function $x \mapsto h(x, \gamma)$ is differentiable on $\mathbb{R}^d \setminus \gamma$. Let f be a function from $\mathbb{R}^d \times \mathscr{C}$ to $\mathbb{R}$ which is differentiable and with compact support with respect to the first variable. Moreover we assume the integrability of both terms below. Then*

$$E\left(\sum_{x \in \Gamma} \nabla_x f(x, \Gamma \setminus \{x\})\right) = \beta E\left(\sum_{x \in \Gamma} f(x, \Gamma \setminus \{x\}) \nabla_x h(x, \Gamma \setminus \{x\})\right). \quad (5.46)$$

Proof By the standard GNZ equation (5.22) applied to the function $\nabla_x f$, we obtain

$$E\left(\sum_{x \in \Gamma} \nabla_x f(x, \Gamma \setminus \{x\})\right) = zE\left(\int_{\mathbb{R}^d} e^{-\beta h(x, \Gamma)} \nabla_x f(x, \Gamma) dx\right).$$

By a standard integration by part formula with respect to the first variable x, we find that

$$E\left(\sum_{x \in \Gamma} \nabla_x f(x, \Gamma \setminus \{x\})\right) = z\beta E\left(\int_{\mathbb{R}^d} \nabla_x h(x, \Gamma) e^{-\beta h(x, \Gamma)} f(x, \Gamma) dx\right).$$

Using again the GNZ equation we finally obtain (5.46).

Note that Eq. (5.46) is a vectorial equation. For convenience it is possible to obtain a real equation by summing each coordinate of the vectorial equation. The gradient operator is simply replaced by the divergence operator.

Remark 5.3 (On the Activity Parameter z) The parameter z does not appear in the variational GNZ equation (5.46). Therefore these equations do not characterize the Gibbs measures as in Proposition 5.5. Actually these variational GNZ equations characterize the mixing of Gibbs measures with random activity (See [11] for details).

Let us now explain how to estimate β^* from these variational equations. When the observation window Λ_n is large enough we identify the expectations of sums in (5.46) by the sums. Then the estimator of β^* is simply defined by

$$\hat{\beta}_n = \frac{\sum_{x \in \Gamma_{\Lambda_n}} \mathrm{div}_x f(x, \Gamma\backslash\{x\})}{\sum_{x \in \Gamma_{\Lambda_n}} f(x, \Gamma\backslash\{x\})\mathrm{div}_x h(x, \Gamma\backslash\{x\})}. \tag{5.47}$$

Note that this estimator is very simple and quick to compute in comparison to the MLE, MPLE or the general Takacs-Fiksel estimators. Indeed, in (5.47), there are only elementary operations (no optimization procedure, no integral to compute).

Let us now finish this section with a consistency result. More general results for consistency, asymptotic normality and practical estimations are available in [1].

Theorem 5.10 *Let Γ be a GPP for a stationary and finite range energy function H, activity z^* and inverse temperature β^*. We assume that, for any $\gamma \in \mathscr{C}$, the function $x \mapsto h(x, \gamma)$ is differentiable on $\mathbb{R}^d\backslash\gamma$. Let f be a stationary function from $\mathbb{R}^d \times \mathscr{C}$ to $\mathbb{R}$, differentiable with respect to the first variable and such that*

$$E\left((|f(0, \Gamma)| + |\nabla_x f(0, \Gamma)| + |f(0, \Gamma)\nabla_x h(0, \Gamma)|)e^{-\beta^* h(0,\Gamma)}\right) < +\infty \tag{5.48}$$

and

$$E\left(f(0, \Gamma)div_x h(0, \Gamma)e^{-\beta^* h(0,\Gamma)}\right) \neq 0. \tag{5.49}$$

Then the estimator $\hat{\beta}_n$ converges almost surely to β^*.

Proof As usual, without loss of generality, we assume that the Gibbs distribution of Γ is ergodic. Then by the ergodic theorem the following limits both hold almost surely

$$\lim_{n \mapsto +\infty} \frac{1}{\lambda^d(\Lambda_n)} \sum_{x \in \Gamma_{\Lambda_n}} \mathrm{div}_x f(x, \Gamma\backslash\{x\}) = E\left(\sum_{x \in \Gamma_{[0,1]^d}} \mathrm{div}_x f(x, \Gamma\backslash\{x\})\right) \tag{5.50}$$

and

$$\lim_{n \mapsto +\infty} \frac{1}{\lambda^d(\Lambda_n)} \sum_{x \in \Gamma_{\Lambda_n}} f(x, \Gamma \backslash \{x\}) \mathrm{div}_x h(x, \Gamma \backslash \{x\})$$

$$= E \left(\sum_{x \in \Gamma_{[0,1]^d}} f(x, \Gamma \backslash \{x\}) \mathrm{div}_x h(x, \Gamma \backslash \{x\}) \right). \tag{5.51}$$

Note that both expectations in (5.50) and (5.51) are finite since by the GNZ equations, the stationarity and assumption (5.48)

$$E \left(\sum_{x \in \Gamma_{[0,1]^d}} |\mathrm{div} f(x, \Gamma \backslash \{x\})| \right) = E \left(|\mathrm{div} f(0, \Gamma)| e^{-\beta^* h(0, \Gamma)} \right) < +\infty$$

and

$$E \left(\sum_{x \in \Gamma_{[0,1]^d}} |f(x, \Gamma \backslash \{x\}) \mathrm{div} h(x, \Gamma \backslash \{x\})| \right) = E \left(|f(0, \Gamma)| \mathrm{div} h(0, \Gamma)| e^{-\beta^* h(0, \Gamma)} \right)$$

$$< +\infty.$$

We deduce that almost surely

$$\lim_{n \mapsto +\infty} \hat{\beta}_n = \frac{E \left(\sum_{x \in \Gamma_{[0,1]^d}} \mathrm{div} f(x, \Gamma \backslash \{x\}) \right)}{E \left(\sum_{x \in \Gamma_{[0,1]^d}} f(x, \Gamma \backslash \{x\}) \mathrm{div} h(x, \Gamma \backslash \{x\}) \right)},$$

where the denominator is not null thanks to assumption (5.49). Therefore it remains to prove the following variational GNZ equation

$$E \left(\sum_{x \in \Gamma_{[0,1]^d}} \nabla_x f(x, \Gamma \backslash \{x\}) \right) = \beta^* E \left(\sum_{x \in \Gamma_{[0,1]^d}} f(x, \Gamma \backslash \{x\}) \nabla_x h(x, \Gamma \backslash \{x\}) \right). \tag{5.52}$$

Note that this equation is not a direct consequence of the variational GNZ equation (5.46) since the function $x \mapsto f(x, \gamma)$ does not have a compact support. We need the following cut-off approximation. Let us consider $(\psi_n)_{n \geq 1}$ any sequence of functions from $\mathbb{R}^d$ to $\mathbb{R}$ such that ψ_n is differentiable, equal to 1 on Λ_n, 0 on Λ_{n+1}^c and such that $|\nabla \psi_n|$ and $|\psi_n|$ are uniformly bounded by a constant C (which does not depend on n). It is not difficult to build such a sequence of functions. Let us now apply the variational GNZ equation (5.46) to the function $(x, \gamma) \mapsto \psi_n(x) f(x, \gamma)$, we obtain

$$E\left(\sum_{x\in\Gamma}\psi_n(x)\nabla_x f(x,\Gamma\backslash\{x\})\right) + E\left(\sum_{x\in\Gamma}\nabla_x\psi_n(x)f(x,\Gamma\backslash\{x\})\right)$$

$$= \beta^* E\left(\sum_{x\in\Gamma}\psi_n(x)f(x,\Gamma\backslash\{x\})\nabla_x h(x,\Gamma\backslash\{x\})\right). \tag{5.53}$$

Thanks to the GNZ equation and the stationarity we get

$$\left| E\left(\sum_{x\in\Gamma}\psi_n(x)\nabla_x f(x,\Gamma\backslash\{x\})\right) - \lambda^d(\Lambda_n)E\left(\sum_{x\in\Gamma_{[0,1]^d}}\nabla_x f(x,\Gamma\backslash\{x\})\right)\right|$$

$$\leq C z^* \lambda^d(\Lambda_{n+1}\backslash\Lambda_n)E\left(|\nabla_x f(0,\Gamma)|e^{-\beta^* h(0,\Gamma)}\right),$$

and

$$\left| E\left(\sum_{x\in\Gamma}\psi_n(x)f(x,\Gamma\backslash\{x\})\nabla_x h(x,\Gamma\backslash\{x\})\right)\right.$$

$$\left. -\lambda^d(\Lambda_n)E\left(\sum_{x\in\Gamma_{[0,1]^d}}f(x,\Gamma\backslash\{x\})\nabla_x h(x,\Gamma\backslash\{x\})\right)\right|$$

$$\leq C z^* \lambda^d(\Lambda_{n+1}\backslash\Lambda_n)E\left(|f(0,\Gamma)\nabla_x h(0,\Gamma)|e^{-\beta^* h(0,\Gamma)}\right),$$

and finally

$$\left| E\left(\sum_{x\in\Gamma}\nabla_x\psi_n(x)f(x,\Gamma\backslash\{x\})\right)\right| \leq C z^* \lambda^d(\Lambda_{n+1}\backslash\Lambda_n)E\left(|f(0,\Gamma)|e^{-\beta^* h(0,\Gamma)}\right).$$

Therefore, dividing Eq. (5.53) by $\lambda^d(\Lambda_n)$, using the previous approximations and letting n go to infinity, we find exactly the variational equation (5.52). The theorem is proved.

Acknowledgements The author thanks P. Houdebert, A. Zass and the anonymous referees for the careful reading and the interesting comments. This work was supported in part by the Labex CEMPI (ANR-11-LABX-0007-01), the CNRS GdR 3477 GeoSto and the ANR project PPP (ANR-16-CE40-0016).

References

1. A. Baddeley, D. Dereudre, Variational estimators for the parameters of Gibbs point process models. Bernoulli **19**(3), 905–930 (2013)
2. A. Baddeley, R. Turner, Practical maximum pseudolikelihood for spatial point patterns (with discussion). Aust. N. Z. J. Stat. **42**(3), 283–322 (2000)
3. A.J. Baddeley, M.N.M. van Lieshout, Area-interaction point processes. Ann. Inst. Stat. Math. **47**(4), 601–619 (1995)
4. A. Baddeley, P. Gregori, J. Mateu, R. Stoica, D. Stoyan, *Case Studies in Spatial Point Process Models*. Lecture Notes in Statistics, vol. 185 (Springer, New York, 2005)
5. J. Besag, Spatial interaction and the statistical analysis of lattice systems. J. R. Stat. Soc. Ser. B **36**, 192–236 (1974). With discussion by D. R. Cox, A. G. Hawkes, P. Clifford, P. Whittle, K. Ord, R. Mead, J. M. Hammersley, and M. S. Bartlett and with a reply by the author
6. J.-M. Billiot, J.-F. Coeurjolly, R. Drouilhet, Maximum pseudolikelihood estimator for exponential family models of marked Gibbs point processes. Electron. J. Stat. **2**, 234–264 (2008)
7. J.T. Chayes, L. Chayes, R. Kotecký, The analysis of the Widom-Rowlinson model by stochastic geometric methods. Commun. Math. Phys. **172**(3), 551–569 (1995)
8. S.N. Chiu, D. Stoyan, W.S. Kendall, J. Mecke, *Stochastic Geometry and Its Applications*, 3rd edn. (Wiley, Chichester, 2013)
9. J.-F. Coeurjolly, D. Dereudre, R. Drouilhet, F. Lavancier, Takacs-Fiksel method for stationary marked Gibbs point processes. Scand. J. Stat. **39**(3), 416–443 (2012)
10. D.J. Daley, D. Vere-Jones, *An Introduction to the Theory of Point Processes. Vol. I. Elementary Theory and Methods*. Probability and Its Applications (New York), 2nd edn. (Springer, New York, 2003).
11. D. Dereudre, *Diffusion infini-dimensionnelles et champs de Gibbs sur l'espace des trajectoires continues*. PhD, Ecole polytechnique Palaiseau (2002)
12. D. Dereudre, The existence of quermass-interaction processes for nonlocally stable interaction and nonbounded convex grains. Adv. Appl. Probab. **41**(3), 664–681 (2009)
13. D. Dereudre, Variational principle for Gibbs point processes with finite range interaction. Electron. Commun. Probab. **21**, Paper No. 10, 11 (2016)
14. D. Dereudre, P. Houdebert, Infinite volume continuum random cluster model. Electron. J. Probab. **20**(125), 24 (2015)
15. D. Dereudre, F. Lavancier, Consistency of likelihood estimation for Gibbs point processes. Ann. Stat. **45**(2), 744–770 (2017)
16. D. Dereudre, R. Drouilhet, H.-O. Georgii, Existence of Gibbsian point processes with geometry-dependent interactions. Probab. Theory Relat. Fields **153**(3–4), 643–670 (2012)
17. D. Dereudre, F. Lavancier, K. Staňková Helisová, Estimation of the intensity parameter of the germ-grain quermass-interaction model when the number of germs is not observed. Scand. J. Stat. **41**(3), 809–829 (2014)
18. R.L. Dobrushin, E.A. Pecherski, A criterion of the uniqueness of Gibbsian fields in the noncompact case, in *Probability Theory and Mathematical Statistics (Tbilisi, 1982)*. Lecture Notes in Mathematics, vol. 1021 (Springer, Berlin, 1983), pp. 97–110
19. T. Fiksel, Estimation of parametrized pair potentials of marked and nonmarked Gibbsian point processes. Elektron. Informationsverarb. Kybernet. **20**(5–6), 270–278 (1984)
20. H.-O. Georgii, *Canonical Gibbs Measures*. Lecture Notes in Mathematics, vol. 760 (Springer, Berlin, 1979). Some extensions of de Finetti's representation theorem for interacting particle systems
21. H.-O. Georgii, Large deviations and the equivalence of ensembles for Gibbsian particle systems with superstable interaction. Probab. Theory Relat. Fields **99**(2), 171–195 (1994)
22. H.-O. Georgii, *Gibbs Measures and Phase Transitions*. de Gruyter Studies in Mathematics, vol. 9, 2nd edn. (Walter de Gruyter, Berlin, 2011)
23. H.-O. Georgii, T. Küneth, Stochastic comparison of point random fields. J. Appl. Probab. **34**(4), 868–881 (1997)

24. H.-O. Georgii, H.J. Yoo, Conditional intensity and Gibbsianness of determinantal point processes. J. Stat. Phys. **118**(1–2), 55–84 (2005)
25. H.-O. Georgii, H. Zessin, Large deviations and the maximum entropy principle for marked point random fields. Probab. Theory Relat. Fields **96**(2), 177–204 (1993)
26. C.J. Geyer, J. Møller, Simulation procedures and likelihood inference for spatial point processes. Scand. J. Stat. **21**(4), 359–373 (1994)
27. X. Guyon, *Random Fields on a Network*. Probability and Its Applications (New York) (Springer, New York, 1995). Modeling, statistics, and applications, Translated from the 1992 French original by Carenne Ludeña
28. P. Hall, On continuum percolation. Ann. Probab. **13**(4), 1250–1266 (1985)
29. J.L. Jensen, Asymptotic normality of estimates in spatial point processes. Scand. J. Stat. **20**(2), 97–109 (1993)
30. J.L. Jensen, H.R. Künsch, On asymptotic normality of pseudo likelihood estimates for pairwise interaction processes. Ann. Inst. Stat. Math. **46**(3), 475–486 (1994)
31. J.L. Jensen, J. Møller, Pseudolikelihood for exponential family models of spatial point processes. Ann. Appl. Probab. **1**(3), 445–461 (1991)
32. W.S. Kendall, J. Møller, Perfect simulation using dominating processes on ordered spaces, with application to locally stable point processes. Adv. Appl. Probab. **32**(3), 844–865 (2000)
33. O.K. Kozlov, Description of a point random field by means of the Gibbs potential. Uspehi Mat. Nauk **30**(6(186)), 175–176 (1975)
34. J.L. Lebowitz, A. Mazel, E. Presutti, Liquid-vapor phase transitions for systems with finite-range interactions. J. Stat. Phys. **94**(5–6), 955–1025 (1999)
35. S. Mase, Uniform LAN condition of planar Gibbsian point processes and optimality of maximum likelihood estimators of soft-core potential functions. Probab. Theory Relat. Fields **92**(1), 51–67 (1992)
36. K. Matthes, J. Kerstan, J. Mecke, *Infinitely Divisible Point Processes* (Wiley, Chichester, 1978). Translated from the German by B. Simon, Wiley Series in Probability and Mathematical Statistics
37. J. Mayer, E. Montroll, Molecular distributions. J. Chem. Phys. **9**, 2–16 (1941)
38. R. Meester, R. Roy, *Continuum Percolation*. Cambridge Tracts in Mathematics, vol. 119 (Cambridge University Press, Cambridge, 1996)
39. I.S. Molchanov, Consistent estimation of the parameters of Boolean models of random closed sets. Teor. Veroyatnost. i Primenen. **36**(3), 580–587 (1991)
40. J. Møller, *Lectures on random Voronoĭ tessellations*. Lecture Notes in Statistics, vol. 87 (Springer, New York, 1994)
41. J. Møller, K. Helisová, Likelihood inference for unions of interacting discs. Scand. J. Stat. **37**(3), 365–381 (2010)
42. J. Møller, R.P. Waagepetersen, *Statistical Inference and Simulation for Spatial Point Processes*. Monographs on Statistics and Applied Probability, vol. 100 (Chapman & Hall/CRC, Boca Raton, 2004)
43. X. Nguyen, H. Zessin, Integral and differential characterizations Gibbs processes. Mathematische Nachrichten, **88**(1), 105–115 (1979)
44. Y. Ogata, M. Tanemura, Likelihood analysis of spatial point patterns. J. R. Stat. Soc. Ser. B **46**(3), 496–518 (1984)
45. S. Poghosyan, D. Ueltschi, Abstract cluster expansion with applications to statistical mechanical systems. J. Math. Phys. **50**(5), 053509, 17 (2009)
46. C. Preston, *Random fields*. Lecture Notes in Mathematics, vol. 534 (Springer, Berlin, 1976)
47. D. Ruelle, *Statistical Mechanics: Rigorous Results* (W. A. Benjamin, Inc., New York, 1969)
48. D. Ruelle, Superstable interactions in classical statistical mechanics. Commun. Math. Phys. **18**, 127–159 (1970)
49. R. Takacs, Estimator for the pair-potential of a Gibbsian point process. Statistics, **17**(3), 429–433 (1986)
50. J. van den Berg, C. Maes, Disagreement percolation in the study of Markov fields. Ann. Probab. **22**(2), 749–763 (1994)

51. M.N.M. van Lieshout, *Markov Point Processes and Their Applications* (Imperial College Press, London, 2000)
52. B. Widom, J.S. Rowlinson, New model for the study of liquid-vapor phase transitions. J. Chem. Phys. **52**, 1670–1684 (1970)

LECTURE NOTES IN MATHEMATICS

 Springer

Editors in Chief: J.-M. Morel, B. Teissier;

Editorial Policy

1. Lecture Notes aim to report new developments in all areas of mathematics and their applications – quickly, informally and at a high level. Mathematical texts analysing new developments in modelling and numerical simulation are welcome.

   Manuscripts should be reasonably self-contained and rounded off. Thus they may, and often will, present not only results of the author but also related work by other people. They may be based on specialised lecture courses. Furthermore, the manuscripts should provide sufficient motivation, examples and applications. This clearly distinguishes Lecture Notes from journal articles or technical reports which normally are very concise. Articles intended for a journal but too long to be accepted by most journals, usually do not have this "lecture notes" character. For similar reasons it is unusual for doctoral theses to be accepted for the Lecture Notes series, though habilitation theses may be appropriate.

2. Besides monographs, multi-author manuscripts resulting from SUMMER SCHOOLS or similar INTENSIVE COURSES are welcome, provided their objective was held to present an active mathematical topic to an audience at the beginning or intermediate graduate level (a list of participants should be provided).

   The resulting manuscript should not be just a collection of course notes, but should require advance planning and coordination among the main lecturers. The subject matter should dictate the structure of the book. This structure should be motivated and explained in a scientific introduction, and the notation, references, index and formulation of results should be, if possible, unified by the editors. Each contribution should have an abstract and an introduction referring to the other contributions. In other words, more preparatory work must go into a multi-authored volume than simply assembling a disparate collection of papers, communicated at the event.

3. Manuscripts should be submitted either online at www.editorialmanager.com/lnm to Springer's mathematics editorial in Heidelberg, or electronically to one of the series editors. Authors should be aware that incomplete or insufficiently close-to-final manuscripts almost always result in longer refereeing times and nevertheless unclear referees' recommendations, making further refereeing of a final draft necessary. The strict minimum amount of material that will be considered should include a detailed outline describing the planned contents of each chapter, a bibliography and several sample chapters. Parallel submission of a manuscript to another publisher while under consideration for LNM is not acceptable and can lead to rejection.

4. In general, **monographs** will be sent out to at least 2 external referees for evaluation.

   A final decision to publish can be made only on the basis of the complete manuscript, however a refereeing process leading to a preliminary decision can be based on a pre-final or incomplete manuscript.

   Volume Editors of **multi-author works** are expected to arrange for the refereeing, to the usual scientific standards, of the individual contributions. If the resulting reports can be forwarded to the LNM Editorial Board, this is very helpful. If no reports are forwarded or if other questions remain unclear in respect of homogeneity etc, the series editors may wish to consult external referees for an overall evaluation of the volume.

5. Manuscripts should in general be submitted in English. Final manuscripts should contain at least 100 pages of mathematical text and should always include

   – a table of contents;
   – an informative introduction, with adequate motivation and perhaps some historical remarks: it should be accessible to a reader not intimately familiar with the topic treated;
   – a subject index: as a rule this is genuinely helpful for the reader.
   – For evaluation purposes, manuscripts should be submitted as pdf files.

6. Careful preparation of the manuscripts will help keep production time short besides ensuring satisfactory appearance of the finished book in print and online. After acceptance of the manuscript authors will be asked to prepare the final LaTeX source files (see LaTeX templates online: https://www.springer.com/gb/authors-editors/book-authors-editors/manuscriptpreparation/5636) plus the corresponding pdf- or zipped ps-file. The LaTeX source files are essential for producing the full-text online version of the book, see http://link.springer.com/bookseries/304 for the existing online volumes of LNM). The technical production of a Lecture Notes volume takes approximately 12 weeks. Additional instructions, if necessary, are available on request from lnm@springer.com.

7. Authors receive a total of 30 free copies of their volume and free access to their book on SpringerLink, but no royalties. They are entitled to a discount of 33.3 % on the price of Springer books purchased for their personal use, if ordering directly from Springer.

8. Commitment to publish is made by a *Publishing Agreement*; contributing authors of multiauthor books are requested to sign a *Consent to Publish form*. Springer-Verlag registers the copyright for each volume. Authors are free to reuse material contained in their LNM volumes in later publications: a brief written (or e-mail) request for formal permission is sufficient.

Addresses:
Professor Jean-Michel Morel, CMLA, École Normale Supérieure de Cachan, France
E-mail: moreljeanmichel@gmail.com

Professor Bernard Teissier, Equipe Géométrie et Dynamique,
Institut de Mathématiques de Jussieu – Paris Rive Gauche, Paris, France
E-mail: bernard.teissier@imj-prg.fr

Springer: Ute McCrory, Mathematics, Heidelberg, Germany,
E-mail: lnm@springer.com

Printed in the United States
By Bookmasters